Advances in Carbohydrate Chemistry

Advances in Carbohydrate Chemistry

Edited by **Allegra Smith**

NY RESEARCH PRESS

New York

Published by NY Research Press,
23 West, 55th Street, Suite 816,
New York, NY 10019, USA
www.nyresearchpress.com

Advances in Carbohydrate Chemistry
Edited by Allegra Smith

International Standard Book Number: 978-1-63238-030-2 (Hardback)

Printed in China.

Contents

Preface IX

Chapter 1 **Application of Bioactive Coatings Based on Chitosan for Soybean Seed Protection** 1
Defang Zeng, Xinrong Luo, and Renjie Tu

Chapter 2 **A New Rapid and Sensitive Spectrophotometric Method for Determination of a Biopolymer Chitosan** 6
Mohamed E. I. Badawy

Chapter 3 **Acetylation of Wood Flour from Four Wood Species Grown in Nigeria Using Vinegar and Acetic Anhydride** 13
Yakubu Azeh, Gabriel Ademola Olatunji, Cheku Mohammed and Paul Andrew Mamza

Chapter 4 **Factors Involved in the *In Vitro* Fermentability of Short Carbohydrates in Static Faecal Batch Cultures** 19
Eva Gietl, Wim Mengerink, Jaap de Slegte, Glenn Gibson, Robert Rastall and Ellen van den Heuvel

Chapter 5 **Synthesis and Characterization of Graft Copolymer of Dextran and 2-Acrylamido-2-methylpropane Sulphonic Acid** 29
Venkanna Azmeera, P. Adhikary, and S. Krishnamoorthi

Chapter 6 **Development and Characterization of Semi-IPN Silver Nanocomposite Hydrogels for Antibacterial Applications** 36
A. Chandra Babu, M. N. Prabhakar, A. Suresh Babu, B. Mallikarjuna, M. C. S. Subha and K. Chowdoji Rao

Chapter 7 **Thermal and Photoresponsive Studies of Starch Modified with 2-(5-(4-Dimethylamino-benzylidine)-4-oxo- 2-thioxo-thiazolidin-3-yl)acetic Acid** 44
Ambily Chandran, Sunny Kuriakose and Tessymol Mathew

Chapter 8 **Scanning Electron Microscopy and Kinetic Studies of Ketene-Acetylated Wood/Cellulose High-Density Polyethylene Blends** 52
Yakubu Azeh, Gabriel Ademola Olatunji and Paul Andrew Mamza

Chapter 9 **Partial Chemical Characterization of Immunomodulatory Polysaccharides from *Plantago palmata* Hook. f. s. Leaves** 59
Gabriel Biringanine, Moustapha Ouedraogo, Bernard Vray, Anne Berit Samuelsen and Pierre Duez

Chapter 10 **The Study of the Reaction of Pectin-Ag(0) Nanocomposites Formation** 66
Nadezhda Viktorovna Ivanova, Natalya Nikolaevna Trofimova, Lydmila Alexandrovna Es'kova, and Vasilyi Anatolievich Babkin

Chapter 11 **Certain Rheological Aspects of Functionalized Guar Gum** 75
Meenu Kapoor, Dhriti Khandal, Ruchi Gupta, Pinklesh Arora, Geetha Seshadri, Saroj Aggarwal and Rakesh Kumar Khandal

Chapter 12 **High-Performance Anion-Exchange Chromatography Coupled with Pulsed Electrochemical Detection as a Powerful Tool to Evaluate Carbohydrates of Food Interest: Principles and Applications** 90
Claudio Corradini, Antonella Cavazza, and Chiara Bignardi

Chapter 13 **Analysis by Vibrational Spectroscopy of Seaweed Polysaccharides with Potential Use in Food, Pharmaceutical, and Cosmetic Industries** 103
Leonel Pereira, Saly F. Gheda and Paulo J. A. Ribeiro-Claro

Chapter 14 **Latent Fingerprint Enhancement Using Tripolyphosphate-Chitosan Microparticles** 110
Issa M. A. Il Dueik and Gordon A. Morris

Chapter 15 **Production Methods for Hyaluronan** 114
Carmen G. Boeriu, Jan Springer, Floor K. Kooy, Lambertus A. M. van den Broek and Gerrit Eggink

Chapter 16 **Identification and Quantification of Protein Glycosylation** 128
Ziv Roth, Galit Yehezkel, and Isam Khalaila

Chapter 17 **The Effect of Tribomechanical Micronization and Activation on Rheological, Thermophysical, and Some Physical Properties of Tapioca Starch** 138
Zoran Herceg, Verica Batur, Anet Režek Jambrak, Tomislava Vukušić, Ines Gmajnički and Igor Špoljarić

Chapter 18 **ZnO Nanoparticles-Chitosan Composite as Antibacterial Finish for Textiles** 145
Asmaa Farouk, Shaaban Moussa, Mathias Ulbricht and Torsten Textor

Chapter 19 **Conformational Search on the Lewis X Structure by Molecular Dynamic: Study of Tri- and Pentasaccharide** 153
N. Khebichat, K. Nekkaz, and S. Ghalem

Chapter 20 **Preparation and Characterization of Chitosan/Zinc Oxide Nanoparticles for Imparting Antimicrobial and UV Protection to Cotton Fabric** 160
M. M. AbdElhady

Chapter 21 **Effect of Molecular Sizes of Chondroitin Sulfate on Interaction with L-Selectin** 166
Naoko Igarashi, Atsuko Takeguchi, Shinobu Sakai, Hiroshi Akiyama, Kyohei Higashi and Toshihiko Toida

Chapter 22 **Synthesis and Characterization of Acrylic Acid-g-(*k*-Carrageenan) Copolymer and Study of Its Application** 175
Arti Srivastava and Rajesh Kumar

Chapter 23 **Miscibility Studies of Hydroxypropyl Cellulose/Poly(Ethylene Glycol) in Dilute Solutions and Solid State** 183
K. Sudharsan Reddy, M. N. Prabhakar, P. Kumara Babu, G. Venkatesulu, U. Sajan Kumarji Rao, K. Chowdoji Rao and M. C. S. Subha

Chapter 24 **Brönsted Acidic Ionic Liquid 1-(1-Propylsulfonic)-3-methylimidazolium-Chloride Catalyzed Hydrolysis of D-Cellobiose in Aqueous Medium** 192
Ananda S. Amarasekara and Bernard Wiredu

Permissions

List of Contributors

Preface

The chemical study of carbohydrates and its various properties is called carbohydrate chemistry. It is a fairly new field and has grown substantially over the past few decades, given the exponential rise in the interest in various food groups. It involves multiple dimensions of study like the synthesis of carbohydrates, glycosidic bond formations, monosaccharides, oligosaccharides, reactions to carbohydrates and functions of carbohydrates.

Glycosidic bond formation is further split into chemical glycosylation, fischer glycosylation, glycosyl halide and Koenigs-Knorr reaction while the reaction of carbohydrates concerns itself with the presence of carbohydrates in organic reactions like Cyanohydrin reaction, Amadori rearrangement, Ferrier rearrangement and Wohl degradation.

The functions of carbohydrates look at energy supply from the brain, mainly in the form of glucose and avoiding both breakdown of amino acids and ketosis from the breakdown of fatty acids. This is a field of science which has found critical applications in biological spheres.

This book documents and sheds light on various aspects of carbohydrate chemistry. It is a compilation of works from some of the most renowned experts in carbohydrate chemistry who have carefully and tirelessly worked to gather information and knowledge in this branch of science. I'd like to extend my gratitude to the team at the publishing house for their constant support at each and every step of the publication process.

Editor

1

Application of Bioactive Coatings Based on Chitosan for Soybean Seed Protection

Defang Zeng, Xinrong Luo, and Renjie Tu

School of Resource & Environmental Engineering and Key Laboratory of Hubei Province for Mineral Resources Processing and Environment, Wuhan University of Technology, Wuhan 430070, China

Correspondence should be addressed to Defang Zeng, turenjie617@163.com

Academic Editor: José A. Lopes da Silva

Soybean seeds suffer attacks of various pests that result in a decreased yield in northeastern China. Until recently, people use pesticides such as insecticides to achieve the goal of controlling pests. Chitosan extracted from deacetylation of chitin is promising candidates as a seed-coating agent to control agrotis ypsilon, soybean pod borer, and soybean aphid effectively. An experimental study on influences of chitosan with different concentrations on pest controlling and soybean growth was made in the paper. Coating based on chitosan was used as a feeding deterrent and for enhancing the germination and quality of soybean seeds. Results indicated that all chitosan coating had a significant effect on antifeeding against pests; with the increasing concentration, antifeedant rate (AR) were increased obviously, especially when in the concentration of 5%, santifeedant rate of agrotis ypsilon, soybean pod borer, and soybean aphid reached 82.89%, 87.24%, and 80.21%, respectively. Also chitosan coating increased seed germination, plant growth, and soybean yield efficiently, especially when, in the concentration of 5%, the yield was increased by about 20% compared with CK. The application of chitosan in soybean seed coated is an appropriate option to control pests replacing high-toxicity pesticides and enhance soybean yield.

1. Introduction

Soybean is one of the major economic crops and famous for its rich plant proteins in China. However, during the soybean growth process, soybean suffers attacks of various pests such as agrotis ypsilon, soybean pod borer, and soybean aphid that result in considerable crop losses in northeastern China. Agrotis ypsilon is a cosmopolitan important underground pest with a wide distribution in China, which can attack many kinds of seedlings of crops and trees. Soybean pod borer is a devastating insect pest of many legume crops in tropical and semitropical regions, such as Africa and southeastern Asia. Larvae are pale-colored and feed on the floral parts and pods of legume plants, rendering them unmarketable. Plants are not killed, but a large proportion of the pods may be damaged and unmarketable. Flowers are also eaten and destroyed. Soybean aphid has a tube-like mouthpart that sucks juices and nutrients from the plant. Plants may yellow and become stunted, reducing pod and seed production if aphid populations become extremely high. The honeydew acts as a substrate for sooty mould to develop which turns the leaves black and rubbery.

To prevent considerable economic losses, many studies [1–3] have developed for crop protection, such as seed-coated technology as a pesticide against crop pests and diseases. Results indicated that it can prevent the seeds from pests in the soil and plant and improve the germination potential: germination percentage and crop yield [4]. Therefore, soybean seeds are usually coated by seed-coating agents before being planted in Northeast China [5]. Results are usually positive since the number of pests is reduced, and the quality of the seeds is improved when conventional treatments are used. However, these conventional seed-coating agents are not the best alternative for the environment due to their accumulative toxicity in the soil. In addition, pests may develop resistance against them, which implies the use of increasingly larger amounts to be efficient [6, 7]. Thus, there is a demand for a safer and more ecofriendly alternative [8, 9].

Chitosan is a carbohydrate biopolymer derived from deacetylation of chitin, which is found in the crustacean's shells, insect's cuticle, and cell wall of fungi. The positive charge of chitosan confers to this polymer numerous and unique physiological and biological properties with great potential in a wide range of industries such as pharmacology, medicine, and agriculture [10–12].

This effect has been shown when seeds were treated by chitosan solutions; these treatments improved the germination of several cultures and contributed to harvests yields. The application of chitosan in agriculture consisted of coating seeds and plants by dipping or spraying. The main objective of this study was to investigate the effect of chitosan as a seed treatment on insect pest control, seed germination, soybean growth, and yield.

2. Materials and Methods

Chitosan (Aokang Biotech Co., China), soybean seeds (Wuyin number nine, Hubei Seed Co., China). Agrotis ypsilon, soybean pod borer, and soybean aphid (Fourth instars larvae, Shandong Agricultural University, China), sodium hydroxide, and acetic acid (analytically pure, Hubei University Chemical Plant).

Chitosan was dissolved at 1%, 2%, 3%, 4%, and 5% (w/v) into aqueous solutions of acetic acid at 1% (w/v). The pH of solution was adjusted to 6.0 using 1% NaOH. Soybean seeds (10 g) were introduced into a seed coater. During the rotation of the coater, 0.2 mL solution treatment was added to the seeds. After coating, seeds were air-dried for 2 h at room temperature.

2.1. Laboratory Method for Antifeeding Test. According to the guideline for laboratory bioassay of pesticides, the antifeeding effect of chitosan was studied in no-choice bioassays with the artificial mixed feeding method and leaf discs method. The artificial mixed feeding method was as follows: 100 g artificial feed was individually dipped in different concentrations of chitosan solutions and then put into each hole with diameter of 1.5 cm and depth of 1.5 cm, where the fourth instars larvae were fed. The leaf discs method was as follows: fresh cotton leaf discs of 4 cm diameter were punched using a cork borer and were individually dipped in different concentrations of chitosan solutions. Each fourth instars larva was introduced into per Petri dish. Artificial feeds and leaf discs treated with water (CK) were used as controls, and each treatment was replicated 3 times. The antifeeding effect was determined at 48 h after breeding in the incubator at temperature of $25 \pm 1^{\circ}$C and 75–85% relative humidity under a day length of 16 L : 8 D. An electronic balance and a leaf area meter were used to measure the consumption. The test was valid only at less than 5% mortality of larvae. The nonselective antifeedant rate (AR) was calculated by the following formula:

$$\text{AR}\ (\%) = \frac{A - B}{A} \times 100\%, \tag{1}$$

where A and B are the consumption of the control group and the treatment group, respectively.

2.2. Laboratory Method for the Germination Abilities Test. According to the rules for seed testing of the International Seed Testing Association [13], 100 seeds taken from each group were arranged on two layers of wet filter paper in each Petri dishes filled with wet sand. Each Petri dish contained 25 seeds, and each treatment was replicated 4 times. All Petri dishes were incubated in the constant temperature and humidity incubator at $28 \pm 1^{\circ}$C and air relative humidity of 85%. Germinability (GE), germination percentage (GP), and germination index (GI) of soybean seeds were investigated on the third day and seventh day, respectively. The calculation formulas were as follows:

$$\begin{aligned} \text{GE} &= \frac{C}{E} \times 100\%, \\ \text{GP} &= \frac{D}{E} \times 100\%, \\ \text{GI}\ (\%) &= \sum \frac{F}{G} \times 100\%, \end{aligned} \tag{2}$$

where C is the number of germinated seeds on the third day, D is the number of germinated seeds on the seventh day, E is the number of total seeds investigated, F is the number of germinated seeds one day, and G is the days of seeds germination.

2.3. Field Trial. The field trial was conducted in the plots of the experimental field in the Liaoning Provincial Seed Group Co., Ltd, China. The soybean seeds were mixed with each of the seed-coating agents at a ratio of 1 : 50 (w/w) and coated by hand uniformly. After spreading and airing for about 30 min, the seeds were sown into the field. The experiments were designed as a randomized block design with each treatment consisting of several plots separated from each other by a row 20 cm wide. Each plot was 4.0 m wide by 6.0 m long, contained 250 soybean seeds, and was sown either with coated seeds or with uncoated seeds to serve as control. Each treatment was replicated three times. A random sample of 100 seedlings was selected to determine the germination percentage (GP), and 100 plants were randomly selected to determine the growth indexes at flowering stage and pod-setting stage, respectively. Total yield of each plot was determined by weighing the seeds of all established plants after threshing.

3. Results

The artificial mixed feeds weight and mean leaf area consumed were significantly less than the control group for every chitosan treatment against agrotis ypsilon, soybean pod borer, and soybean aphid. For all chitosan treatments, the artificial mixed feeds weight and leaf area consumed declined with increasing concentration. It can be seen from Table 1 that all natural polysaccharide coating had a significant effect on antifeeding against pests; with the increasing concentration, antifeedant rates (AR) were increased obviously; especially when in the concentration of 5%, antifeedant rate of agrotis ypsilon, soybean pod borer, and soybean aphid reached 82.89%, 87.24%, and 80.21%, respectively, and the best to soybean pod borer.

TABLE 1: Effects of different treatments on the consumption of artificial mixed feeds weight and cucumber leaf area.

T	AR (%) Agrotis ypsilon	Soybeanpod borer	Soybean aphid
1% NP	53.51 ± 1.1c	55.56 ± 1.2c	29.16 ± 1.3c
2% NP	61.40 ± 0.9c	62.14 ± 1.0c	41.55 ± 1.1c
3% NP	67.98 ± 1.5b	69.55 ± 1.5b	50.06 ± 1.4b
4% NP	74.12 ± 1.3b	72.02 ± 1.4b	65.33 ± 1.5
5% NP	82.89 ± 1.2a	87.24 ± 1.0a	80.21 ± 1.1b

Values are expressed as mean ± standard error. Means with different letters is significance different at $P < 0.05$.

TABLE 2: Effects of different treatments on the emergence of soybean pests in the field trials.

Treatment	Soybean aphid emergence percentage (%) Weeks after planting		No. of soybean aphid (no./plant) Weeks after planting		Leaf roll emergence percentage (%) Weeks after planting	
	3	6	3	6	3	6
1% NP	73.14	93.28	5875	37564	6.98.	15.34
2% NP	67.45	91.37	3276	35867	6.87	14.86
3% NP	56.65	89.92	2094	26755	6.35	12.56
4% NP	42.76	86.57	1889	19353	4.89	9.84
5% NP	35.16	84.46	1574	14735	3.72	8.45
CK	87.78	100	15650	58646	36.56	55.46

TABLE 3: Effects of different treatments on germination abilities.

Treatments	Germination abilities GE (%)	GP (%)	GI (%)
1% NP	74.1 ± 1.5c	82.3 ± 1.3c	44.3 ± 0.5c
2% NP	75.9 ± 1.6c	82.9 ± 1.2c	45.8 ± 0.6c
3% NP	78.3 ± 1.8b	88.3 ± 1.2b	46.9 ± 0.8b
4% NP	81.2 ± 1.7a	89.5 ± 1.1a	48.1 ± 0.7c
5% NP	81.9 ± 1.4a	92.5 ± 1.0a	49.7 ± 0.4c
CK	71.9 ± 2.1	79.8 ± 1.5	41.9 ± 0.6

Values are expressed as mean ± standard error. Means with different letters is significance different at $P < 0.05$.

The field trial results indicated that chitosan coating protected soybean plant against pests efficiently (Table 2). Compared with CK, soybean aphid emergence and leaf-roll emergence percentage of soybean plants treated with chitosan were significantly reduced. With the increasing concentration of chitosan, soybean aphid emergence percentage, no. of soybean aphid, and leaf roll emergence percentage were reduced, it was observed that there were significant differences between 1% and 5% chitosan treated.

Results related to germinability (GE), germination percentage (GP), and germination index (GI) were shown in Table 3. The chitosan coating treatment resulted in a significant increase in GE, GP, and GI as compared to the blank control group (CK). The formulation with the highest concentration of chitosan (5%) was the best treatment reaching over 90%. However, it was observed that there were not significant differences between 4% and 5% chitosan treated.

The main performance indexes of soybean growth such as germination percentage, plant height, and dry weight were improved under chitosan treated. From Figure 1, there were obvious differences between coated seedlings and noncoated ones. The roots were longer and more well developed; the stems were longer and thicker than those of CK. It can be seen from Tables 4 to 5 that different concentrations had different degrees of improvement. And pest damage of seedlings and leaves greatly bated. Especially when in the concentration of 5%, the yield was increased by about 20% compared with CK.

4. Discussion

4.1. The Mechanism of Antifeedant Effect. Our research indicated that chitosan acted as a seed-coating agent to control agrotis ypsilon, soybean pod borer and soybean aphid effectively. Chitosan stimulates plants to produce systematic antibodies, which can produce repellant effects to deter insect pests. In addition, chitosan with natural antifungal property can produce significant repellent action to crop pests [14, 15]. Chitosan as signal molecule plays a barrier function in feeding behavior, which causes the unusual discharge of the nervous system and prevents animals from getting correct information of taste [16, 17].

4.2. The Mechanism of the Yield Increase Effect. Figure 2 has showed clearly that the face of seed coated with chitosan 2(a) forms a layer of chunky membrane, but this 2(b) not coated does not. Chitosan has excellent film-forming property, making it easy to form a semipermeable film on the seed surface which can maintain the seed moisture and absorb the soil moisture, and thus it can promote seed germination.

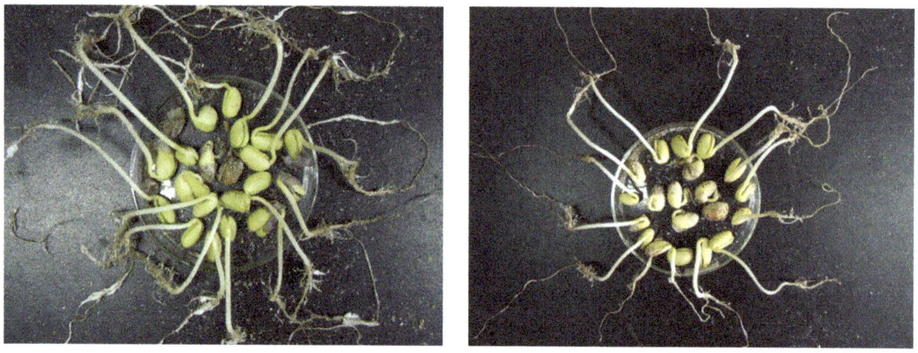

(a) Coated with NP (b) CK

FIGURE 1: Contrast section photos of soybean seedling growth coated with NP and CK.

TABLE 4: Effects of different treatments on germination percentage, plant height, and dry weight in different growth period.

Treatment	Germination Percentage (%)	Seedling stage		Full-bloom stage		Pod bearing period	
		Plant height (cm)	Dry weight (g)	Plant height (cm)	Dry weight (g)	Plant height (cm)	Dry weight (g)
1% NP	80.72	8.4	0.42	21.8	1.84	38.8	7.53
2% NP	81.21	8.6	0.44	22.4	1.84	39.1	7.84
3% NP	85.53	8.6	0.45	22.4	1.91	39.5	7.86
4% NP	89.76	8.7	0.45	22.8	1.94	41.2	7.96
5% NP	94.45	9.4	0.49	24.6	2.18	43.3	8.87
CK	79.84	8.2	0.41	21.5	1.79	38.4	7.32

TABLE 5: Effects of different treatments on yield components of soybean in the field trials.

Treatments	No. of nodes (no./plant)	No. of pods (no./plant)	No. of grains (no./plant)	100-seed weight/g	Yield (10^3 kg/ha)
1% NP	11.8 ± 1.4c	32.5 ± 3.2c	63.2 ± 5.5c	24.42 ± 0.3	2.63 ± 0.025c
2% NP	12.2 ± 1.6c	33.7 ± 2.8c	65.8 ± 5.8c	25.26 ± 0.5c	2.89 ± 0.024c
3% NP	12.5 ± 1.5c	34.1 ± 3.0c	67.2 ± 5.2b	25.76 ± 0.2c	2.95 ± 0.021b
4% NP	12.7 ± 1.3c	34.6 ± 2.7c	68.4 ± 5.6b	26.35 ± 0.4b	3.05 ± 0.023a
5% NP	12.9 ± 1.2c	35.8 ± 3.3c	75.2 ± 5.1a	27.85 ± 0.3a	3.12 ± 0.022a
CK	11.2 ± 1.7	31.9 ± 3.1	60.6 ± 5.7	23.13 ± 0.5	2.58 ± 0.024

Values are expressed as mean ± standard error. Means within a column designated with different letters are significance different at $P < 0.05$.

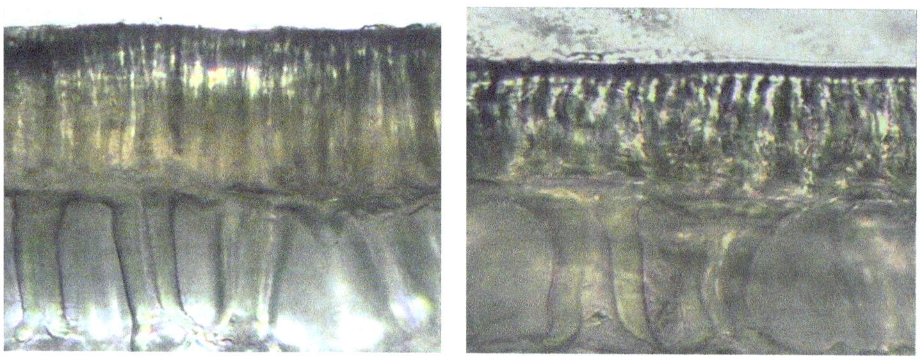

(a) Coated with NP (b) CK

FIGURE 2: Contrast section photos of coated-seeds' and uncoated-seeds' surface membrane.

In contrast, it can cut off excessive soil moisture to prevent the seed from corrupting. Furthermore, chitosan film is also considered to have a good selective permeability, which can prevent oxygen from entering the film, restrict loss of CO_2, and maintain a high concentration of CO_2 in the film, so as to restrain the seed respiration and thus to make the internal nutrient consumption of seeds fall to the lowest possible level [18]. This kind of semi-permeable film is believed to be able to maintain the seed moisture and absorb the soil moisture, and thus it can promote seed germination [19]. Chitosan also can increase soluble sugar content and enhance the activity of protease conversion to protein and increasing free amino acid content, which has obvious inhibiting effect for many plant pathogenic fungi. Chitosan as a novel plant disease inhibitor can induce and improve the disease resistance of plants thus has a repellent effect to the pests in the soil [20]. Chitosan possesses the natural antifungal role, which increases the permeability of the outer membrane and inner membrane and ultimately disrupts bacterial cell membranes, with the release of cellular contents [21, 22].

Acknowledgments

The authors express the appreciation to the Wuhan Science and Technology Bureau of China for financial support (no. 201120922303). They also thank the State Key Laboratory of Crop Biology of Shandong Agricultural University. A special acknowledgment is given to the Wuhan University of Technology for experimental conditions and technical support.

References

[1] R. A. Omolehin, T. O. Ogunfiditimi, and O. B. Adeniji, "Factors influencing adoption of chemical pest control in cowpea production among rural farmers in Makarfi Local Government Area of Kaduna State, Nigeria," *Asian Journal of Plant Sciences*, vol. 3, no. 4, pp. 225–228, 2005.

[2] E. Asare-Bediako, A. A. Addo-Quaye, and A. Mohammed, "Control of diamondback moth (*Plutella xylostella*) on cabbage (*Brassica oleracea* var *capitata*) using intercropping with non-host crops," *American Journal of Food Technology*, vol. 5, no. 4, pp. 269–274, 2010.

[3] M. M. Rahman and M. M. Hossain, "Plant density effects on growth, yield and yield components of two soybean varieties under equidistant planting arrangement," *Asian Journal of Plant Sciences*, vol. 5, no. 10, pp. 278–286, 2011.

[4] N. E. Ahmed, H. O. Kanan, S. Inanaga, Y. Q. Ma, and Y. Sugimoto, "Impact of pesticide seed treatments on aphid control and yield of wheat in the Sudan," *Crop Protection*, vol. 20, no. 10, pp. 929–934, 2001.

[5] N. L. Brooker, C. D. Lagalle, A. Zlatanic, I. Javni, and Z. Petrovic, "Soy polyol formulations as novel seed treatments for the management of soil-borne diseases of soybean.," *Communications in Agricultural and Applied Biological Sciences*, vol. 72, no. 2, pp. 35–43, 2007.

[6] X. Honglu and X. Guomei, "Suspension property of gemini surfactant in seed coating agent," *Journal of Dispersion Science and Technology*, vol. 29, no. 4, pp. 496–501, 2008.

[7] K. Ziani, B. Ursúa, and J. I. Maté, "Application of bioactive coatings based on chitosan for artichoke seed protection," *Crop Protection*, vol. 29, no. 8, pp. 853–859, 2010.

[8] M. E. Arias, J. A. González-Pérez, F. J. González-Vila, and A. S. Ball, "Soil health—a new challenge for microbiologists and chemists," *International Microbiology*, vol. 8, no. 1, pp. 13–21, 2005.

[9] H. F. Huiting and A. Ester, "Effects of seed coatings with thiamethoxam on germination and flea beetle control in flax," *Communications in Agricultural and Applied Biological Sciences*, vol. 72, no. 3, pp. 595–601, 2007.

[10] S. Bautista-Banos, M. Hernandez-Lopez, and E. Bosquez-Molina, "Growth inhibition of select fungi by chitosan and plant extracts," *Mexican Journal of Phytopathology*, vol. 22, pp. 178–186, 2004.

[11] M. Rinaudo, "Chitin and chitosan: properties and applications," *Progress in Polymer Science*, vol. 31, no. 7, pp. 603–632, 2006.

[12] H. Tang, P. Zhang, T. L. Kieft et al., "Antibacterial action of a novel functionalized chitosan-arginine against Gram-negative bacteria," *Acta Biomaterialia*, vol. 6, no. 7, pp. 2562–2571, 2010.

[13] X. H. Wu, W. H. Zhang, and P. F. Liu, "Research and development trend of the seed coating agent in China," *Plant Protection and Promotion of Technology*, vol. 10, pp. 36–38, 2003.

[14] N. Shibuya and E. Minami, "Oligosaccharide signalling for defence responses in plant," *Physiological and Molecular Plant Pathology*, vol. 59, no. 5, pp. 223–233, 2001.

[15] P. P. Ma and L. Q. He, "Progress of chitosan in suppression of plant diseases," *Natural Product Research and Development*, vol. 13, no. 6, pp. 82–86, 2001.

[16] T. Tokunaga, N. Takada, and M. Ueda, "Mechanism of antifeedant activity of plumbagin, a compound concerning the chemical defense in carnivorous plant," *Tetrahedron Letters*, vol. 45, no. 38, pp. 7115–7119, 2004.

[17] D. R. Perera, G. Armstrong, and N. Senanayake, "Effect of antifeedants on the diamondback moth (*Plutella xylostella*) and its parasitoid *Cotesia plutellae*," *Pest Management Science*, vol. 56, no. 5, pp. 486–490, 2002.

[18] R. T. Furbank, R. White, J. A. Palta, and N. C. Turner, "Internal recycling of respiratory CO_2 in pods of chickpea (*Cicer arietinum* L.): the role of pod wall, seed coat, and embryo," *Journal of Experimental Botany*, vol. 55, no. 403, pp. 1687–1696, 2004.

[19] W. V. Sigler and R. F. Turco, "The impact of chlorothalonil application on soil bacterial and fungal populations as assessed by denaturing gradient gel electrophoresis," *Applied Soil Ecology*, vol. 21, no. 2, pp. 107–118, 2002.

[20] H. P. Chen and L. L. Xu, "Progress of study on chitosan in regulating plants' growth and eliciting plants' defense responses," *Acta Botanica Yunnanica*, vol. 27, no. 6, pp. 613–619, 2005.

[21] H. Liu, Y. Du, X. Wang, and L. Sun, "Chitosan kills bacteria through cell membrane damage," *International Journal of Food Microbiology*, vol. 95, no. 2, pp. 147–155, 2004.

[22] K. K. Nielsen, P. Jorgensen, and J. D. Mikkelsen, "Antifungal activity of sugar beet chitinase against *Cercospora beticola*: an autoradiographic study on cell wall degradation," *Plant Pathology*, vol. 43, no. 6, pp. 979–986, 1994.

A New Rapid and Sensitive Spectrophotometric Method for Determination of a Biopolymer Chitosan

Mohamed E. I. Badawy

Department of Pesticide Chemistry and Technology, Faculty of Agriculture, Alexandria University El-Shatby, Alexandria 21545, Egypt

Correspondence should be addressed to Mohamed E. I. Badawy, m_eltaher@yahoo.com

Academic Editor: R. J. Linhardt

A novel approach of spectrophotometric quantification of chitosan based on one-step depolymerization with sodium nitrite followed by reaction of the end product with thiobarbituric acid has been proposed, optimized, and validated. In this process, chitosan is converted into 2,5-anhydro-D-mannose that reacts with thiobarbituric acid to form pink color. The color that resulted from the reaction was stabilized and measured at 555 nm. The method optimization was essential as many procedural parameters influenced the accuracy of the determination including hydrolysis conditions, thiobarbituric acid concentration, reaction time, pH, reaction temperature, and color stability period. Under given optimized conditions that appeared to be critical, chitosan was quantitatively analyzed and the calibration graph was linear over the range of 10–50 μg/mL ($r^2 = 0.999$). This approach was applied for determination of chitosan in pharmaceutical formulation (chitocal) and had a recovery rate of higher than 96%. The developed method is easy to use and highly accurate.

1. Introduction

Recently, much attention has been paid to chitosan as a potential polysaccharide resource. Chitosan is a linear amino polysaccharide of glucosamine and *N*-acetylglucosamine units and is obtained by alkaline deacetylation of chitin [1]. The following major characteristics of chitosan make this polymer advantageous for numerous applications: (1) it has a defined chemical structure; (2) it can be chemically and enzymatically modified; (3) it is physically and biologically functional; (4) it is biodegradable and biocompatible with many organs, tissues, and cells; (5) it can be processed into several products including flakes, fine powders, beads, membranes, sponges, cottons, fibers, and gels. Therefore, chitosan becomes of great interest as a new functional material of high potential in various fields and the discovery or development of methods for chitosan determination is imperative [2–5].

Although research activities dealing with chitosan are numerous, a generally accepted simple method for direct quantitative analysis is lacking. Chitosan can be degraded to glucosamine monomer by hydrolysis, and there are several reports on the determination of glucosamine by chromatographic, colorimetric, and fluorimetric techniques, or a combination of these [6–8]. Eikenes et al. [6] developed a method for determination of chitosan in wood and water samples based on acidic hydrolysis of chitosan to glucosamine followed by online derivatization by *o*-phthalaldehyde, chromatographic separation, and fluorescent detection. Roseman and Daffner [9] determined the concentration of glucosamine by acetylating followed by photometric detection of *N*-acetyl glucosamine.

A novel approach of colorimetric quantification of chitosan based on the derivatization reaction of amino groups in chitosan with *o*-phthalaldehyde and a thiol-*N*-acetyl-L-cysteine has been developed by Larionova co-workers [10]. The procedure conditions have been optimized for chitosan assay in the presence of polyanionic electrolyte dextran sulphate. The method has proven to be convenient and reliable for quantitative determination of either the concentrations of chitosan of various molecular weights or their degree of deacetylation.

Based on the fact that, in dilute solutions, the amino group is at its most accessible form for optimum quantification, several spectrophotometric approaches have been reported. The determination of aminoglucose by the indole reaction [11] and the formation of a complex of chitosan with ninhydrin are the most used protocols [12, 13]. These methods did not involve any pretreatment (i.e., hydrolysis) of the solution, but the method was highly dependent on the degree of deacetylation of chitosan. More reliable analytical methods are colorimetric assay of chitosan with various anionic dyes, for instance, C.I. Acid Orange 7, C.I. Acid Red 13, C.I. Acid Red 27, Orange II, Alizarin S, Alizarin GG, Congo Red [14], Cibacron Brilliant Red 3B-A [15, 16]. However, these techniques based on the reactions between free amino groups of chitosan and acting agents are ineligible for determination of chitosan in presence of substances bearing primary amino groups, including peptides [17]. Recently, an electrochemical method based on the oxidation-reduction peaks produced by chitosan solubilisation in an acidic buffer has been reported as a sensitive measurement for this polymer [18]. However, the application of this analysis to complex mixtures would be difficult.

Most quantitative analysis of polysaccharides such as chitosan entails total hydrolysis to monosaccharides followed by subsequent characterization of the monomers [6, 19, 20]. However, the glycosidic linkages of chitosan are resistant to acid hydrolysis such as the anthrone or phenol-sulphuric acid methods [21]. In addition, such hydrolysis methods are difficult to apply. Chitosanases may be used for depolymerization of chitosan, but availability and costs limit the use of such enzymes [22].

In the present work, a selective and sensitive method for determination of chitosan was developed using sodium nitrite hydrolysis and online derivatization with thiobarbituric acid in basic medium to form pink color. Sodium nitrite causes depolymerization and deamination of chitosan and converts it to 2,5-anhydro-D-mannose [19, 20]. The procedure conditions have been optimized and evaluated for chitosan assay in chitocal as a pharmaceutical formulation.

2. Materials and Methods

2.1. Principle of the New Assay. The method is based on determination of chitosan after deaminative cleavage by sodium nitrite ($NaNO_2$) and coupling of the 2,5-anhydro-D-mannose thus produced with thiobarbituric acid to give pink product. 2,5-Anhydro-D-mannose represents the sum of glucosamine (GlcN) and *N*-acetyl glucosamine (GlcNAc). Depolymerization of chitosan by using of $NaNO_2$ is selective, rapid, easily controlled, no side reactions have been observed with chitosan, and, as a consequence, the stoichiometry and reaction products are well established [23, 24]. Furthermore, because chitosan is soluble in dilute acids, the reaction with $NaNO_2$ is homogeneous. The depolymerization mechanism involves a deamination of (1 → 4)-linked 2-acetamido-2-deoxy-β-D-glucopyranose unit forming 2,5-anhydro-D-mannose at the new reducing end as shown in Figure 1(a) [24–27]. The end product contains an aldehyde which is determined colorimetrically with thiobarbituric acid. Thiobarbituric acid undergoes a Knoevenagel condensation reaction with aldehydes to give the color-condensed product according to Tietze and Beifuss [28] (Figure 1(b)). The observed level of pink product was related to the chitosan present in the sample.

2.2. Assay Reagents. All experiments were performed with analytical reagent grade chemicals. Low molecular weight of acid-soluble chitosan (made from coarse ground crab with 89% degree of deacetylation) was purchased from Sigma-Aldrich Co. (USA). Chitosan stock solution was made by dissolving 100 mg of chitosan in 100 mL of 1% (v/v) aqueous acetic acid; the chitosan dissolves readily on shaking. From this solution, suitable dilutions can be made when required. An aqueous solution of 0.5 M $NaNO_2$ was prepared, and the solution should be kept between 0 and 4°C. 2-Thiobarbituric acid standard solution (0.04 M) was prepared by dissolving 0.576 g of reagent (≥98%, Sigma-Aldrich Co., USA) in 100 mL of distilled water with warming for 15 min.

2.3. Procedure. Chitosan solution was pipetted into the test tube, and the $NaNO_2$ reagent was added from a micropipette (0.1 mL of 0.5 M). The mixture was shaken briefly. The tubes were kept at 80°C for 30 min in a water bath to complete the depolymerization-deamination reaction. After depolymerization, the pH was raised to 8 by adding 0.2 mL NaOH (0.1 M) and the solution was then shaken, and 0.5 mL of the thiobarbituric acid solution (0.04 M) was finally added. The tubes were incubated a second time at 80°C for 10 min in a water bath. The solution was briefly cooled, and the absorbance was measured at 555 nm by Spectrophotometer (Unico 1200-Spectrophotometer, USA) against blank.

2.4. Assay of Chitosan in Pharmaceutical Formulation. In Chitocal capsules (500 mg + 100 mg ascorbic acid + 80 mg gymnema sylvestre, manufactured by Aldebeiky Group-DBK Pharma, Cairo, Egypt), the well-mixed powdered content of 10 capsules was used in the assay. A weight of 100 mg of the powder was transferred into volumetric flask and dissolved in 50 mL aqueous acetic acid solution (1%, v/v) then adjusted to volume. Different aliquots (20, 40, 60, 80, and 100 μL) of this solution were analyzed according to the proposed procedure as previously mentioned. For the recovery study, the standard addition method has been used. Different amounts of the corresponding pure chitosan were added to a fixed amount of dosage form (capsule) where chitosan had been determined. Results are the average of triplicate readings of all determinations in the solutions.

2.5. Statistical Methods. Linear regression analysis of the absorbance at 555 nm versus chitosan concentration, determination of confidence intervals, and other statistical tests were carried out with SPSS 12.0 software program (Statistical Package for Social Sciences, USA).

3. Results and Discussion

3.1. Spectral Analysis. Firstly, we checked the use of acetic acid to dissolve chitosan because this is the usual solvent for

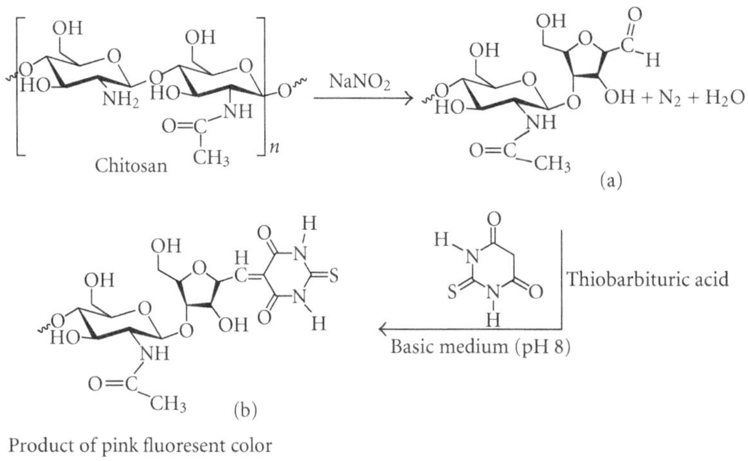

FIGURE 1: The mechanism by which the nitrous acid reaction leads to chain cleavage resulting in a 2,5-anhydro-D-mannose-reducing end (**a**) and proposed reaction of the reducing end in basic medium with thiobarbituric acid to generate a pink fluorescent dye (**b**).

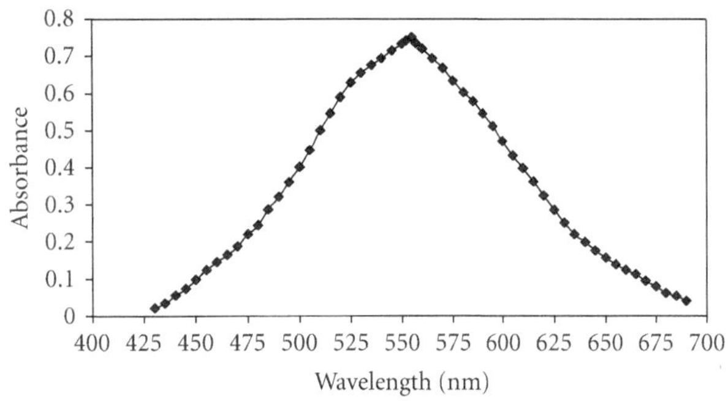

FIGURE 2: Absorbance spectrum of the pink-fluorescent product from the reaction between reducing end of chitosan and thiobarbituric acid. 400 to 700 nm spectrum is plotted and clearly shows the maximum wavelength is 555 nm for the pink color formed.

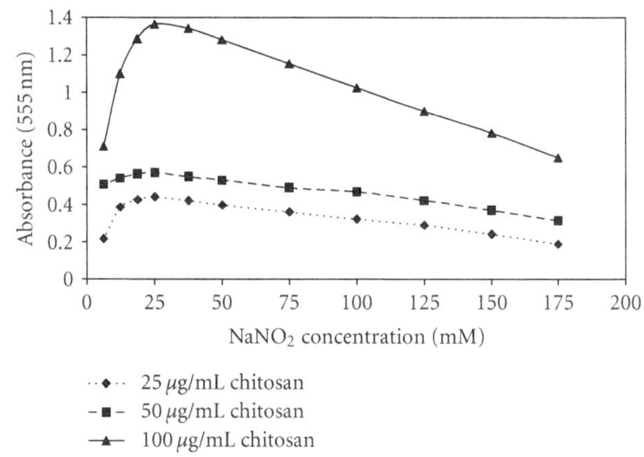

FIGURE 3: Effect of sodium nitrite concentration that used as a depolymerized reagent of chitosan for 30 min on the rate and intensity of the color development.

chitosan solubility. Absorption scans in the range of 400 to 700 nm were performed on 25 μg/mL chitosan solution in 1% (v/v) aqueous acetic acid solution. The same solution without chitosan was used as blank. The color developed within seconds of the mixing of thiobarbituric acid solution with the reagents and reached stability after 10 min. The spectral absorbance of this color is shown in Figure 2. The product reacts with thiobarbituric acid giving rise to a pink-colored derivative with excitation maximum located at 555 nm. Therefore, all absorbance measurements were performed at this wavelength. To obtain high sensitivity, it is necessary to investigate the effect of all parameters that could influence the chemical reactions and the performance of the colored complex.

3.2. Optimization of the Reaction Conditions. For survey of the optimal reaction conditions, seven factors were considered including the sample size, the amount of NaNO$_2$, the pH, the temperature, the depolymerization time, the amount of the thiobarbituric acid reagent, and the stability of the color formed. Three chitosan concentrations (10, 25 and 50 μg/mL) were selected to allow study the effect of these factors. Therefore, the sequence was adopted as follows: chitosan, NaNO$_2$, pH increase to 8, thiobarbituric acid, and then the absorbance measured. Using this sequence, when the 25 mM NaNO$_2$ was added to chitosan solution, incubation at 80°C for 30 min, and 10.8 mM thiobarbituric acid as added immediately after the pH adjusted to 8, the maximal absorbance was obtained.

3.2.1. Effect of the Sodium Nitrite Concentration. The effect of NaNO$_2$ concentration was investigated within the range of 6.25 to 175 mM. The results revealed that the absorbance increased by increasing the reagent concentration up to 25 mM and decreased at higher concentrations (Figure 3). Therefore, 25 mM NaNO$_2$ was the optimum concentration and was applied in the proposed method. It could be deduced from the curves that 25 ug/mL chitosan was more sensitive to NaNO$_2$ degradation compared with that of 50 and 100 ug/mL chitosan. It seems that the depolymerization of chitosan was influenced not only by the concentration of

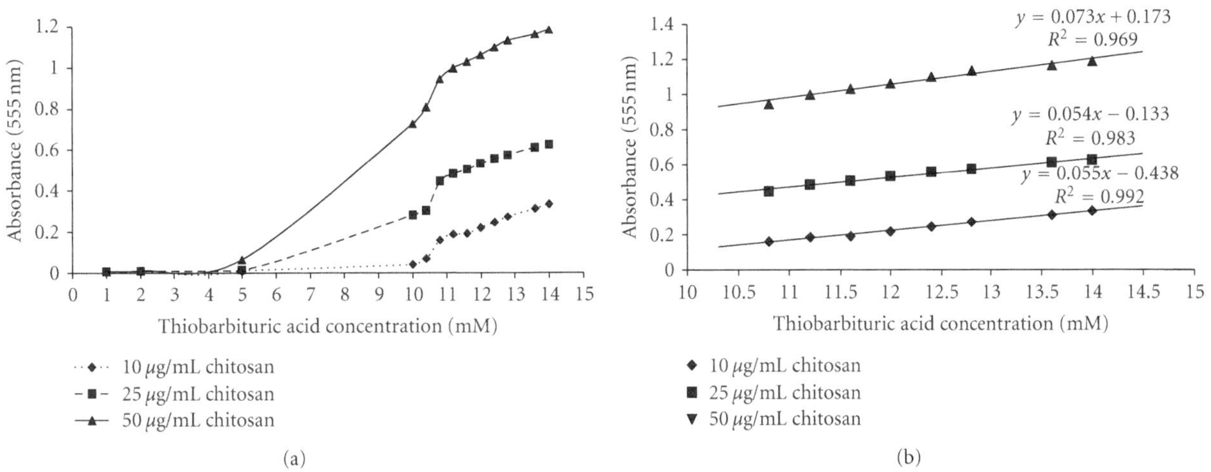

FIGURE 4: Influence of the concentration of thiobarbituric acid on the intensity of the color development (a) and the linear correlation between thiobarbituric acid and color development (b).

$NaNO_2$ but also by the initial concentration of chitosan. This phenomenon is related to the structure of chitosan in the solution. Normally chitosan molecule takes the shape of an extended random coil in the solution. When the concentration is high, due to the larger viscosity of the solution and strong intermolecular interactions, accessible chain segments can only stretch in a limited area, decreasing the contact probability with $NaNO_2$, thus resulting in a lower degradation rate. In this case, low chitosan concentration would be preferred to yield end product in a short time.

3.2.2. Effect of the TBA Concentration on Development of the Color. In order to study the effect of thiobarbituric acid, three chitosan concentrations (10, 25, and 50 ug/mL) were tested then the absorbencies at 555 nm were then plotted against serial concentrations of the thiobarbituric acid (Figure 4(a)). The influence of the thiobarbituric acid concentration was studied in the range between 1 to 14 mM. The rate of the reaction was found to increase continuously (presumably as reducing ends are produced) with a concentration of the thiobarbituric acid reagent up to 10 mM and then remain constant between 11.2 to 14 mM. A concentration of 10.8 mM of thiobarbituric acid was selected as the optimum concentration for the three chitosan concentrations. Moreover, regression analysis of the thiobarbituric acid concentrations ranged from 10.8 to 14 mM yielded a linear correlation with correlation coefficients of 0.992, 0.983, and 0.969 for 10, 25, and 50 μg/mL chitosan, respectively (Figure 4(b)).

3.2.3. Effect of the Reaction Time on the Color Development. In order to investigate the effect of the reaction time, keeping $NaNO_2$ and thiobarbituric acid at 25 and 10.8 mM, respectively, chitosan concentrations from 5 to 100 μg/mL were degraded by $NaNO_2$ in 1% acetic acid solution (v/v) and the reaction was examined after 15, 30, and 45 min, and the corresponding samples were processed in the same way as that described before. The absorbance values of the samples were determined at 555 nm, and the results are illustrated in Figure 5(a). It will be observed from the readings given in this figure that the maximum color intensity developed after the solutions have been heated for 30 min. This means that complete depolymerization or total acid hydrolysis of all chitosan concentrations occurred mainly in the first 30 min. It has been found that the colors obtained after heating of the chitosan solutions with $NaNO_2$ reagent for 45 min or longer are of a slightly different tint from those obtained with solutions which have been heated for 30 min. Therefore, this time was selected as the optimum for the determination reaction.

3.2.4. Effect of the pH. Another possible reason for the low color yields could be differences in the pH optimum for the reaction. The thiobarbituric acid reaction was carried out in the pH that ranged from 5 to 12 (Figure 5(b)). The pink complex gave its maximal absorbance only within a narrow pH interval, 7-8. The three chitosan concentrations gave highest color yield at pH $\approx$ 8. Therefore, this pH value was selected as the optimum for the determination reaction.

3.2.5. Calibration Curve and Final Optimized Assay. Different concentrations of chitosan ranged from 5 to 60 μg/mL were mixed in test tubes with 100 μL of $NaNO_2$ (0.5 M). After incubating for 30 min at 80°C in a static water bath, the pH was increased to 8 by adding 200 μL NaOH (0.1 M). 540 μL of thiobarbituric acid (0.04 M) was added, and the tubes were incubated a second time at 80°C for 10 min in a water bath and then cooled to room temperature. The absorbance was measured at 555 nm by spectrophotometer. The plot of absorbencies against serial concentrations of chitosan gave straight line as shown in Figure 6.

3.2.6. Stability of the Color Complex. The color of complex formed (Table 1) remained stable between 5 and 180 min after color development. Fading of the color with time was found to be very slow, and no fading occurred during the

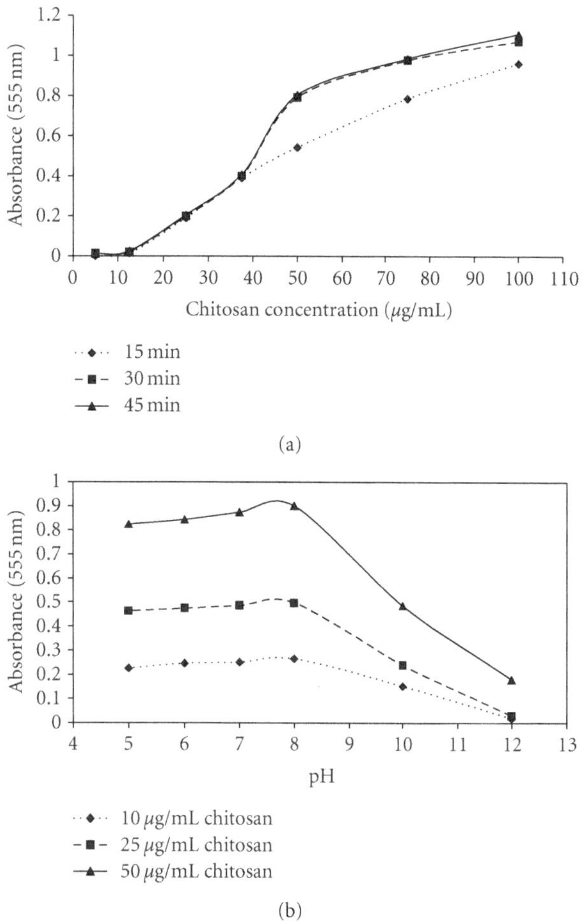

FIGURE 5: The influence of the time of heating with the sodium nitrite reagent on the subsequent color development (a) and the function of pH for reaction of the reducing end of chitosan with thiobarbituric acid after 30 min reaction time (b).

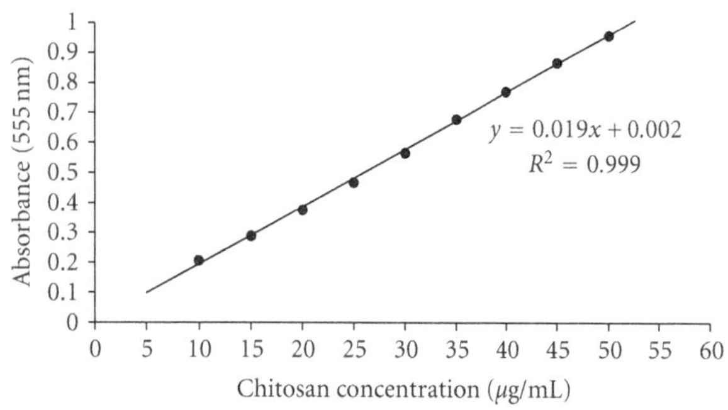

FIGURE 6: Standard analytical curve for chitosan at optimum conditions in the proposed method at 555 nm. Each value is the mean of four determinations.

first 180 min, even when exposed to indoors illumination as long as direct sunlight was avoided. After 12 h, decrease in absorbance was recorded, not exceeding 1% per hour.

3.3. Analytical Sensitivity. As for the slope of the calibration line, analytical sensitivity for chitosan determination was found to be 0.0019 absorbance/amount (0.1 μg/mL chitosan).

TABLE 1: Stability of color complex by the proposed method.

Time after color development (min)	Absorbance (555 nm) with the following amount of chitosan (μg/mL)		
	5	10	25
5	0.086	0.151	0.410
10	0.086	0.152	0.412
15	0.091	0.153	0.415
20	0.095	0.154	0.411
25	0.098	0.152	0.410
30	0.098	0.152	0.411
35	0.101	0.154	0.412
40	0.100	0.152	0.411
45	0.100	0.154	0.409
50	0.102	0.157	0.409
55	0.097	0.155	0.404
60	0.100	0.157	0.405
90	0.092	0.153	0.402
120	0.092	0.153	0.409
150	0.092	0.150	0.407
180	0.092	0.150	0.414
1080 (18 h)	0.046	0.080	0.231

The values are means of four replicates.

3.4. Linearity of the Color Responses. A linear relationship between the absorbance and the chitosan quantity was observed within the range of 10–50 μg/mL. The upper limit of the linearity in the assay was 50 μg/mL chitosan. In the regression analysis, the r value was 0.999, $P < 0.001$, the slope was 0.0191, and the intercept was 0.0028.

3.5. Lower Detection Limit. The detection limit of the method was determined by evaluating the zero calibrator 10 times. The limit of detection (LoD), defined as the mean chitosan value of the zero calibrator (limit of blank, LoB) + 1.645 (SD low concentration sample) according to the method of Armbruster and Pry [29], was found to be 0.185, where LoB = mean blank + 1.645 (SD blank).

3.6. Applications. To evaluate the performance of the method for analysis of real samples, determination of chitosan in pharmaceutical formulation of Chitocal capsules was investigated. The results presented in Table 2 revealed that the capsules product presented contents of chitosan in the range of 73.82 to 80.81% of that reported on the label with average of 76.11%. In addition, the recoveries for the addition of different concentrations of chitosan to Chitocal samples are in the range of 96.67 to 98.46% with average of 97.68% and indicate that the proposed method is suitable for determination of trace amounts of chitosan in the real samples.

Table 2: Determination of chitosan in pharmaceutical formulation by the proposed method.

Sample	Chitosan content declared in label (mg)	Sample taken (μL)	Chitosan calculated (mg)	Chitosan found (mg)	% of chitosan in formulation ± SE	Recovery (%)[a] ± SE
		20	0.029	0.022	73.82 ± 1.54	96.67 ± 1.75
		40	0.059	0.044	74.86 ± 1.87	97.28 ± 0.88
Chitocal capsule	500 mg	60	0.088	0.067	75.52 ± 1.84	97.89 ± 0.56
		80	0.118	0.089	75.59 ± 2.19	98.11 ± 1.15
		100	0.147	0.119	80.81 ± 2.51	98.46 ± 0.82
Mean ± SE					76.11 ± 1.55	97.68 ± 0.32

[a]Recovery (%) values were calculated from the amount of chitosan that had been determined after addition of its different amounts to a fixed amount of dosage form (capsule). The data are average of three determintaions.

In spite of the availability of several methods for quantitative determination of chitosan, only isolated examples have been found on the estimation of the content of this polymer in different samples. In the present study, a method for quantifying the chitosan in pure and commercial products was developed. This method is based on the previous reports on its unique degradability in nitrous acid solution. In previous reports, hydrochloric acid was used for the hydrolysis of chitin and chitosan prior to glucosamine analysis [19, 30–33]. However, hydrolysis by HCl is usually a time-consuming process to achieve complete depolymerization and high yield. Bosworth and Scott [19] reported that the maximum yield of hydrolysis of chitosan with 5.5 N HCl at 102°C was not reached in less than 23 h. Chen and Chiou [30] recovered only 63.7% of chitin as glucosamine after 60 min of hydrolysis with 2.5 M HCl at 140°C. Wu et al. [34] used 6 M HCl at 110°C and recovered 93.0% of chitosan after 3 h and only 68.7% of chitin after 12 h as glucosamine. Zhu et al. [33] got 85.3% recovery of chitin at 110°C after 4 h. In all of these works, the sum of free and acetylated glucosamine was reported. The recoveries of chitosan from the commercial preparation with the present method were >98% (Table 2), indicating high performance of the method for the analysis of materials containing this biopolymer.

In earlier methods, glucosamine has been quantified by hydrolysis in HCl solutions. However, part of the sample that was not depolymerized by HCl and stayed in the solid form was not included in the measurement [19, 30, 33, 34]. However, in the present method, chitosan samples dissolved in aqueous acetic acid solution and became completely soluble and the chitosan is not precipitated from the solution. In this case, the whole initial sample is considered for the estimation of chitosan content.

4. Conclusions

A colorimetric method is described for the determination of chitosan in pure and pharmaceutical formulation. This method depends on the acid hydrolysis of chitosan molecule by one step with sodium nitrite treatment to convert it into 2,5-anhydro-D-mannose at the new reducing end which reacts with thiobarbituric acid in basic medium to give rise very stable pink-colored solution. The procedure has been applied for estimation of the chitosan in Chitocal capsules as a pharmaceutical formulation with high recoveries of 96.67 to 98.46%, indicating that the proposed method is suitable for determination of trace amounts of chitosan in the real samples.

References

[1] H. K. No and S. P. Meyers, "Preparation of chitin and chitosan," in *Chitin Handbook*, R. A. A. Muzzarelli and M. G. Peter, Eds., pp. 475–489, European Chitin Society, 1997.

[2] S. Hirano, "Chitin biotechnology applications," *Biotechnology Annual Review*, vol. 2, pp. 237–258, 1996.

[3] Y. Shigemasa and S. Minami, "Applications of chitin and chitosan for biomaterials," *Biotechnology and Genetic Engineering Reviews*, vol. 13, pp. 383–420, 1996.

[4] E. I. Rabea, M. E. T. Badawy, C. V. Stevens, G. Smagghe, and W. Steurbaut, "Chitosan as antimicrobial agent: applications and mode of action," *Biomacromolecules*, vol. 4, no. 6, pp. 1457–1465, 2003.

[5] F. Shahidi, J. K. V. Arachchi, and Y. J. Jeon, "Food applications of chitin and chitosans," *Trends in Food Science and Technology*, vol. 10, no. 2, pp. 37–51, 1999.

[6] M. Eikenes, M. Fongen, L. Roed, and Y. Stenstrøm, "Determination of chitosan in wood and water samples by acidic hydrolysis and liquid chromatography with online fluorescence derivatization," *Carbohydrate Polymers*, vol. 61, no. 1, pp. 29–38, 2005.

[7] A. Tsuji, T. Kinoshita, and M. Hoshino, "Analytical chemical studies on amino sugars. II. Determination of hexosamines using 3-methyl-2-benzothiazolone hydrazone hydrochloride," *Chemical and Pharmaceutical Bulletin*, vol. 17, no. 7, pp. 1505–1510, 1969.

[8] A. Tsuji, T. Kinoshita, and M. Hoshino, "Microdetermination of hexosamines," *Chemical and Pharmaceutical Bulletin*, vol. 17, no. 1, pp. 217–218, 1969.

[9] S. Roseman and I. Daffner, "Colorimetric method for determination of glucosamine and galactosamine," *Analytical Chemistry*, vol. 28, no. 11, pp. 1743–1746, 1956.

[10] N. I. Larionova, D. K. Zubaerova, D. T. Guranda, M. A. Pechyonkin, and N. G. Balabushevich, "Colorimetric assay of chitosan in presence of proteins and polyelectrolytes by using *o*-phthalaldehyde," *Carbohydrate Polymers*, vol. 75, no. 4, pp. 724–727, 2009.

[11] Z. Dische and E. Borenfreund, "A spectrophotometric method for the microdetermination of hexosamines," *The Journal of Biological Chemistry*, vol. 184, no. 2, pp. 517–522, 1950.

[12] E. Curotto and F. Aros, "Quantitative determination of chitosan and the percentage of free amino groups," *Analytical Biochemistry*, vol. 211, no. 2, pp. 240–241, 1993.
[13] S. Prochazkova, K. M. Vårum, and K. Østgaard, "Quantitative determination of chitosans by ninhydrin," *Carbohydrate Polymers*, vol. 38, no. 2, pp. 115–122, 1999.
[14] C. Wischke and H. H. Borchert, "Increased sensitivity of chitosan determination by a dye binding method," *Carbohydrate Research*, vol. 341, no. 18, pp. 2978–2979, 2006.
[15] B. Miralles, M. Mengíbar, R. Harris, and A. Heras, "Suitability of a colorimetric method for the selective determination of chitosan in dietary supplements," *Food Chemistry*, vol. 126, no. 4, pp. 1836–1839, 2011.
[16] R. A. A. Muzzarelli, "Colorimetric determination of chitosan," *Analytical Biochemistry*, vol. 260, no. 2, pp. 255–257, 1998.
[17] D. Sekmokiene, V. Speiciene, A. Salaseviciene, and G. Garmiene, "Determination of chitosan in food products," *Veterinarija ir Zootechnika*, vol. 31, pp. 87–92, 2005.
[18] G. Lu, L. Wang, R. Wang, Y. Zeng, and X. Huang, "Determination of chitosan by cathodic stripping voltammetry," *Analytical Sciences*, vol. 22, no. 4, pp. 575–578, 2006.
[19] T. R. Bosworth and J. E. Scott, "A specific fluorometric assay for hexosamines in glycosaminoglycans, based on deaminative cleavage with nitrous acid," *Analytical Biochemistry*, vol. 223, no. 2, pp. 266–273, 1994.
[20] A. Zamani, A. Jeihanipour, L. Edebo, C. Nlklasson, and M. J. Taherzadeh, "Determination of glucosamine and *N*-acetyl glucosamine in fungal cell walls," *Journal of Agricultural and Food Chemistry*, vol. 56, no. 18, pp. 8314–8318, 2008.
[21] L. Daniels, R. S. Hanson, and J. A. Phillips, "Chemical analysis," in *Methods for General and Molecular Bacteriology*, P. Gerhardt, R. G. E. Murray, W. A. Wood, and N. R. Krieg, Eds., pp. 518–519, American Society for Microbiology, Washington, DC, USA, 1994.
[22] M. Izume and A. Ohtakara, "Preparation of D-glucosamine oligosaccharides by the enzymatic hydrolysis of chitosan," *Agricultural and Biological Chemistry*, vol. 51, pp. 1989–11991, 1987.
[23] S. Mao, X. Shuai, F. Unger, M. Simon, D. Bi, and T. Kissel, "The depolymerization of chitosan: effects on physicochemical and biological properties," *International Journal of Pharmaceutics*, vol. 281, no. 1-2, pp. 45–54, 2004.
[24] H. Sashiwa, H. Saimoto, Y. Shigemasa, and S. Tokura, "*N*-acetyl group distribution in partially deacetylated chitins prepared under homogeneous conditions," *Carbohydrate Research*, vol. 242, pp. 167–172, 1993.
[25] G. G. Allan and M. Peyron, "Molecular weight manipulation of chitosan. I: kinetics of depolymerization by nitrous acid," *Carbohydrate Research*, vol. 277, no. 2, pp. 257–272, 1995.
[26] G. G. Allan and M. Peyron, "Molecular weight manipulation of chitosan. II: prediction and control of extent of depolymerization by nitrous acid," *Carbohydrate Research*, vol. 277, no. 2, pp. 273–282, 1995.
[27] K. Tømmeraas, K. M. Vårum, B. E. Christensen, and O. Smidsrød, "Preparation and characterisation of oligosaccharides produced by nitrous acid depolymerisation of chitosans," *Carbohydrate Research*, vol. 333, no. 2, pp. 137–144, 2001.
[28] L. F. Tietze and U. Beifuss, "Sequential transformations in organic chemistry: a synthetic strategy with a future," *Angewandte Chemie International Edition in English*, vol. 32, no. 2, pp. 131–163, 1993.
[29] D. A. Armbruster and T. Pry, "Limit of blank, limit of detection and limit of quantitation," *Clinical Biochemistry Review*, vol. 29, supplement i, pp. S49–S52, 2008.
[30] W. Chen and R. Y. Y. Chiou, "A modified chemical procedure for rapid determination of glucosamine and its application for estimation of mold growth in peanut kernels and koji," *Journal of Agricultural and Food Chemistry*, vol. 47, no. 5, pp. 1999–2004, 1999.
[31] C. H. Ng, S. Hein, S. Chandrkrachang, and W. F. Stevens, "Evaluation of an improved acid hydrolysis-HPLC assay for the acetyl content in chitin and chitosan," *Journal of Biomedical Materials Research—Part B*, vol. 76, no. 1, pp. 155–160, 2006.
[32] F. Niola, N. Basora, E. Chornet, and P. F. Vidal, "A rapid method for the determination of the degree of N-acetylation of chitin-chitosan samples by acid hydrolysis and HPLC," *Carbohydrate Research*, vol. 238, pp. 1–9, 1993.
[33] X. Zhu, J. Cai, J. Yang, and Q. Su, "Determination of glucosamine in impure chitin samples by high-performance liquid chromatography," *Carbohydrate Research*, vol. 340, no. 10, pp. 1732–1738, 2005.
[34] T. Wu, S. Zivanovic, F. A. Draughon, and C. E. Sams, "Chitin and chitosan-value-added products from mushroom waste," *Journal of Agricultural and Food Chemistry*, vol. 52, no. 26, pp. 7905–7910, 2004.

Acetylation of Wood Flour from Four Wood Species Grown in Nigeria Using Vinegar and Acetic Anhydride

Yakubu Azeh,[1] Gabriel Ademola Olatunji,[2] Cheku Mohammed,[1] and Paul Andrew Mamza[1,3]

[1] *Department of Chemistry, Ibrahim Badamasi Babangida University, Lapai, Nigeria*
[2] *Department of Chemistry, University of Ilorin, Ilorin, Nigeria*
[3] *Department of Chemistry, Ahmadu Bello University, Zaria, Nigeria*

Correspondence should be addressed to Yakubu Azeh; azehyakubu@yahoo.com

Academic Editor: R. J. Linhardt

Effect of acetylation on pretreated wood flour of four different wood species, Boabab (*Adansonia digitata*), Mahoganny (*Daniella oliveri*), African locust bean (*Parkia biglobosa*) and Beech wood (*Gmelina arborea*), had been investigated. The first batch of wood species were acetylated using acetic anhydride while the second batch were acetylated with commercial vinegar. Both experiments were conducted in the presence of varying amount of $CaCl_2$ as catalyst and at temperature of 120°C for 3 h. The success of acetylation was determined based on Weight Percent Gain for each sample treated with either chemicals used. FT-IR, a veritable tool was used for the analysis of both treated and untreated samples to further investigate the success of acetylation. The results showed the presence of important band such as carbonyl absorptions at 1743, 1744, 1746, 1731, 1718 and 1696 cm^{-1} as appeared separately in the spectra of acetylated samples, confirming esterification occurred. The purpose of this work was to investigate the applicability of vinegar for acetylation of lignocellulosic fibers. Blends/composites were prepared by solution casting and their kinetics investigated in distilled water. The results indicated they could be used in outdoor applications such as, decking and packaging.

1. Introduction

Composites have been described as materials composed of a hard material with discontinuous reinforcement that is embedded in a weaker, continuous matrix. Where the reinforcement matrix [1] maintains the position and orientation of the reinforcement. The constituents of the composites retain their individual, physical and chemical properties. Composite give a combination of qualities that are very different from the individual constituents that constituted the composite. Several reports on thermoplastic composites have been documented. Different types of modified and unmodified natural fibers such as wood fibers and flour, kenaf fibers, sago, rice starch, cornstarch, henequen fibers, and pineapple-leaf fibers, have been used as fillers in polymer matrices [1, 2].

Dimensional stability and strength of unmodified wood flour polyethylene composites was reported to have improved by increased in fiber loading(s). Unmodified starch have been used to produce composites of low density polyethylene (LDPE). Composites of unmodified starch have been reported to exhibit low mechanical properties, though with improved biodegradation. The introduction of ester groups unto starch surface, manipulate its properties and support the blending mechanism [3]. Wood-Plastic composites have been described as products form by combining treated or untreated wood in the form of fiber or particles and a thermoplastic polymer such as polyethylene, polyvinyl chloride, or polypropylene [4]. Wood-plastic composites have significant applications in various fields such as construction/engineering. For wood as a construction material, its use is impaired under certain degradative agents due to its hydrophilic nature, presence of cellulose, a food substrate for microorganisms. For this reason, ecofriendly methods to improve the performance of wood are been developed. All over the world, efforts has been made to move away from toxic to nontoxic means of wood modification.

Chemical modification of wood is a very good alternative to conventional wood preservation methods. Acetylation has received considerable attention more than any other chemical modification technique known [5]. This method not only protects wood/wood-based products against degradative agents, but also changes the hydrophilic nature of the wood components (cellulose, hemicellulose, and lignin) to hydrophobicity by increasing the acetyl contents in the polymers while decreasing their –OH groups. This, therefore, enhances the surface energy and interfacial interaction between acetylated wood-dust and the hydrophobic polymers. It has been reported that, for every acetyl gain, one –OH group is eliminated. Implying that there is a reduction in hydrogen bonding within the wood reactive polymers, this can increase the interfacial bonding of PE with the acetylated wood surface [6]. Since the surface qualities plays a significant role to satisfactory interfacial performance, we, therefore, acetylate four different wood flours of four wood species grown in Nigeria using acetic anhydride and as well investigate the feasibility to use commercially available vinegar for acetylation of cellulose in lignocellulosic materials because of its low cost, availability, and renewability. Acetylation of cellulose in wood flour was carried out in the presence of varying amounts of $CaCl_2$ as a catalyst to manipulate its properties in order to support the blending mechanism. The objectives of this work were to investigate the feasibility of vinegar for acetylation, preparation of blends/composite, and investigation of their sorption kinetics in distilled water for application in industry.

2. Materials and Methods

Wood flour used was from four wood species Baobab (*Adansonia digitata*), mahogany (*Daniella oliveri*), African locust bean (*Parkia biglobosa*), and Beechwood (*Gmelina arborea*) collected from a saw mill located in the Niger state Nigeria. Reagents and chemicals used were obtained from May & Baker, and these include commercial vinegar, acetic anhydride, NaOH, ethanol, toluene, sulphuric acid, hypochlorite, and $CaCl_2$. All reagents used were of analytical grade.

2.1. Soxhlet Extraction. To reduce the influence of wood extractives on the acetylation, 5 g of 2 mm size wood particles was extracted with a mixture of ethanol-toluene (2 : 1 v/v) for 3 h. After extraction, samples were rinsed with ethanol followed by hot water and then oven-dried at 105°C for 24 h to reach a constant weight. The extractible content was calculated as a percentage of oven-dried test samples.

2.2. Sample Pretreatment by Mercerization. Mercerization of the 2 g oven-dried extracted sample was mercerized using 20 mL of 10% NaOH solution. The mixture was placed on a shaker and shaked at 75 rpm for 1 h. Afterwards, the sample was washed with distilled water until it was neutral. The residue was oven-dried at 105°C for 1 h. The procedure was repeated for each wood flour sample.

2.3. Chemical Modification

2.3.1. Acetylation. Acetylation was carried out in batches. The first batch was conducted using acetic anhydride while the second batch was carried out using commercial vinegar as acetylating agents in the presence of varying amounts of $CaCl_2$ as a catalyst. 2 g of oven-dried samples of wood species oven-dried at 105°C with a constant weight (W_{unt}) were treated with the acetylating chemicals each with 0.5 and 1 g $CaCl_2$ catalyst at 120°C during 3 h refluxing. After modification, the residues were thoroughly washed with distilled water until neutral and then oven-dried for 3 h. The extent of acetylation was calculated as weight percent gains (WPGs) based on the differences in oven-dried weight of the samples before and after modification (W_{unt}) and after modification (W_{trt}) according to the equation

$$\text{WPG} = \frac{M_{trt} - M_{unt}}{M_{unt}} \times 100. \tag{1}$$

2.3.2. Blending Procedures. The dissolution of high-density polyethylene (packaged water sachets) follows the method described by [7]. In brief, 1 g pieces of nonprinted portion were placed in a reaction flask and 30 mL of toluene was added. The contents was placed on a hot plate at 100°C, and a thermometer was inserted. The sachets swelled at 60°C and then rapidly dissolved at 80°C to give a clear solution. Blending of high-density polyethylene and acetylated wood flour is thus described. Into the clear solution of high-density polyethylene sachets was introduced 0.5/1.0 g of each vinegar/acetic anhydride treated or untreated wood flour while stirring for 5 min and then spread on clean metal plates to form a film. The thin films formed were peeled off the metal plates to afford acetylated wood flour/polyethylene blends [2].

2.4. Water Absorption Test of Acetylated and Untreated Blends. Three sets of specimens were prepared. Acetylated blends and untreated blends were subjected to water absorption test. The blends were soaked for 1, 2, and 16 h, respectively. After each immersion, the blends were removed and wiped off using filter paper and then weighed and weights were noted, using the formula

$$\frac{H_2 - H_1}{H_1} \times 100\%, \tag{2}$$

where H_1 = initial weight of blend before immersion in water and H_2 = weight of blend after immersion in water.

2.5. Fourier Transform Infrared Spectroscopy Analysis. The properties of acetylated and untreated samples were characterized using FT-IR, Perkin-ELMER-8400S Spectrophotometer in the range 4000 cm^{-1} to 400 cm^{-1}. Samples were run using the KBr pellet technique at the National Research Institute for Chemical Technology (NARICT), Zaria, Kaduna-Nigeria.

Table 1: Assignment of the IR spectra bands of functional groups in acetylated wood flour treated with acetic anhydride.

Band position (cm^{-1})	Functional group
3973–3352	–OH bonded stretching vibrations
1746–1731	C=O stretching vibrations of acetate group
2938–2917	C–H methyl and methylene group
1373–1369	($-C-CH_3-$) in acetate group due to acetylation
1285–1241	(ν_{C-O}) stretching band vibrations of the acetyl moieties
1628–1618	–OH of absorbed water
1527–1511	(C=C), aromatic skeletal absorption caused by lignin
1044–1042	C–O stretching vibrations in cellulose, hemicelluloses, and primary alcohol
1052	C–O stretching in acetyl (covalence vibration)
620–603	C–H out-of-plane stretching absorption of aromatic ring vibrations caused by lignin

Table 2: Assignment of the IR spectra bands of functional groups in acetylated wood flour treated with Vinegar based on acetic anhydride and ketene related work.

Band position (cm^{-1})	Functional group
3978–3111	–OH stretching (bonded) vibrations
1718	C=O stretching vibrations of acetate group
1696	C=O stretching in acetyl groups
2723 and 2820	C–H stretch of methylene in celluloses and hemicelluloses
2928–2904	CH_3 symmetric stretch of methyl groups of aliphatic
1350–1340	($-C-CH_3-$) deformation of acetate group in cellulose and hemicelluloses
1254–1247	(ν_{C-O}) stretching band vibrations of the acetyl moieties and (C=O) deformation in the ester bond during acetylation
1629–1624	C=C stretching of aromatic ring of syringyl in lignin
1654	C=O stretching of (COOH) in glucuronic acid
1527–1509	C=C Stretching of the aromatic ring caused by lignin
1044–1042	C–O stretching vibrations in cellulose, hemicelluloses, and primary alcohol
1439–1428	CH_2 deformation and stretching in cellulose, lignin, and xylan
625–603	C–H out-of-plane stretching absorption of aromatic ring vibrations caused by lignin

3. Results and Discussion

The IR spectra of acetic anhydride or vinegar treated wood flour showing the major absorption bands are shown in Tables 1 and 2. The following strong absorption bands as reflected separately on the IR spectra of vinegar treated wood flour are characteristic absorptions of bonded –OH, and these include 3978–3339, 3477–3375, 3426–3360, 3907–3372, 3429–3387, and 3874–3111 cm^{-1} [2, 8, 9]. The following peak intensities at 3029, 3104, 3345, 3602, 3352, 3852, and 3938 cm^{-1} as reflected separately on the IR spectra of acetic anhydride treated wood flour are characteristic absorptions of bonded –OH group stretching vibrations in cellulose and hemicelluloses. Acetic anhydride or vinegar treated wood flour showed reductions in –OH absorption bands. The –OH absorption bands of untreated wood flour were predominantly detected at 4008, 4055, 4107, 4290, 4394, 4413, 4567, 4570, 4685, and 4696 cm^{-1}. The intensity of the –OH absorption bands in the treated samples decreased. This decrease in the intensity of –OH band is an indication that the hydroxyl group contents in wood flour were reduced during reaction. Indicating that some level of acetylation had taken place [2, 9, 10]. The presence of –OH absorption in vinegar/acetic anhydride treated wood flour has been attributed to hydroxyl functionalities that are not accessible to chemical reagents [11]. The band at 1746–1731 cm^{-1} as reflected on acetic anhydride treated wood flour is characteristic absorption of carbonyl (C=O ester) stretching vibration of acetate group in cellulose and uronic ester in hemicelluloses. This band showed evidence of acetylation [9, 12–15]. The peak absorption increased in acetic anhydride treated samples. Indicating high level of acetyl gain as shown in Table 3. While vinegar treated samples showed this band at 1718 cm^{-1} and 1696 cm^{-1}. The low level of acetylation could be attributed to low-percent acetyl content in vinegar and shorter reaction time adopted for acetylation. Since vinegar treated Baobab sample gave 11.61% (WPG) during four-hour reaction unlike the other samples who gave less than 11.61% during 3 h reaction as reflected in Table 3. This implies that longer reaction times are required for acetylation using vinegar. 1696 cm^{-1} carbonyl (C=O) absorption peak has been reported in ketene treated wood cellulose [2]. The peak here has been attributed to the low penetration of acetyl, consequently leading to lower percent gains. The absorption bands at 2723 cm^{-1} and 2820 cm^{-1} observed in vinegar treated samples have been assigned to C–H stretching vibrations of $-CH_3$ and $-CH_2$

Table 3: Average weight percent gain (WPG) in vinegar and acetic anhydride treated wood flour.

Sample	Acetic anhydride (%)	Sample	Vinegar (%)
Gmelina	17	Gmelina	10.8
Mahogany	12	Mahogany	8.5
Baobab	19	Baobab	11.61
Locust bean	7	Locust bean	10.5

Table 4: Percent of water absorption after soaking for 1, 2 and 16 hours.

S/N	Blend sample	Dry weight (g)	Weight after soaking/h			Percentage of water absorption (%)/h		
			1 h	2 h	16 h	1 h	2 h	16 h
1	Acetic anhydride treated blends	1.00	1.08	1.16	1.38	8	16	38
2	Acetic anhydride untreated blends	1.00	1.34	1.39	1.49	34	39	49
3	Untreated blends	1.00			2.70			70
	Vinegar treated blends	1.00			1.52			52

groups in cellulose and hemicelluloses [2, 15]. The presence of absorption peaks at both 2938–2091 cm^{-1} reflected on acetic anhydride and 2928–2904 cm^{-1} in vinegar treated samples has been assigned to asymmetric stretching vibration of aliphatic $-CH_3$ group which is evidence of acetylation [2, 8, 12, 16]. The small bands at 1628–1618 and 1629–1624 cm^{-1} in acetic anhydride or vinegar treated samples are assigned to adsorbed water, β-glycosidic linkages (ether -C–O–C-) in the cellulose chain or sugar units or could be due to aromatic ring (C=C) vibrations in lignin (syringyl) [2, 9–12]. The absorption bands at 1527–1509 cm^{-1} are characteristic to absorption of aromatic (C=C) stretching vibrations caused by lignin (guaiacyl) [2, 8–12]. This suggests the presence of lignin [9, 11]. The bands at 1439–1428 cm^{-1} are due to C–H deformations and bending vibrations of $-CH_2$ (methylene) in cellulose and hemicelluloses [2, 14, 15, 17, 18]. These bands were not found in acetic anhydride treated wood flour. The reason could be that longer reaction time adopted led to product (cellulose/hemicelluloses) degradation and subsequently their extraction by acetylating chemical occurred and was probably washed alongside with the byproducts. An important band which showed that some level of acetylation was achieved by the two reagents used for acetylation was observed at 1330–1373 cm^{-1} for acetic anhydride and 1340–1348 for vinegar. This important band was attributed to aliphatic C–H deformation/bending vibration of $-CH_3$ in the acetyl groups and this is evidence of the formation of ester bond due to acetylation in cellulose and hemicelluloses [2, 9, 10, 13]. This absorption peak was missing in the control samples. A prominent absorption band which further gave evidence of acetylation appeared at 1285–1242 cm^{-1} for acetic anhydride while 1254–1247 cm^{-1} peak absorption was detected for vinegar as alternative acetylation agent. This band has been assigned for the stretching of C–O and the deformation of C=O in the acetate bond formed during acetylation in xylan and lignin [8, 9, 13, 15]. The absence of absorption band at 1800–1760 cm^{-1} in all the spectra of acetic anhydride treated samples was an indication that the acetylated products were free of unreacted acetic anhydride, implying that the reagent was used up during the reaction, while the absence of absorption peak at 1700 cm^{-1} in all spectra is evidence of the absence of unreacted carboxylic acid in vinegar treated samples and as byproducts in acetic anhydride treated wood flour, respectively. The bands at 1044–1042 cm^{-1} are assigned to C–O stretching vibrations in cellulose, hemicelluloses, and that of primary alcohol [2, 9, 10, 12]. The medium band at 1052 cm^{-1} ascribed to C–O stretching vibration of the acetyl group in acetate [2, 12, 14] is evidence of acetylation.

3.1. Water Absorption Kinetics of Acetylated and Control Wood Flour Samples. Three sets of specimens were prepared, as shown in Table 4. Blends prepared from acetylated wood flour were flexible, brittle, and very smooth and absorbed less water compared to untreated blends which were coarse with rough surface. The modification of fiber surface really helps to manipulate its properties and supports the blending mechanism as was observed on blends made from treated wood flour. It has been established that when the accessible hydroxyl groups in the cell wall polymers have been substituted by acetyl group, reduction in water and moisture sorption are observed, and this depends on the level of acetylation too. The results also showed that water absorption of blends is dependent on fibers in the matrix. It is expected that the sorption rate for 16 h soaking duration should give about 16 times the results obtained for 1/2 h soaking periods. However, this was not observed, indicating that the fibers in the polymer matrix have reached their saturation point and thereby cannot absorb much water molecules. Acetylated blends reached the fiber saturation point faster compared to untreated blends as indicated by the percent water uptake/sorption rate in Table 4. This is attributed to the modification of the hydrophilic fiber surface by acetyl

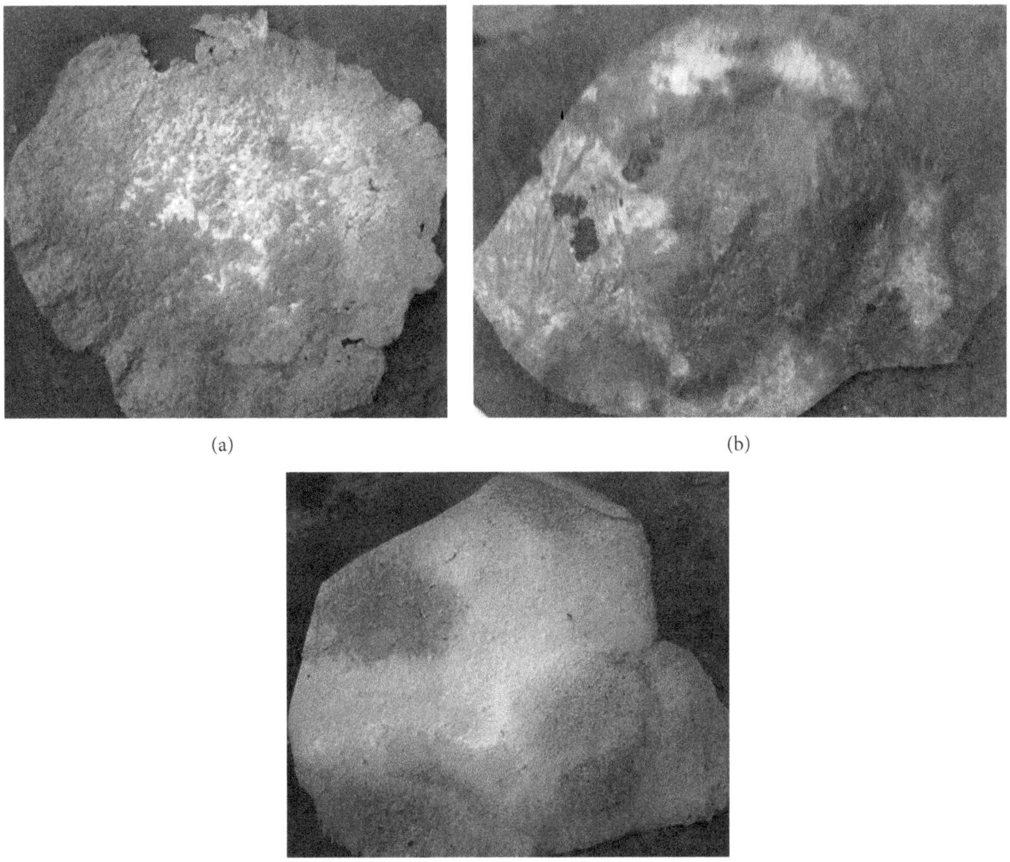

(a) (b) (c)

Figure 1: (a) photographic plate of acetylated blend (1 g wood flour/1 g polyethylene loading), (b) photographic plate of untreated blend (0.5 g wood flour/1 g polyethylene loading), and (c) photographic plate of acetylated blends (0.5 g wood flour/1 g polyethylene loading).

moieties and subsequent hydrophobicity improvement of fiber surface which allowed for proper interfacial bonding with polyethylene.

3.2. Results of the Formation of Blends. Blends obtained from the incorporation of acetylated wood flour with polyethylene were flexible and had smooth surface (Figures 1(a) and 1(b)). These blends were easily peeled off the film forming plate while those obtained from untreated wood flour were porous, coarse, and difficult to peel off from the film forming plate. For this reason, blends had rough surface (Figure 1(c)). Poor blend formation exhibited by untreated wood flour could be attributed to facial differences between the two materials—hydrophilic biopolymers in wood flour and the hydrophobic polyethylene, causing very poor interfacial interaction between the materials [2]. Treated wood flour resulted in the formation of flexible blends with smooth surface after incorporation and solution casting on metal plates. The properties exhibited by these blends was an indication that acetylation using vinegar and acetic anhydride successfully transformed accessible hydrophilic hydroxyl surfaces of the biopolymers in wood flour to hydrophobic surface (acetyls) which causes proper interfacial interaction with polyethylene as reported by [2, 19, 20].

4. Conclusion

In this work we have been able to carry out acetylation of cellulose in wood flour using vinegar successfully. Results of the kinetic studies of blends indicated that treated blends had lower water sorption than untreated blends, implying that treated blends had lower hygroscopic surface with a high hydrophobic energy surface that facilitated its interaction with polyethylene during the blending process, while untreated blends had high water sorption, due to the presence of accessible hydrophilic hydroxyl groups in cellulose and other biopolymers in wood which allowed for strong formation of hydrogen bonds with water molecules. Treated wood flour blends could be very useful in outdoor applications such as decking and packaging.

References

[1] A. G. Supri and B. Y. Lim, "Effect of treated and untreated filler loading on the mechanical, morphological, and water absorption properties of water hyacinth fibers-low density polyethylene composites," *Journal of Physical Science*, vol. 20, no. 2, pp. 85–96, 2009.

[2] A. Yakubu, G. A. Olatunji, O. Sunday, and A. Olubunmi, "Ketene acetylated wood cellulose for industrial applications

in wood-base and polymer industry," *Journal of Environmental Science and Technology*, vol. 5, no. 3, pp. 168–176, 2012.

[3] K. Sriroth, R. Chollakup, K. Piyachomkwan, and C. G. Oates, "Biodegradable Plastics from Cassava Starch in Thailand," Department of Biotechnology, Kasetsart University, Bangkok, Thailand.

[4] B. K. Segerholm, R. E. Ibach, and M. E. P. Walinder, "Moisture Sorption of artificially aged wood-plastic composite," *BioResources*, vol. 7, no. 1, pp. 1283–1293, 2012.

[5] R. M. Rowell, R. Simonson, S. Hess, D. V. Plackett, and E. Dunningham, "Acetyl distribution in acetylated whole wood and reactivity of isolated wood cell wall components to acetic anhydride," *Wood Fibre Science*, vol. 26, no. 1, p. 11, 1994.

[6] C. R. Frihart, R. Brandon, and R. E. Ibach, "Selectivity of bonding for modified wood. USDA forest service, forest products laboratory," in *Proceedings of the 27th Annual Meeting of the Adhesion Society, Inco*, From Molecules and Mechanic to Optimization and Design of Adhesive Joints, pp. 15329–18332, Adhesion Society, Madison, Wis, USA, 2004.

[7] S. A. Abdulkareem and B. Garba, "Novel application of polymer dissolution technique," *Nigerian Journal of Pure and Applied Science*, vol. 20, pp. 1799–1485, 2005.

[8] V. M. Tuong and J. Li, "Effect of heat treatment on the change in color and dimensional stability of acacia hybrid wood," *BioResources*, vol. 5, no. 2, pp. 1257–1267, 2010.

[9] B. Mohebby, "Application of ATR infrared spectroscopy in wood acetylation," *Journal of Agricultural Science and Technology*, vol. 10, no. 3, pp. 253–259, 2008.

[10] R. Bodîrlău and C. A. Teacă, "Fourier transform infrared spectroscopy and thermal analysis of lignocellulose fillers treated with organic anhydrides," *Romanian Journal in Physics*, vol. 54, no. 1-2, pp. 93–104, 2009.

[11] A. S. H. Callum, *How does the chemical modification of wood provide protection against decay fungi, school of agriculture and forest science*, Presentation for COST E22, University of Wales Bangor, 2002.

[12] M. Nasar, A. Emam, M. Sultan, and A. A. Abdel-Hakim, "Optimization of characterization of sugar cane bagasse liquefaction process," *Indian Journal of Science and Technology*, vol. 13, no. 12, pp. 207–212, 2010.

[13] Y. Indrayan, S. Yusuf, Y. S. Hadi, D. Nandika, and R. E. Ibach, "Dry wood termite resistance of Acetylated and Polymerized Tributyltin Acrylate (TBTA) indonesian and USA wood," in *Proceedings of the 3rd International Wood Science Symposium*, Core University Program in the Field of Wood Science, November 2000.

[14] R. T. Morrison and R. N. Boyd, *Organic Chemistry*, 6th edition, 1961.

[15] M. O. Adebajo and R. L. Frost, "Infrared and ^{13}C MAS nuclear magnetic resonance spectroscopic study of acetylation of cotton," *Spectrochimica Acta A*, vol. 60, no. 1-2, pp. 449–453, 2004.

[16] N. S. Cetin and N. Ozmen, "Acetylation of wood components and fourier transform infra-red spectroscopy studies," *African Journal of Biotechnology*, vol. 10, no. 16, pp. 3091–3096, 2011.

[17] A. Sandak, J. Sandak, W. Pradzynski, M. Zborowska, and M. Negri, "Near infrared spectroscopy as a tool for characterization of wood surface," *Folia Forestalia Polonica*, vol. 40, pp. 31–40, 2009.

[18] P. Sikorski, M. Wada, L. Heux, H. Shintani, and B. T. Stokke, "Crystal structure of cellulose triacetate I," *Macromolecules*, vol. 37, no. 12, pp. 4547–4553, 2004.

[19] J. George, S. S. Bhagawan, and S. Thomas, "Effects of environment on the properties of low-density polyethylene composites reinforced with pineapple-leaf fibre," *Composites Science and Technology*, vol. 58, no. 9, pp. 1471–1485, 1998.

[20] Y. Azeh, G. A. Olatunji, and P. A. Mamza, "Scanning electron microscopy and kinetic studies of ketene-acetylated wood/cellulose high-density polyethylene blends," *International Journal of Carbohydrate Chemistry*, vol. 2012, Article ID 456491, 7 pages, 2012.

4

Factors Involved in the *In Vitro* Fermentability of Short Carbohydrates in Static Faecal Batch Cultures

Eva Gietl,[1,2] Wim Mengerink,[1] Jaap de Slegte,[1] Glenn Gibson,[2] Robert Rastall,[2] and Ellen van den Heuvel[1]

[1] *FrieslandCampina Research, FrieslandCampina, 7400 AB Deventer, The Netherlands*
[2] *Department of Food and Nutritional Sciences, The University of Reading, Whiteknights, P.O. Box 226, Berkshire, Reading RG6 6AP, UK*

Correspondence should be addressed to Eva Gietl, eva.gietl@aevotis.com

Academic Editor: R. J. Linhardt

In recent years, research has focused on the positive effects of prebiotics on intestinal health and gut microbiota. The relationship between their chemical structure and their fermentation pattern by human intestinal microbiota is still not well understood. The aim of this study was to improve understanding of this relationship and identify factors that may be used to design galactooligosaccharides that reach more distal regions than commercial prebiotics which mainly target the proximal colon. The following factors were investigated: monomer type, linkage, substitution, and degree of polymerisation. Total organic acid production from sugars by faecal bacteria was fitted to a model which allowed an estimate of the time when half of the maximal organic acid concentration was reached (T_{50}) in static faecal batch cultures. The different factors can be grouped by their effectiveness at prolonging fermentation time as follows: substitution is most effective, with methylgalactose, β-galactose-pentaacetate, D-fucose, and galactitol fermented more slowly than D-galactose. Monomers and linkage also influence fermentation time, with L rhamnose, arabinose, melezitose, and xylose being fermented significantly slower than D-glucose ($P < 0.05$), maltose, isomaltose, cellobiose, and gentiobiose showing that Glcα1-6Glc and Glcβ1-4Glc were utilised slowest. Chain length had the smallest effect on fermentation time.

1. Introduction

Epidemiological data [1–4] indicate that a diet high in dietary fibres from vegetables and grains is correlated to a reduced risk of certain diseases in humans, among them colon cancer, ulcerative colitis, and cardiovascular disease. Experimental evidence suggests that the short-chain fatty acids (SCFA), butyric, acetic, and propionic acids [3–5], play a role in this risk reduction. They are produced by the intestinal microbiota as a result of dietary fibre fermentation.

Dietary fibres are plant materials, mostly storage carbohydrates or cell wall components, that cannot be digested in the human small intestine and reach the colon intact. They are available as substrates for the resident microbiota [6]. Some dietary fibres may also have prebiotic potential. Prebiotics are "non-digestible food ingredients that beneficially affect the host by selectively stimulating the growth and/or activity of one or a limited number of bacterial species already resident in the colon, and thus attempt to improve host health" [7]. Bacterial fermentation of carbohydrates in the colon results in the production of SCFA, mainly acetic, propionic, and butyric acid. Other organic acids also occur, such as lactic acid. The SCFA pattern, for example, the ratio of acetate, propionate, and butyrate to each other, is strongly determined by the fermented fiber/carbohydrate [8]. Each of them has positive effects on the health of the host. Butyrate has anticarcinogenic properties and contributes towards mucosal integrity by influencing the cell cycle progression of colonic epithelial cells [9–11]. Propionate may decrease cardiovascular disease risk by its influence on cholesterol synthesis [12], while acetate may improve cognitive function [13, 14]. *In vitro* and *in vivo* data suggest that fermentable prebiotics, such as fructooligosaccharides (FOS) and galactooligosaccharides

(GOS), are rapidly fermented in the proximal colon [15, 16]. Therefore, their beneficial effects may be limited to the proximal part of the colon and of short duration.

Because the intestinal microflora uses mainly the available carbohydrates as energy source in the proximal region of the colon, carbohydrates become depleted in more distal areas and the microflora switches to proteolytic fermentation [17]. Protein breakdown and amino acid dissimilation generate harmful putrefactive substances such as ammonia, amines, phenols, and indoles [18]. Van Nuenen et al. [19] showed a higher *in vitro* production of ammonia when faecal samples from patients with inflammatory bowel disease (IBD: Crohn's disease as well as ulcerative colitis) were used as inocula. This may indicate that individuals with IBD host a more proteolytic microbiota and/or have a lower supply of carbohydrates to the colon.

By prolonging the availability of carbohydrates towards the distal part of the colon, increased saccharolytic fermentation might occur throughout the colon, and the production of harmful compounds of protein breakdown is reduced. This is underscored by data of McIntyre et al. [20], who demonstrated, in rats, that slowly fermented fibres resulted in increased concentrations of SCFA in the distal large bowel compared to more rapidly fermented carbohydrates.

There are many properties that influence the fermentation time of fibres. Olano-Martin et al. [21] showed that degree of polymerisation (DP) and methylation influenced fermentation time by comparing the metabolism of high and low methylated pectins and their oligosaccharides *in vitro*. Oligosaccharides with lower chain length were fermented preferentially which also has been demonstrated *in vitro* by other researchers [22, 23]. Van Laere et al. [24] showed that linear oligosaccharides were fermented quicker and by a wider variety of species than branched chain oligosaccharides *in vitro*. While it is also generally assumed that solubility predicts fermentability [6], other factors that could influence fermentation time have not been studied in this respect, such as monosaccharide composition, and the linkage between the monosaccharides.

In the present study we focus on factors that can be modified in the synthesis of GOS; GOS is enzymatically produced by β-galactosidase from lactose in a reversible reaction. The primary objective here was to investigate the relationship between carbohydrate structure and fermentability in order to identify factors that might prolong the fermentation time of GOS: (1) type of monomers (i.e., acceptor substrates different from lactose), (2) isomery of the monomers (e.g., L-sugars as acceptor substrates), (3) linkages between the monomers (α1-4, α1-6, β1-4, and β1-6), (4) the substitution of the side-groups/chains (reduction, acetylation, a methylation), and (5) variation in chain length (DP < 8). A further outcome was to determine if and how these factors would affect the SCFA pattern produced by human faecal microbiota which may provide benefits that exceed gut health.

2. Materials and Methods

If not otherwise specified, materials were obtained from Sigma.

2.1. Nonstirred, Non-pH-Controlled In Vitro Fermentations. Glass culture tubes containing 9 mL medium (pH 7) per litre: peptone 0.78 g (Oxoid), NaCl 0.49 g, $Na_2HPO_4 \cdot 12H_2O$ 7.1 g, KH_2PO_4 1.6 g, Bile salts no. 3 (Oxoid) 0.4 g, cysteine·HCl 0.5 g, $K_2HPO_4 \cdot 3H_2O$ 0.5 g, $MgSO_4$ 0.01 g, $CaCl_2 \cdot 2H_2O$ 0.007 g, haemin-solution (0.5 g/L;) 10 mL, Tween 80 2 mL, vitamin K1 10 μL, resazurin-solution 4 mL. The medium was heat-sterilised and degassed by boiling, and the hot tubes were placed to cool in an anaerobic cabinet (90% N_2, 5% CO_2, 5% H_2) to maintain anaerobicity. To each tube, 0.5 mL faecal slurry and 0.5 mL carbohydrate-PBS solution (10% w/v) were added. The fermentations were carried out in three parallel runs, one per faecal donor, for each carbohydrate. A glucose-PBS-solution was treated in the same way and used as a positive control. Incubation and sample collection were carried out in the anaerobic cabinet at 37°C. The initial experiments were conducted over a period of 24 h. Subsequent incubations were prolonged to first 48, then 72 h, as indicated below.

In vivo, acetate, propionate, and butyrate do not accumulate in the colon but are absorbed by the colonic mucosa [25, 26] while in static batch culture tests these short chain acids accumulate and could possibly inhibit microorganisms by the drop in pH. To prevent the inactivation of microorganisms due to this accumulation the buffer concentration of the batch culture medium was increased such that the pH did not fall below pH 5.

2.2. Faecal Slurry Preparation. Healthy adults who had not used antibiotics for at least three months donated stool samples (three males, one female; aged 27 to 60 years) prior to each experiment. None of the volunteers had any history of gastrointestinal disorder. All volunteers were eating a mixed Dutch diet. Within 30 min following defaecation, faecal slurries were prepared from 1 part freshly voided faecal sample and 9 parts prereduced phosphate-buffered saline (PBS) pH 7 (10% w/v). For every experiment, three fresh faecal samples were obtained and immediately used.

2.3. Factors Studied. Table 1 states the manufacturers/suppliers of the test substrates.

2.4. Composition of Monomers, D-/L-Isomery, and Linkage. In separate experiments, substrates were tested at a starting concentration of 0.5% (w/v): monosaccharides, di- and trisaccharides, D- and L-isomers of arabinose and xylose, and the glucopyranosyl-D-glucoses D-maltose (α1-4), isomaltose (α1-6), cellobiose (β1-4) and gentiobiose (β1-6) (Table 1).

2.5. Substitution/Derivatisation. Methyl-β-galactopyranoside (methylgalactose; see Table 1 for manufacturers of the substrates used), β-galactose pentaacetate, D-fucose (deoxygalactose), and galactitol were studied at a final concentration of 1% (w/v) to examine the influence of different substitutions on fermentation times. Instead of D-glucose, D-galactose was used as a positive control. In this experiment, the concentration of substrate was doubled, as compared to the other experiments, to make differences

Table 1: Substrates used in faecal batch culture experiments to determine influence of chain length, composition, and linkage on fermentation time.

Sugar	DP	Monomer(s)	Linkage	Substitution/derivatisation	Manufacturer
D-glucose	1	D-glucose			Sigma
D-mannose	1	D-mannose			Sigma
L-rhamnose	1	L-rhamnose		6-Deoxy-D-mannose	Sigma
D-arabinose	1	D-arabinose			Sigma
L-arabinose	1	L-arabinose			Merck
D-xylose	1	D-xylose			Sigma
L-xylose	1	L-xylose			Fluka
D-fucose	1	D-galactose		Reduction (6-Deoxy-D-galactose)	Fluka
D-galactose	1	D-galactose			Merck
Lactose	2	D-glucose, D-galactose	β1-4		Sigma
Lactulose	2	D-galactose, D-fructose	β1-4		Sigma
Sucrose	2	D-glucose, D-fructose	α1-2		Sigma
D-maltose	2	D-glucose	α1-4		Sigma
Isomaltose	2	D-glucose	α1-6		Sigma
Cellobiose	2	D-glucose	β1-4		Sigma
Gentiobiose	2	D-glucose	β1-6		Sigma
Raffinose	3	D-galactose, D-glucose, D-fructose	α1-6 α2-1		Sigma
Melezitose	3	D-glucose, D-fructose, D-glucose	α1-3 β2-1		Sigma
Vivinal-GOS	1–8	D-galactose, D-glucose	β1-4, β1-6		Royal Friesland Foods
Glucuronic acid	1	D-glucose		Uronic acid (oxidation at C6)	Sigma
Galacturonic acid	1	D-galactose, D-fructose		Uronic acid (oxidation at C6)	Sigma
Glucosamine	1	D-glucose		amine	Sigma
N-acetylated glucosamine	1	D-glucose		Acetylated amine	Sigma
Methyl-β-Galactopyranoside	1	D-galactose		Methylation	Fluka
β-galactose pentaacetate	1	D-galactose		Acetylation	Fluka
Galactitol	1	D-galactose		Reduction (reduction at C1)	Fluka

between the test substances more pronounced as β-galactose pentaacetate, and galactitol did not dissolve in the growth medium.

The sugar derivatives glucuronic acid, galacturonic acid, glucosamine, and N-acetylated glucosamine (GlcNAc) were tested at a concentration of 0.5% (w/v, Table 1).

2.6. Chain Length. To determine the influence of chain length on fermentation time, different fractions of GOS (Vivinal-GOS, FrieslandCampina Domo, Zwolle, The Netherlands) were tested with a starting concentration of 0.5% (w/v) total carbohydrates; Vivinal-GOS consisting mainly of mono-, di-, tetra-, and pentasaccharides, GOS^0 (purified GOS) which was treated with lactase (Maxilact) and contained no monosaccharides (removed by ion-exchange chromatography) or lactose, and pools of purified GOS (Vivinal-GOS) which were fractionated on a Mitsubishi UBK-530 cation exchange column (Pharmacia XP 260, 1 m length) in aqueous liquid chromatography. The latter resulted in GOS^1 (mostly tetra- and pentasaccharides), GOS^2 (tri-, tetra-, and pentasaccharides), and GOS^3 (disaccharides, not including lactose, tri-, and tetrasaccharides).

2.7. Sample Collection. The incubation time lasted 72 h in most experiments, except for tests on the substitution of oligosaccharides (48 h), the composition of monomers, the D-/L-isomery and chain length (24 h), as explained above. At incubation times of 0, 2, 4, 6, 8, 24, 48, and 72 h, samples of 1 mL were taken. Samples were centrifuged immediately (5 min, 16,110 ×g, room temperature) and the supernatants stored at −20°C.

2.8. SCFA Analysis. Supernatants were thawed at room temperature. Twenty μL $K_4Fe(CN)_6 \cdot 3H_2O$, 20 μL $ZnSO_4 \cdot 7H_2O$ and 10 μL 0.25 mM NaOH were added to 450 μL supernatant. The samples were mixed after the addition of every reagent and centrifuged again (5 min, 16 110 ×g); supernatants were transferred into glass vials. A high performance liquid chromatography (HPLC) method, according to Guerrant [27], was used with an ion-exchange column (HPX-87H), 0.006 M H_2SO_4 as eluent, and UV-detection at 205 nm. An organic acid mixture, containing lactate and the SCFA formate, acetate, propionate, butyrate, isobutyrate, valerate, and isovalerate, was employed in making a calibration curve.

2.9. Carbohydrate Recovery. Samples taken at 24 h were thawed and filter-sterilised. They were then diluted 500 times with 3% (w/v) acetonitrile. Determinations of the test carbohydrates were then carried out on a DIONEX HPLC, with a PA1 (Ion exchange) column and pulsed amperometric detection (PAD). The eluents (12.5 mM NaOH, 125 mM NaOH, 125 mM NaOH + 500 mM Na-acetate) formed a gradient from low-to-high molarity NaOH.

This method allows the quantification of water-soluble sugars and their derivatives, but not of the oligosaccharide mixture of GOS, as this is a complex mixture of various sugars. Therefore, results for GOS were analysed semiquantitatively by calculating the percentage of the area under every peak compared to total area under the curve.

2.10. Data Analysis. Total SCFA and lactate production, that is, the sum of succinate, lactate, formate, acetate, propionate, butyrate, isobutyrate, valerate, and isovalerate, was fitted to a logistic growth model [28] using the advanced linear/nonlinear models-linear estimation of STATISTICA (StatSoft Benelux, Groningen).

The logistic model

$$F = \frac{MP}{1 + e^{-k(T - T_{50})}}, \tag{1}$$

where F is the amount of total organic acids at time t, MP is the maximal production of total organic acids, k is the fractional rate constant (which determines the speed of the organic acid production), T is time of fermentation, and T_{50} is time when total organic acid production reaches half of the maximum. MP, k, and T_{50} were estimated for every sample. We chose the T_{50} value for comparing sugars on time of fermentation because it is independent of the total organic acid production. Means were compared by t-test within an experiment. For interexperimental comparisons, values were corrected for D-glucose and then compared by t-test. The resulting t-values were corrected for false discovery rate to avoid false significances [29]. Differences were considered as significant if $P < 0.05$.

The organic acid concentrations were treated as the T_{50} values above.

3. Results

A correlation was found between total organic acid production and carbohydrate recovery, as measured during the first 24 h fermentation of 0.5% glucose and 1% (w/v) galactose ($R^2 = 0.994$ and 0.995, resp., Figure 1). This indicates that the rate of production of organic acids is a good reflection of the rate of fermentation. Quantitative analysis of recovery was not possible for the different fractions of GOS using Dionex HPLC because it is a complex mixture of oligosaccharides. Therefore, organic acid production was used to compare the rate of fermentation of test substances differing in structural factors.

3.1. Composition of Monomers, D-/L-Isomery. Fermentation of the di- and trisaccharides resulted in a concentration of total organic acids of 29 to 41 mM at 24 h. In the other experiments regarding D-/L-isomery and monomers, total organic acid concentrations between 25 and 60 mM were found (Table 2). Compared to D-glucose, fermentation of D-arabinose, L-xylose, and D-mannose resulted in a significantly lower amount of lactate at 24 h ($P = 0.002$, 0.000, 0.003, resp., Table 2). Fermentation of D-glucose resulted in an acetate : propionate : butyrate (A : P : B) ratio of 57 : 20 : 23; lower acetate and higher butyrate proportions than in the other experiments. D-mannose and L-rhamnose fermentation resulted in an A : P : B ratio of 45 : 29 : 26 and 46 : 34 : 20, respectively, while the D-glucose control in this experiment yielded mainly acetate (82 : 8 : 10).

Fermentation of D-mannose, L-arabinose, L-rhamnose, melezitose, D-xylose, D-arabinose, and L-xylose resulted in significantly higher T_{50} values than did the fermentation of D-glucose (Table 3). Based on HPLC analyses, only L-xylose was not completely utilised by 24 h of incubation (data not shown).

3.2. Linkage. Isomaltose yielded 32 mM organic acids as maximal concentration, while the other substrates reached between 50 and 60 mM at 24 h. The fermentation products of isomaltose were mainly acetate and propionate (A : P : B ratio of 76 : 14 : 6), whereas fermentation of cellobiose, maltose, and gentiobiose resulted mainly in lactate (30, 54, and 44% of total organic acid production, resp.) and acetate (A : P : B ratio of 54 : 3 : 1, 48 : 4 : 1, and 49 : 2.5 : 1, resp.) production (Table 2).

D-maltose (Glcα1-4Glc), gentiobiose (Glcβ1-6Glc), cellobiose (Glcβ1-4Glc), and isomaltose (Glcα1-6Glc) had significantly higher T_{50} values than D-glucose (Table 3).

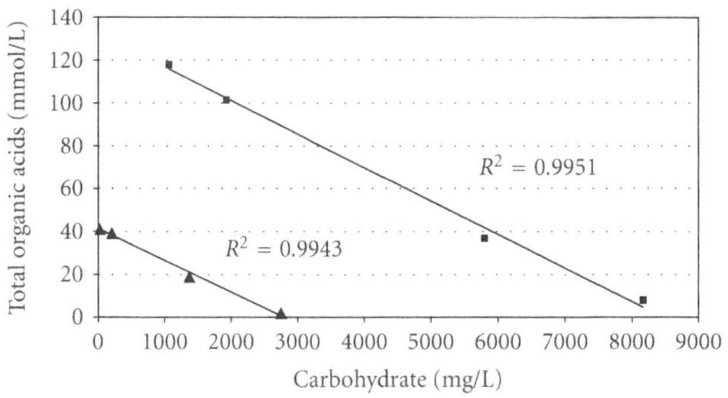

Figure 1: Correlation between total organic acid production, that is, the sum of acetate, propionate, isobutyrate, butyrate, isovalerate, and valerate, and carbohydrate consumption for 1% (w/v) galactose (squares) and 0.5% (w/v) glucose (triangles) in static faecal batch culture experiments ($n = 3$).

Nevertheless, less than 1% (w/v) of the original substrate was recovered after 24 h fermentation (data not shown).

3.3. Substitution/Derivatisation. GlcNAc, glucosamine, galacturonate, and glucuronate fermentations resulted in significantly higher T_{50} values than D-glucose fermentation (Table 3). Their total organic acid yields after 24 h were markedly lower, due to low lactate yields (Table 2). Concentrations of the branched-chain fatty acids, valerate and formate, were significantly lower in the glucuronic and galacturonic acid fermentations than in the positive control (data not shown). GlcNAc and glucosamine resulted in high concentrations of formate and lactate during the first 8 h of incubation (data not shown), which disappeared after 8 h.

Galactitol and β-galactose pentaacetate did not dissolve in the growth medium, but remained as a precipitate. Fermentation of methylgalactose closely resembled the pattern of D-galactose fermentation while D-fucose, galactose pentaacetate, and galactitol yielded significantly lower lactate, and in the case of galactitol and D-fucose, acetate concentrations and reached significantly higher T_{50} values (Tables 2 and 3).

3.4. Chain Length. Vivinal-GOS and D-glucose had almost identical organic acid production values. Vivinal-GOS (mainly DP 1 to 5) and GOS^0 (mainly DP 2 to 5) had the same estimated T_{50} value as D-glucose. GOS^1 (mainly DP 4 and 5), GOS^2 (DP 3 to 5), and GOS^3 (DP 2 to 4) had longer fermentation times. GOS^0 through GOS^3 fermentation yielded significantly less lactate and GOS^1 through GOS^3 significantly more acetate than D-glucose and Vivinal-GOS fermentations.

GOS consist of galactose chains with a terminal glucose molecule. During fermentative hydrolysis, D-galactose and some D-glucose are released (Figure 2, 4 hours: D-galactose and D-glucose peaks appear) and subsequently consumed by the bacteria (Figure 2, 8 hours: D-galactose and D-glucose have disappeared again). As quantitative measurement of all peaks of GOS was not possible, the utilisation of substrate over time could only be shown semiquantitatively. Following 24 h incubation, only 6% of the initial Vivinal-GOS could be recovered and 8 to 12% of GOS^0 and the pools (data not shown). However, in all samples independent of substrate, peak 22 of the chromatogram was the most resistant to degradation (Figure 2).

4. Discussion

The present study was designed to investigate the different factors that affect the chemical structure of GOS and that can be changed in the synthesis of GOS. The approach was to use the isolated factors instead of synthesising new GOS and test these. The advantage of this approach was that more structure elements could be tested independent of each other. However, it has to be kept in mind that *in vivo* most of the mono- and disaccharides used would be digested and absorbed in the small intestine and not reach the colon.

This study has shown a structure-function relationship between factors, such as the substitution of side-chains, monomer composition and linkage, and fermentation time of carbohydrates in static faecal batch culture experiments. Furthermore, these factors not only affected the total amount of SCFA from which the T_{50} was calculated, but also the SCFA ratios. Static batch culture experiments have widely been used to compare substrates in faecal fermentations, using either SCFA production or gas production [28, 30, 31] to characterise the rate of fermentation. In the present study, organic acid production was used for curve fitting and estimation of fermentation time because of a good correlation with carbohydrate recovery. Total organic acid production for 0.5% (w/v) D-glucose varied between 40 and 70 mM at 24 h, which seems high but is comparable to literature data [32, 33]. This variation may be due to a variation in the microbiota and subsequently its fermentative pathways [34–36], leading to different ratios of SCFA to gas and ethanol production. The experiments were performed over several months, and the faecal samples were provided by the same four donors throughout the study. However, the microbial composition of the faecal samples was not analysed during the study so that changes in the faecal microbiota were not monitored. The monosaccharides D-arabinose, D-xylose, D-mannose, and L-rhamnose as well as glucosamine increased the concentration of propionate significantly compared to D-glucose ($P < 0.05$). This is comparable to Mortensen et al. [37] who also observed an increase in propionate for L-rhamnose, D-arabinose, and D-xylose. Hove et al. [38] report that SCFA pattern is influenced mostly by the substrate, and not by the bacterial composition of the faecal fermentation. Therefore, these sugars should be considered in the design of a potential new prebiotic oligosaccharide targeted at cardiovascular disease [12]. No differences were observed between D-glucose and D-galactose regarding total organic acid production over time. Table 2 shows that 0.5% D-glucose yielded from 40 to 70 mM total organic acids in 24 h, while 1% D-galactose yielded 101 mM total organic acids in the same time. Therefore, D-galactose was chosen as a positive control in the substitution experiment. By using a positive control within each experiment, fermentation times

TABLE 2: Concentration of lactate and SCFA after 24 h of fermentation. Substrates (0.5% (w/v) except substitution group: 1% (w/v)) were fermented anaerobically in static faecal batch culture incubations ($n = 3$).

		Lactate	±Stdev	Acetate	±Stdev	Propionate	±Stdev	Butyrate	±Stdev	Total organic acids	±Stdev
		mmol/L									
Monomer composition	Glucose	**16.09**^{a}	1.14	**20.08**	1.36	**3.61**^{a}	1.66	**0.00**	0.00	**41.83**^{ab}	3.64
	L-arabinose	**18.65**^{abc}	5.00	**24.25**	1.76	**6.19**^{ab}	2.61	**0.00**	0.00	**51.46**^{b}	3.63
	D-arabinose	**0.23**^{b}	0.40	**19.78**	4.89	**16.91**^{cd}	1.79	**0.00**	0.00	**38.18**^{a}	5.12
	L-xylose	**0.53**^{b}	0.56	**19.20**	11.34	**2.87**^{a}	0.23	**0.95**	0.82	**24.90**^{ab}	13.00
	D-xylose	**4.79**^{abc}	4.19	**19.92**	2.48	**13.89**^{ce}	1.52	**0.00**	0.00	**40.03**^{a}	5.65
	Glucose	**0.00**^{a}	0.00	**16.95**	0.60	**6.07**^{a}	1.57	**6.92**	0.37	**32.16**^{ab}	4.77
	Lactose	**6.99**^{abc}	12.11	**19.72**	2.06	**3.98**^{a}	0.02	**8.76**	3.92	**41.06**^{ab}	9.11
	Lactulose	**3.49**^{ac}	6.05	**23.75**	3.20	**3.39**^{a}	0.29	**4.06**	2.86	**37.39**^{a}	4.13
	Melezitose	**1.73**^{a}	3.00	**20.50**	1.51	**5.55**^{a}	1.43	**5.02**	1.34	**36.47**^{ab}	3.60
	Raffinose	**1.76**^{a}	3.04	**19.10**	2.22	**4.55**^{a}	0.90	**6.60**	3.41	**33.53**^{ab}	0.70
	Sucrose	**0.00**^{a}	0.00	**0.95**	0.92	**14.22**^{aef}	4.10	**1.93**	1.90	**29.93**^{b}	4.69
	Glucose	**29.62**^{a}	7.18	**27.73**	5.17	**3.21**^{a}	1.80	**2.82**	1.66	**67.69**	4.73
	D-mannose	**16.84**^{cb}	6.88	**15.78**	2.02	**10.18**^{bdf}	1.35	**9.33**	1.22	**52.20**	11.96
	L-rhamnose	**0.10**^{abc}	0.17	**25.61**	5.13	**19.09**^{c}	1.26	**10.99**	4.34	**60.74**	1.98
Linkage	Glucose	**25.51**^{a}	1.11	**23.26**^{abc}	6.20	**1.30**^{a}	0.69	**0.32**	0.14	**52.99**^{ac}	4.13
	Maltose	**15.60**^{b}	2.24	**28.53**^{b}	2.64	**1.79**^{b}	0.70	**0.75**	0.64	**52.35**^{a}	1.38
	Isomaltose	**0.00**^{c}	0.00	**24.78**^{c}	2.28	**4.67**^{ab}	3.04	**1.34**	1.30	**32.05**^{b}	2.68
	Cellobiose	**26.92**^{a}	2.56	**14.10**^{a}	1.69	**1.93**^{ab}	0.69	**0.59**	0.55	**49.95**^{ac}	4.75
	Gentiobiose	**26.66**^{a}	0.44	**29.76**^{b}	2.25	**1.48**^{ab}	0.39	**0.56**	0.14	**60.12**^{c}	1.25
Substitution	Galactose	**38.11**^{a}	2.73	**56.75**^{a}	2.37	**4.39**^{ab}	1.44	**0.45**	0.15	**101.44**^{a}	6.01
	Methylgalactose	**34.76**^{a}	3.09	**50.89**^{b}	0.36	**3.51**^{a}	0.95	**0.51**	0.53	**91.03**^{a}	4.74
	Galactose pentaacetate	**0.10**^{b}	0.17	**40.66**^{abc}	7.82	**4.28**^{ab}	1.23	**0.58**	1.01	**47.88**^{b}	9.53
	Galactitol	**2.21**^{b}	2.48	**15.28**^{cd}	13.23	**4.41**^{ab}	1.24	**0.63**	1.09	**24.52**^{bc}	12.78
	Fucose	**0.00**^{b}	0.00	**25.61**^{d}	4.85	**11.75**^{b}	4.04	**1.56**	1.45	**40.45**^{c}	9.6
Derivatisation	Glucose	**27.48**^{a}	4.04	**27.80**^{ac}	0.25	**1.89**^{a}	0.19	**0.51**	0.45	**62.24**^{a}	4.17
	Glucuronic acid	**3.41**^{b}	2.81	**23.71**^{abc}	6.40	**3.13**^{ab}	0.93	**5.51**		**33.74**^{b}	6.00
	Galacturonic acid	**0.67**^{ab†}	0.19	**20.31**^{b}	0.74	**4.11**^{ab}	1.54	**0.00**	0.00	**32.86**^{b}	8.48
	Glucosamine	**0.33**^{b}	0.47	**15.68**^{ab}	6.14	**4.99**^{b}	0.45	**3.11**	0.82	**27.08**^{b}	10.54
	GlcNAc	**0.73**^{b}	0.64	**32.43**^{c}	3.97	**4.89**^{ab}	1.93	**5.08**†	4.77	**45.90**^{b}	3.03
Chain length	Glucose	**18.68**^{ab}	3.06	**20.74**^{a}	2.17	**1.57**	0.25	**0.27**	0.26	**41.56**	4.00
	Vivinal-GOS	**18.14**^{a}	2.25	**21.67**^{a}	0.97	**1.41**	0.28	**0.22**	0.22	**41.87**	1.73
	GOS^0	**13.49**^{bc}	1.25	**22.81**^{ab}	1.10	**1.30**	0.25	**0.36**	0.51	**38.28**	1.05
	GOS^1	**10.91**^{c}	0.64	**25.13**^{b}	2.28	**1.42**	0.30	**0.40**	0.56	**37.91**	2.12
	GOS^2	**14.94**^{c}	2.34	**25.64**^{b}	1.96	**1.61**	0.21	**0.31**	0.35	**42.61**	3.37
	GOS^3	**13.51**^{ac}	0.83	**24.28**^{b}	1.33	**1.31**	0.11	**0.25**	0.26	**39.48**	1.02

GOS: galactooligosaccharide.
Different superscript letters indicate significant differences within each group ($P < 0.05$).
$^{\dagger}n = 2$.

of substrates can be compared within each data set, and, with adjustment to the control, across data sets.

Fermentation time of different substrates varied considerably. Sugars such as melezitose, L-rhamnose, L-arabinose, D-arabinose, L-xylose, D-xylose, galacturonate, and glucuronate were fermented significantly more slowly than the digestible sugars D-glucose and sucrose and, as expected, more slowly than the indigestible (by human small intestinal enzymes) sugars raffinose and lactulose. A prolonged fermentation of L-rhamnose as compared

Table 3: Halftime of fermentation, expressed as T_{50} values (hours), of substrates (0.5% (w/v) except substitution group: 1% (w/v)) in static faecal batch culture incubations ($n = 3$). Production of total organic acids over 24, 48 and 72 h was used to calculate T_{50} values.

		T_{50} [hours]	SE
Monomer composition	Glucose	3.46^{a}	±0.46
	Raffinose	3.53^{ab}	±0.28
	Sucrose	3.94^{ab}	±0.28
	Lactulose	4.08^{ab}	±0.36
	Lactose	4.18^{ab}	±0.08
	Melezitose†	8.65^{d}	±0.90
	Glucose	4.42^{a}	±0.66
	D-mannose	5.51^{b}	±0.33
	L-rhamnose	8.25^{d}	±0.91
	Glucose	5.51^{a}	±0.44
	L-arabinose	7.83^{c}	±0.31
	D-xylose	10.27^{d}	±0.51
	D-arabinose	11.31^{d}	±0.45
	L-xylose	11.72^{d}	±0.54
Linkage	Glucose	5.21^{a}	±0.26
	Maltose	5.93^{b}	±0.45
	Gentiobiose	6.23^{c}	±0.17
	Cellobiose	7.33^{c}	±0.29
	Isomaltose	7.45^{d}	±0.88
Substitution	Galactose	11.09^{ac}	±1.46
	Methylgalactose	15.93^{ab}	±0.24
	Galactose penta-acetate	20.65^{d}	±0.62
	Galactitol	24.35^{bd}	±1.61
	D-Fucose	19.49^{c}	±1.14
Derivatisation	Glucose	3.90^{a}	±0.22
	GlcNAc	4.87^{b}	±0.14
	Glucosamine	5.77^{b}	±0.98
	Galacturonic acid	10.85^{c}	±1.26
	Glucuronic acid†	11.63^{c}	±0.75
Chain length	Glucose	4.19^{a}	±0.12
	Vivinal-GOS	4.17^{a}	±0.16
	GOS^{0}	4.37^{ac}	±0.18
	GOS^{1}	4.92^{b}	±0.14
	GOS^{2}	5.02^{b}	±0.12
	GOS^{3}	4.82^{bc}	±0.16

GOS: galactooligosaccharide.
Different superscript letters indicate significant differences within each group ($P < 0.05$).
$^{\dagger}n = 2$.

to L-mannose had been reported previously [37], which confirms the results of the present study.

Not in all instances di- or trisaccharides were more slowly fermented than monosaccharides. This was the case for lactose, lactulose, raffinose, and sucrose. Mortensen et al. [37] did not show a further increase in total SCFA after 6 h fermentation of D-glucose, D-galactose, lactose (Galβ1-4Glc), and lactulose (Galβ1-4Fru), either, confirming our results that disaccharides are not *per se* fermented more slowly than monosaccharides. In the present study, melezitose (Glcβ1-3Fruα2-1Glc) was the only trisaccharide that was metabolised significantly slower than lactulose. This

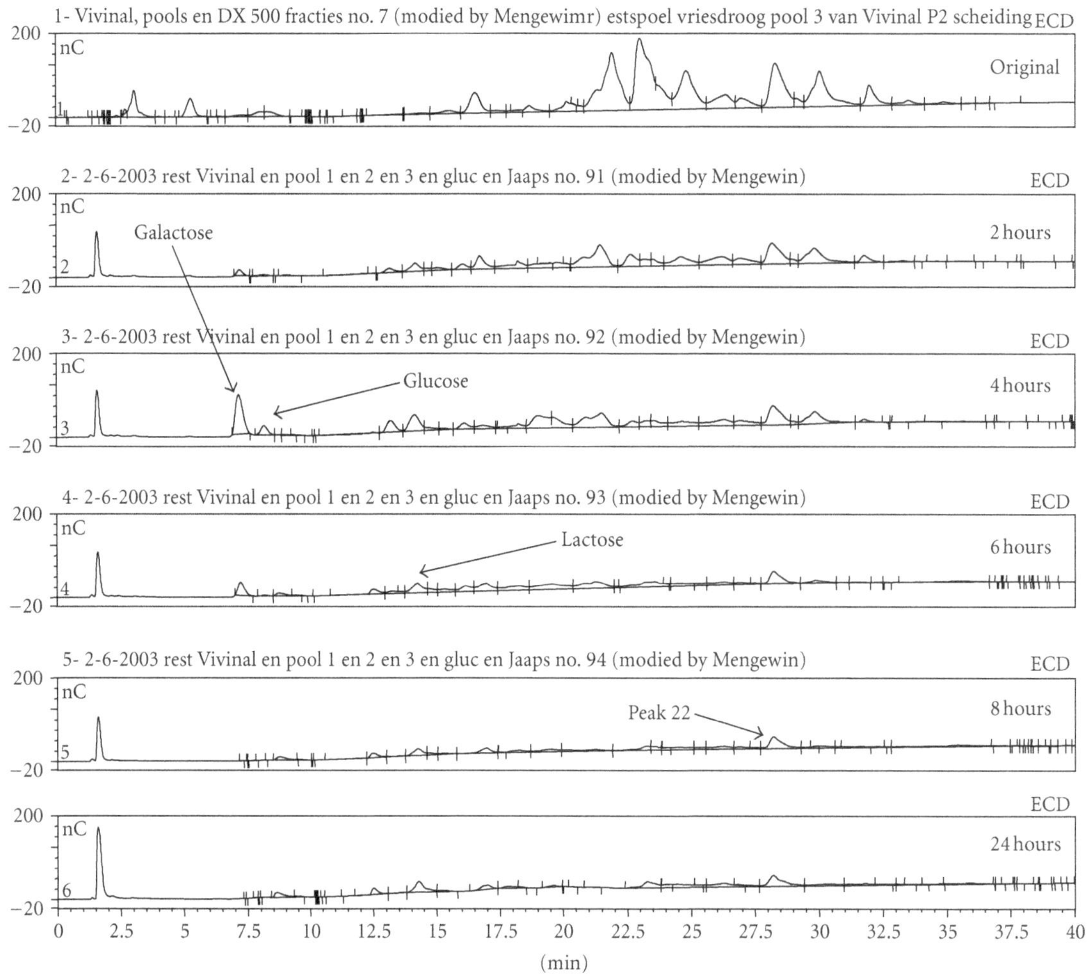

FIGURE 2: Dionex-HPLC chromatogram of the galactooligosaccharide mixture GOS[3] breakdown over 24 h measured in cell-free fermentation supernatants. Static faecal batch culture fermentations were carried out in an anaerobic cabinet with a starting concentration of GOS[3] of 0.5% (w/v). Samples were taken after 2, 4, 6, 8, and 24 h of incubation. The oligosaccharide which is represented by peak 22 seemed to be resistant to fermentation.

may be due to the Glcβ1-3Fru- linkage, as 1–3 bonds occur less often naturally than 1–4 and 1–6 bonds [39].

D-xylose and L- and D-arabinose can all be found in natural polysaccharides, whereas L-xylose is less common. The deduction that L-xylose might therefore be more slowly fermented than D-xylose seemed to be confirmed by the presence of L-xylose after 24 h of fermentation. However, this was not confirmed by the half-time of SCFA production for L-xylose. All isomers were fermented significantly more slowly than D-glucose. It might be possible to design a more distally fermented prebiotic, by synthesising GOS with arabinose or xylose as acceptor substrates, especially because xylan and arabinogalactan degrading enzymes are found in bifidobacteria [36] which are common target organisms of prebiotics because of their beneficial effects on gut health [40, 41]. This is underscored by the results of Van Craeyveld et al. showing that short arabinoxylooligosaccharides (DP 3) increased caecal bifidobacteria numbers, while long arabinoxylooligosaccharides suppressed protein fermentation [42]. The latter indicates distal fermentation.

Another design for distally fermented galacto-oligosaccharides would be to include the linkages Glcα1-6Glc (i.e., isomaltose) and Glcβ1-4Glc (i.e., cellobiose) which were more slowly fermented. Sanz et al. [43] showed that isomaltose and cellobiose also favoured the growth of desirable bacteria (e.g., bifidobacteria) in relation to the growth of undesirable bacteria compared to FOS.

In addition to different acceptor substrates for the design of a distally fermented GOS, derivatisation, and substitution of GOS seems promising. The slower fermentation of β-galactose pentaacetate was not surprising because, in general, solubility and fermentability are correlated, at least when polysaccharides are tested [44]. Fucose and galactitol were both fermented more slowly. Since D-fucose and galactitol are each reduced forms of D-galactose, it might be interesting to reduce GOS to make it less fermentable. While glucosamine was fermented as quickly as D-glucose, GlcNAc, galacturonate, and glucuronate were utilized more slowly. This is in agreement with results from Mortensen et al. [37], who still showed an increase in SCFA concentrations between 6 and 24 h of fermentation for glucuronate and galacturonate.

The comparison between different fractions of GOS did not confirm the hypothesis that a GOS mixture with a higher DP would be fermented more slowly than one with a lower DP. However, it must be kept in mind that the DP of the GOS mixture was less than eight; higher DP oligosaccharides may still display increased fermentation time. Interestingly, there was a difference in the lactate and acetate production from different GOS fractions. GOS consist of galactose chains with a terminal glucose molecule. Different metabolic end-products due to different chain lengths suggest that short-chain oligosaccharides of different DPs modulate the activity of the microbiota via a change in the bacterial composition. In batch culture experiments investigating the effect of DP of short-chain oligosaccharides on the faecal microflora, Sanz et al. observed that lower DP oligosaccharides (DP 2-3) generally score higher on the prebiotic index, indicating a beneficial modulation of the bacterial community, than higher DP oligosaccharides (DP 6–10) [45, 46].

From a multiple comparison (corrected for D-glucose when possible), the different factors can be grouped by their effectiveness at prolonging fermentation as follows: substitution is most effective, followed by the composition of monomers and linkage type. The chain length (within a DP ≤ 8) is the least effective in prolonging fermentation time. It must be remembered, however, that these studies were performed using static batch cultures, which still need to be confirmed *in vivo*.

These factors could be used in the design of novel GOS to create a more distally fermentable prebiotic. The effectiveness of these new carbohydrates needs to be tested *in vitro* and *in vivo*.

Acknowledgments

This work was supported by the European Commission under the Marie Curie Intra-European Fellowship Program (Project QLK1-GH-01-60056-03) and by Friesland Coberco Dairy Foods (now FrieslandCampina). There were no conflicts of interest. Further, we want to thank Kieran Tuohy, Cordula Lindner, Veronika Nieder for their suggestions and sound advice; Alfred Bonte for his great help with the HPLC; Roy Beernink for his help with the fermentations; Rudy de Wit for his help with the curve-fitting. E. Gietl, E. van den Heuvel, J. de Slegte, G. Gibson, and R. Rastall designed research; E. Gietl and W. Mengerink conducted research; E. Gietl analyzed the data; E. Gietl, E. van den Heuvel, R. Rastall, and G. Gibson wrote the paper; E. Gietl had primary responsibility for final content. All authors read and approved the final document.

References

[1] D. P. Burkitt, "Related disease—related cause?" *The Lancet*, vol. 2, no. 7632, pp. 1229–1231, 1969.

[2] F. Levi, C. Pasche, C. La Vecchia, F. Lucchini, and S. Franceschi, "Food groups and colorectal cancer risk," *British Journal of Cancer*, vol. 79, no. 7-8, pp. 1283–1287, 1999.

[3] W. Scheppach, H. Sommer, T. Kirchner et al., "Effect of butyrate enemas on the colonic mucosa in distal ulcerative colitis," *Gastroenterology*, vol. 103, no. 1, pp. 51–56, 1992.

[4] B. F. Hinnebusch, S. Meng, J. T. Wu, S. Y. Archer, and R. A. Hodin, "The effects of short-chain fatty acids on human colon cancer cell phenotype are associated with histone hyperacetylation," *Journal of Nutrition*, vol. 132, no. 5, pp. 1012–1017, 2002.

[5] S. Tedelind, F. Westberg, M. Kjerrulf, and A. Vidal, "Anti-inflammatory properties of the short-chain fatty acids acetate and propionate: a study with relevance to inflammatory bowel disease," *World Journal of Gastroenterology*, vol. 13, no. 20, pp. 2826–2832, 2007.

[6] A. D. Blackwood, J. Salter, P. W. Dettmar, and M. F. Chaplin, "Dietary fibre, physicochemical properties and their relationship to health," *Journal of The Royal Society for the Promotion of Health*, vol. 120, no. 4, pp. 242–247, 2000.

[7] G. R. Gibson and M. B. Roberfroid, "Dietary modulation of the human colonic microbiota: introducing the concept of prebiotics," *Journal of Nutrition*, vol. 125, no. 6, pp. 1401–1412, 1995.

[8] I. R. Rowland, C. J. Rumney, J. T. Coutts, and L. C. Lievense, "Effect of Bifidobacterium longum and inulin on gut bacterial metabolism and carcinogen-induced aberrant crypt foci in rats," *Carcinogenesis*, vol. 19, no. 2, pp. 281–285, 1998.

[9] B. L. Pool-Zobel, S. L. Abrahamse, and G. Rechkemmer, "Protective effects of short-chain fatty acids on early events of carcinogenesis: antigenotoxic effects of butyrate in rat and human colon cells," in *Proceedings of COST Action 92th Workshop on Dietary Fibre and Fermentation in the Colon*, pp. 350–357, Espoo, Finland, 1996.

[10] H. K. Biesalski, P. Fürst, H. Kasper et al., *Ernährungsmedizin. 2. Überarbeitete Auflage*, Georg Thieme, Stuttgart, Germany, 1999.

[11] T. R. Licht, M. Hansen, M. Poulsen, and L. O. Dragsted, "Dietary carbohydrate source influences molecular fingerprints of the rat faecal microbiota," *BMC Microbiology*, vol. 6, article 98, 2006.

[12] J. M. W. Wong, R. De Souza, C. W. C. Kendall, A. Emam, and D. J. A. Jenkins, "Colonic health: fermentation and short chain fatty acids," *Journal of Clinical Gastroenterology*, vol. 40, no. 3, pp. 235–243, 2006.

[13] G. A. Dienel and N. F. Cruz, "Astrocyte activation in working brain: energy supplied by minor substrates," *Neurochemistry International*, vol. 48, no. 6-7, pp. 586–595, 2006.

[14] M. T. Wyss, B. Weber, V. Treyer et al., "Stimulation-induced increases of astrocytic oxidative metabolism in rats and humans investigated with 1-11C-acetate," *Journal of Cerebral Blood Flow and Metabolism*, vol. 29, no. 1, pp. 44–56, 2009.

[15] A. J. McBain and G. T. Macfarlane, "Modulation of genotoxic enzyme activities by non-digestible oligosaccharide metabolism in in-vitro human gut bacterial ecosystems," *Journal of Medical Microbiology*, vol. 50, no. 9, pp. 833–842, 2001.

[16] G. Tzortzis, A. K. Goulas, J. M. Gee, and G. R. Gibson, "A novel galactooligosaccharide mixture increases the bifidobacterial population numbers in a continuous in vitro fermentation system and in the proximal colonic contents of pigs in vivo," *Journal of Nutrition*, vol. 135, no. 7, pp. 1726–1731, 2005.

[17] G. T. Macfarlane, G. R. Gibson, and J. H. Cummings, "Comparison of fermentation reactions in different regions of the human colon," *Journal of Applied Bacteriology*, vol. 72, no. 1, pp. 57–64, 1992.

[18] G. T. Macfarlane and A. J. McBain, "The human colonic microbiota," in *Colonic Microbiota, Nutrition and Health*, G.

R. Gibson MB Roberfroid, Ed., pp. 1–25, Kluwer Academic Publishers, Dordrecht, The Netherlands, 1999.

[19] M. H. M. C. Van Nuenen, K. Venema, J. C. J. Van Der Woude, and E. J. Kuipers, "The metabolic activity of fecal microbiota from healthy individuals and patients with inflammatory bowel disease," *Digestive Diseases and Sciences*, vol. 49, no. 3, pp. 485–491, 2004.

[20] A. McIntyre, G. P. Young, T. Taranto, P. R. Gibson, and P. B. Ward, "Different fibers have different regional effects on luminal contents of rat colon," *Gastroenterology*, vol. 101, no. 5, pp. 1274–1281, 1991.

[21] E. Olano-Martin, K. C. Mountzouris, G. R. Gibson, and R. A. Rastall, "In vitro fermentability of dextran, oligodextran and maltodextrin by human gut bacteria," *British Journal of Nutrition*, vol. 83, no. 3, pp. 247–255, 2000.

[22] E. Olano-Martin, G. R. Gibson, and R. A. Rastall, "Comparison of the in vitro bifidogenic properties of pectins and pectic-oligosaccharides," *Journal of Applied Microbiology*, vol. 93, no. 3, pp. 505–511, 2002.

[23] S. Perrin, C. Fougnies, J. P. Grill, H. Jacobs, and F. Schneider, "Fermentation of chicory fructo-oligosaccharides in mixtures of different degrees of polymerization by three strains of bifidobacteria," *Canadian Journal of Microbiology*, vol. 48, no. 8, pp. 759–763, 2002.

[24] K. M. J. Van Laere, M. Bosveld, H. A. Schols et al., "Fermentative degradation of plant cell wall derived oligosaccharides by intestinal bacteria," in *Proceedings of the International Symposium on "Non-Digestible Oligosaccharides: Healthy Food for the Colon"*, R. Hartemink, Ed., pp. 37–46, Wageningen Graduate School VLAG, 1997.

[25] H. Ruppin, S. Bar-Meir, and K. H. Soergel, "Absorption of short-chain fatty acids by the colon," *Gastroenterology*, vol. 78, no. 6, pp. 1500–1507, 1980.

[26] J. A. Vogt and T. M. S. Wolever, "Fecal acetate is inversely related to acetate absorption from the human rectum and distal colon," *Journal of Nutrition*, vol. 133, no. 10, pp. 3145–3148, 2003.

[27] G. O. Guerrant, M. A. Lambert, and C. W. Moss, "Analysis of short-chain acids from anaerobic bacteria by high-performance liquid chromatography," *Journal of Clinical Microbiology*, vol. 16, no. 2, pp. 355–360, 1982.

[28] J. L. Campbell, C. V. Williams, and J. H. Eisemann, "Fecal inoculum can be used to determine the rate and extent of in vitro fermentation of dietary fiber sources across three lemur species that differ in dietary profile: varecia variegata, Eulemur fulvus and Hapalemur griseus," *Journal of Nutrition*, vol. 132, no. 10, pp. 3073–3080, 2002.

[29] Y. Benjamini and Y. Hochberg, "Controlling the false discovery rate: a practical and powerful approach to multiple testing," *Journal of the Royal Statistical Society B*, vol. 57, pp. 289–300, 1995.

[30] E. Bauer, B. A. Williams, C. Voigt, R. Mosenthin, and M. W. A. Verstegen, "Microbial activities of faeces from unweaned and adult pigs, in relation to selected fermentable carbohydrates," *Animal Science*, vol. 73, no. 2, pp. 313–322, 2001.

[31] M. R. Smiricky-Tjardes, E. A. Flickinger, C. M. Grieshop, L. L. Bauer, M. R. Murphy, and G. C. Fahey, "In vitro fermentation characteristics of selected oligosaccharides by swine fecal microflora," *Journal of Animal Science*, vol. 81, no. 10, pp. 2505–2514, 2003.

[32] A. J. Vince, N. I. McNeil, J. D. Wager, and O. M. Wrong, "The effect of lactulose, pectin, arabinogalactan and cellulose on the production of organic acids and metabolism of ammonia by intestinal bacteria in a faecal incubation system," *British Journal of Nutrition*, vol. 63, no. 1, pp. 17–26, 1990.

[33] P. J. Wood, E. Arrigoni, S. Shea Miller, and R. Amadò, "Fermentability of oat and wheat fractions enriched in β-glucan using human fecal inoculation," *Cereal Chemistry*, vol. 79, no. 3, pp. 445–454, 2002.

[34] N. Peekhaus and T. Conway, "What's for dinner?: entner-Doudoroff metabolism in Escherichia coli," *Journal of Bacteriology*, vol. 180, no. 14, pp. 3495–3502, 1998.

[35] T. L. Miller and M. J. Wolin, "Fermentations by saccharolytic intestinal bacteria," *American Journal of Clinical Nutrition*, vol. 32, no. 1, pp. 164–172, 1979.

[36] G. T. Macfarlane and G. R. Gibson, "Carbohydrate fermentation, energy transduction and gas metabolism in the human large intestine," in *Gastrointestinal Microbiology*, R. I. Mackie and B. A. White, Eds., vol. 1, pp. 269–318, Chapman & Hall, New York, NY, USA, 1997.

[37] P. B. Mortensen, K. Holtug, and H. S. Rasmussen, "Short-chain fatty acid production from mono- and disaccharides in a fecal incubation system: Implications for colonic fermentation of dietary fiber in humans," *Journal of Nutrition*, vol. 118, no. 3, pp. 321–325, 1988.

[38] H. Hove, I. Nordgaard-Andersen, and P. B. Mortensen, "Effect of lactic acid bacteria on the intestinal production of lactate and short-chain fatty acids, and the absorption of lactose," *American Journal of Clinical Nutrition*, vol. 59, no. 1, pp. 74–79, 1994.

[39] S. A. Brooks, M. V. Dwek, and U. Schumacher, *Functional and Molecular Glycobiology*, BIOS Scientific, Oxford, UK, 2002.

[40] F. Depeint, G. Tzortzis, J. Vulevic, K. I'Anson, and G. R. Gibson, "Prebiotic evaluation of a novel galactooligosaccharide mixture produced by the enzymatic activity of Bifidobacterium bifidum NCIMB 41171, in healthy humans: a randomized, double-blind, crossover, placebo-controlled intervention study," *American Journal of Clinical Nutrition*, vol. 87, no. 3, pp. 785–791, 2008.

[41] A. Bernalier, J. Dore, and M. Durand, "Biochemistry of fermentation," in *Colonic Microbiota, Nutrition and Health*, G. R. Gibson and M. B. Roberfroid, Eds., pp. 37–53, Kluwer Academic Publishers, Dordrecht, The Netherlands, 1999.

[42] V. Van Craeyveld, K. Swennen, E. Dornez et al., "Structurally different wheat-derived arabinoxylooligosaccharides have different prebiotic and fermentation properties in rats," *Journal of Nutrition*, vol. 138, no. 12, pp. 2348–2355, 2008.

[43] M. L. Sanz, G. R. Gibson, and R. A. Rastall, "Influence of disaccharide structure on prebiotic selectivity in vitro," *Journal of Agricultural and Food Chemistry*, vol. 53, no. 13, pp. 5192–5199, 2005.

[44] P. Brobech Mortensen and I. Nordgaard-Andersen, "The dependence of the in vitro fermentation of dietary fibre to short-chain fatty acids on the contents of soluble non-starch polysaccharides," *Scandinavian Journal of Gastroenterology*, vol. 28, no. 5, pp. 418–422, 1993.

[45] M. L. Sanz, G. L. Côté, G. R. Gibson, and R. A. Rastall, "Prebiotic properties of alternansucrase maltose-acceptor oligosaccharides," *Journal of Agricultural and Food Chemistry*, vol. 53, no. 15, pp. 5911–5916, 2005.

[46] M. L. Sanz, G. L. Côté, G. R. Gibson, and R. A. Rastall, "Selective fermentation of gentiobiose-derived oligosaccharides by human gut bacteria and influence of molecular weight," *FEMS Microbiology Ecology*, vol. 56, no. 3, pp. 383–388, 2006.

5

Synthesis and Characterization of Graft Copolymer of Dextran and 2-Acrylamido-2-methylpropane Sulphonic Acid

Venkanna Azmeera, P. Adhikary, and S. Krishnamoorthi

Department of Chemistry, Banaras Hindu University, Varanasi 221005, India

Correspondence should be addressed to S. Krishnamoorthi, dr.skmoorthi@gmail.com

Academic Editor: R. J. Linhardt

A novel biodegradable graft copolymer of dextran (Dx) and 2-acrylamido-2-methyl-1-propane sulphonic acid (AMPS) was synthesized by grafting poly-AMPS chains onto dextran backbone by free radical polymerization using ceric ammonium nitrate (CAN) as an initiator. Different amounts of AMPS were used to synthesize four different grades of graft copolymers with different side chain lengths. These grafted polymers were characterized by elemental analysis, FTIR, ^{1}HNMR, rheological technique, scanning electron microscopy (SEM), thermogravimetric analysis (TGA), and X-ray diffractometry (XRD). They exhibited efficient flocculation performance in kaolin suspension.

1. Introduction

Water soluble polymers have received a great attention in recent years. Carbohydrate polymers are often preferred over synthetic polymers due to their nontoxic, low-cost, ease of availability, and biodegradability characteristics [1–3]. They are used in pharmaceuticals, detergents, cosmetics, textiles production, and in waste-water treatment (flocculation) [4].

The choice of dextran relies on several aspects [5–8]. First, dextran of various molecular weights is commercially available. Second, the degree of branching of the dextran can be as low as 0.5%, making them ideal molecular model for scientific studies. Third, dextran possesses high water solubility, is stable under mild acid and basic conditions, and contains a large number of hydroxyl groups for conjugation [9–11]. Dextran (Dx) $(C_6H_{10}O_5)_n$ has been successfully used in the medical and biomedical fields for more than four decades. Dextran is the generic term for a family of neutral water soluble polysaccharides consisting (95%) of α-$(1 \rightarrow 6)$ linked D-glucose main chain with varying branches, and 5% branches may arise from α-$(1 \rightarrow 2)$, α-$(1 \rightarrow 3)$, and α-$(1 \rightarrow 4)$ glycosidic linkages. It has common solubility in water and dipolar aprotic solvents [12]. It is biocompatible and biodegradable and can lead to biopolymers derivatives with balanced hydrophobic/hydrophilic character [13].

Among polymeric flocculants, the synthetic polymers can be tailor-made by controlling molecular weight, molecular weight distribution, the structure of polymers, and the nature and percentage of ionic groups. Thus, the synthetic polymers are very efficient flocculants [14–16]. It has been observed from rheological studies that grafting of shear degradable polymers onto the rigid polysaccharides backbone provides fairly shear stable systems [17].

In this paper, we aim to report the synthesis and characterization of graft copolymers based on poly 2-acylamido-2-methyl-1-propane sulphonic acid (PAMPS) and dextran by ceric ammonium nitrate (CAN) induced solution polymerization technique [18–20]. Four grades of graft copolymers (Dx-g-PAMPS1 to Dx-g-PAMPS4) were synthesized by varying the amount of AMPS in the reaction feed [21]. The graft copolymers were characterized by elemental analysis, FT-IR, FT-^{1}H-NMR, rheological technique, scanning electron microscopy (SEM), thermal analysis (TGA), and X-ray diffractometry (XRD). The flocculation efficiencies of graft

Table 1: Synthetic details of the graft copolymers of Dx and PAMPS.

Serial number	Graft copolymers	Dextran (g)	AMPS (g)	CAN (mg)	GE (%)	Intrinsic viscosity (dL/g)
1	Dx-g-PAMPS1	1	5	20	75.66	13.5
2	Dx-g-PAMPS2	1	6	20	83.42	14.2
3	Dx-g-PAMPS3	1	7	20	96.56	15.3
4	Dx-g-PAMPS4	1	8	20	54.88	14.6

Percentage of grafting efficiency (GE) = weight of pure graft copolymer/weight of product mixture × 100.

Table 2: Elemental analysis and flocculation performance of Dx and its graft copolymers.

Polymer	Elemental analysis			Flocculation performance	
	Carbon (%)	Hydrogen (%)	Nitrogen (%)	Optimum dose (ppm)	Turbidity (NTU)
Dx	38.7	6.2	0.36	4	96
Dx-g-PAMPS1	36.9	6.8	4.6	8	70
Dx-g-PAMPS2	38.9	6.9	4.7	10	51
Dx-g-PAMPS3	37.1	6.5	5.1	4	23
Dx-g-PAMPS4	38.7	6.9	4.5	16	24

copolymers were estimated in kaolin suspension. It was observed that the graft copolymers are efficient flocculants.

2. Experimental

2.1. Materials. Dextran (average M_W = 482000 Da) was procured from Sigma-Aldrich chemicals, USA, 2-acrylamido-2-methyl-1-propane sulphonic acid (AMPS) was procured from Sigma-Aldrich chemicals, Germany. CAN was procured from Loba Chemie, Mumbai. Acetone and hydroquinone were procured from E. Merck, Mumbai. Kaolin is a gift from Rajmahal Quartz-Sand, Kolkata, India.

2.2. Synthesis of Graft Copolymers. Four grades of graft copolymers (Dx-g-PAMPS1 to Dx-g-PAMPS4) of Dx and PAMPS were synthesized by Ce (IV) induced solution polymerization technique. In a 50 mL conical flask, 1 g of dextran was taken and it was dissolved in 25 mL of water and after a clear solution chuck it downs the amount of AMPS. The mixture was stirred continuously for 30 min under nitrogen atmosphere and heating the reaction mixture up to 50°C, then 20 mg of CAN was added. The reaction was carried out for 3 hrs until the solution become viscous. The graft copolymers were precipitated with acetone and washed 10 times by acetone for removal of unreacted AMPS. The resultant graft copolymer was dried under vacuum at 50°C. The synthesis details of the graft copolymers are summarized in Table 1.

2.3. Characterization

2.3.1. Elemental Analysis. The elemental analysis of dextran and all the graft copolymers was performed by using CHN analyzer, CE-440 Elemental analyzer, Mexico. The estimated analysis of three different elements, carbon, hydrogen and nitrogen, was undertaken. The results are shown in Table 2.

2.3.2. Intrinsic Viscosity Measurement. Intrinsic viscosity measurements of the aqueous solution of all Dx and graft copolymers were carried out with the help of Ubbelohde viscometer (CS/S: 0.00386) 25 ± 0.1°C. Intrinsic viscosity of all the samples was determined from the point of intersection of two extrapolated (to zero concentration) plots, that is, inherent viscosity versus concentration (η_{inh} versus C) and reduced viscosity versus concentration (η_{red} versus C). The standard method used to evaluate intrinsic viscosity has been described earlier [7]. The intrinsic viscosities of all the graft copolymers are reported in Table 1.

2.3.3. FTIR Spectroscopy. A Thermo Nicolet FT-IR spectrophotometer (Model JASCO FT-IR-5300) was used to record the IR spectra within the range of 4000–400 cm^{-1}. The IR spectra (Figure 1) of dextran, PAMPS, and graft copolymer (Dx-g-PAMPS3) were recorded in solid state using a KBr pellet method.

2.3.4. ^1H FTNMR Spectroscopy. ^1H NMR spectra (Figure 2) of samples were recorded by JEOL AL 300 FT-NMR spectrometer with sample solution in dimethyl sulphoxide-d_6 (DMSO-d_6) and tetramethylsilane (TMS) was used as an internal reference [22, 23].

2.3.5. Rheological Measurements. The rheological (viscosity versus shear rate) measurements (Figure 3) of aqueous solutions of dextran and graft copolymers (Dx-g-PAM1 to Dx-g-PAMPS4) were carried out using controlled stress Brookfield DV-III Ultra, programmable rheometer. A 2° cone of diameter 4 cm and with truncation of 46 μm was used for the measurements.

2.3.6. Scanning Electron Microscopy (SEM). For this study, dextran was used in powder form, whereas the PAMPS and graft copolymer (Dx-g-PAMPS3) were in the form of small granules (Figure 4). A HRSEM SUPRA 40, ZEISS (Germany) was used for the above study.

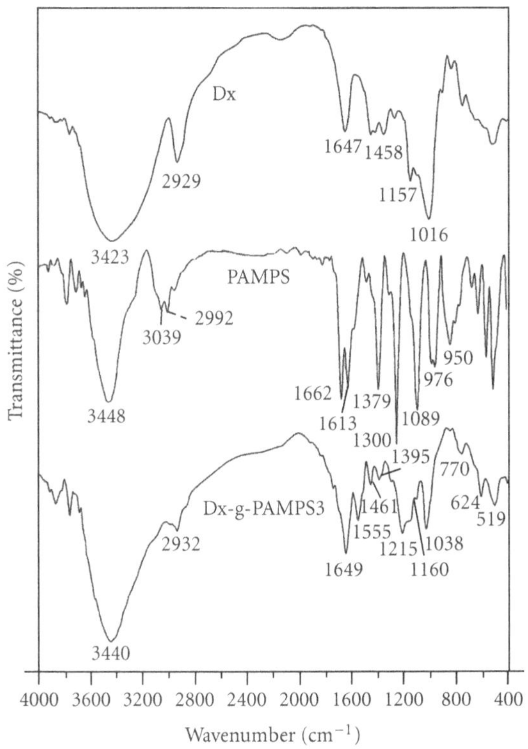

FIGURE 1: FT-IR spectra of Dx, PAMPS, and Dx-g-PAMPS3.

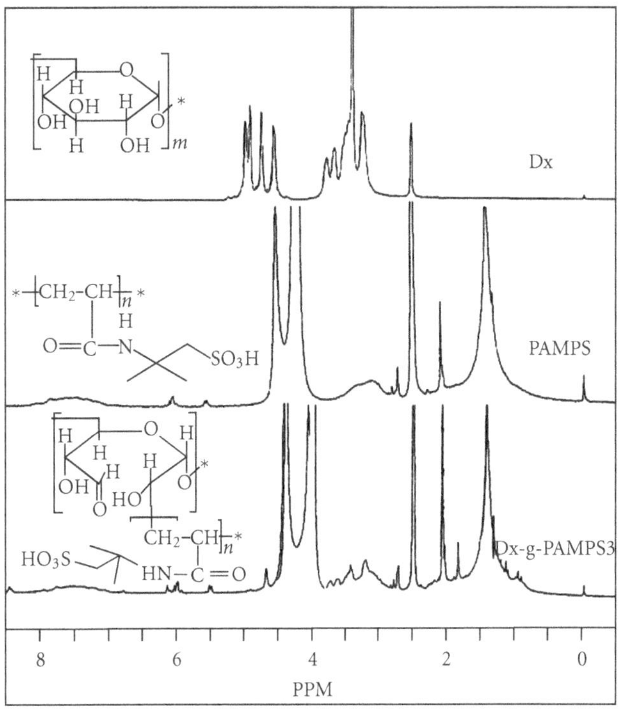

FIGURE 2: FT-^{1}H NMR of Dx, PAMPS, and Dx-g-PAMPS3.

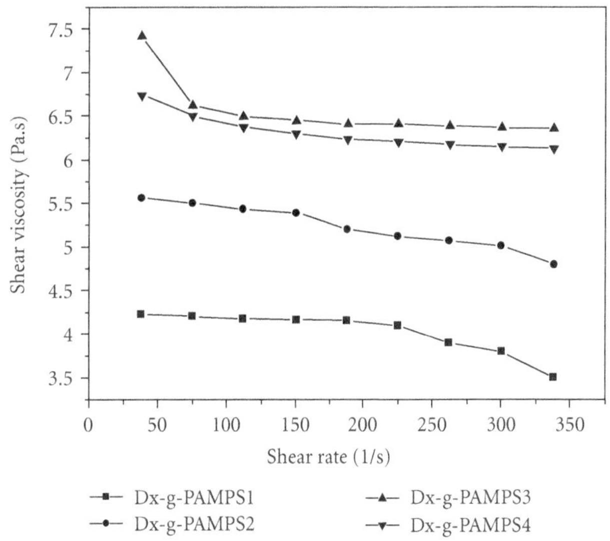

FIGURE 3: Viscosity versus shear rate curves of graft copolymers.

2.3.7. Thermal Analysis. The thermogravimetric analysis (TGA) of Dx, PAMPS, and its graft copolymer (Dx-g-PAMPS3) was carried out with a TGA-DTA, METTER-TOLEDO (Germany) instrument. Analysis of the samples (Figure 5) was performed up to a temperature of 600°C, starting from 50°C in nitrogen atmosphere [24, 25]. The heating rate was uniform in all cases at 10°C per minute.

2.3.8. X-Ray Diffractometry. Dextran, PAMPS, and the graft copolymer (Dx-g-PAMPS3) were subjected to XRD analysis (Figure 6). An 18-KW Cu-rotating anode Rigaku (Tokyo, Japan) X-ray powder diffractometer was used.

3. Flocculation Study

Jar test method was employed to estimate flocculation performance of synthesized graft copolymers. In 1 L beaker, 1 wt% kaolin suspension was taken. The beaker was placed on flocculator dipping the stirrer blade in the suspension. Under a low stirring condition, required quantity of polymer solution was added to beaker to make predetermined dose with respect of suspension volume. After the addition of polymer solution, the suspension was stirred at 185 RPM for 2 min followed by 100 RPM for 5 min. The flocs were allowed to settle down for 10 minutes. Clean supernatant liquid was drawn from a depth of 1.0 cm and its turbidity was measured using a digital nephelometer (digital nephelometer model, Bellstone Hitech International) to express the turbidity in nephelometric unit (NTU). A comparison of flocculation efficiency of Dx-g-PAMPS1 to Dx-g-PAMPS4 is shown in Table 2.

4. Results and Discussion

4.1. Synthesis of Graft Copolymers. Table 1 shows the synthetic details of the graft copolymers based on dextran and PAMPS by radical polymerization technique in aqueous medium using a ceric ammonium nitrate initiation system. The mechanism of ceric-ion-induced initiation involves the formation of chelate complex that decomposes to

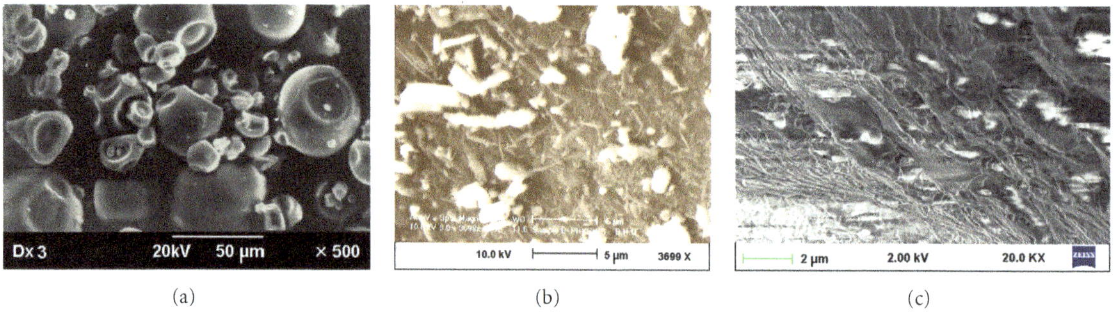

Figure 4: Scanning electron microscope of Dx, PAMPS, and Dx-g-PAMPS3.

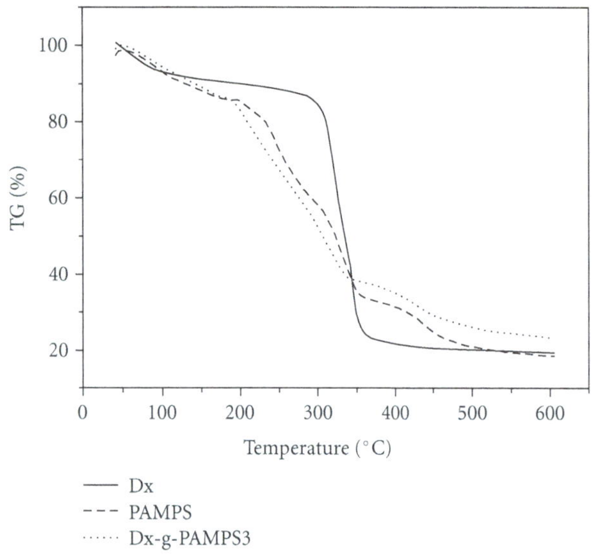

Figure 5: Thermogravimetric analysis (TGA) of Dx, PAMPS, and Dx-g-PAMPS3.

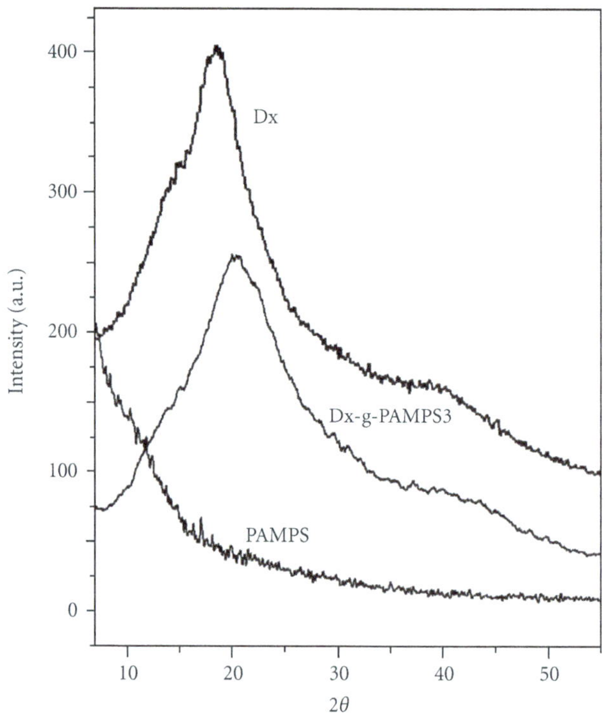

Figure 6: X-ray diffractometry of Dextran, PAMPS, and Dx-g-PAMPS3.

generate free radical sites on the polysaccharide backbone. Four grades of graft copolymers (Dx-g-PAMPS1 to Dx-g-PAMPS4) were synthesized by varying the amount of 2-acrylamido-2-methyl-1-propane sulphonic acid. The amount of AMPS was varied in order to observe the effect with varying chain length of grafted PAMPS. At fixed concentration of CAN, fixed number of grafting sites on the polysaccharide backbone was expected. Thus, difference in AMPS concentration will lead to different chain length of PAMPS. The average length of PAMPS depends on the concentration of AMPS and dextran. At high concentration of AMPS, longer PAMPS chains were expected, whereas shorter chains were expected at lower concentration of AMPS. At optimum concentration of AMPS, longest PAMPS chains were obtained. Therefore, it was determined by intrinsic viscosity. In this case, Dx-g-PAMPS3 is the grade showing highest intrinsic viscosity of 15.3 dL/g having maximum chain length.

4.2. Elemental Analysis. The results of elemental analysis of Dx and its graft copolymers are given in Table 2. Dextran does not show any significant presence of nitrogen. In case of graft copolymers, it was found that there is considerable percentage of nitrogen proving the grafting of PAMPS chains on the polysaccharides backbone. In the series of graft copolymers with Dx, highest percentage of nitrogen was observed in case of Dx-g-PAMPS3. Since this graft copolymer showed highest intrinsic viscosity and maximum grafting efficiency (GE), the highest percentage of nitrogen can be explained on the basis of longer branches of PAMPS on the dextran backbone.

4.3. Intrinsic Viscosity Measurement. The intrinsic viscosity of a polymer is a measure of its hydrodynamic volume in solution, which in turn is a function of the polymer molecular weight, its structure, nature of solvent, and the temperature of the medium. Keeping the other factors constant, for two polymers of approximately similar molecular weight, a branched polymer will have lower hydrodynamic volume and hence a lower intrinsic viscosity as compared to its linear counterpart. Further, along a series of branched polymers, the longer the branches, the higher will be the intrinsic viscosity and vice versa. This has been observed in practice.

In the series of graft copolymers (Dx-g-PAMPS1 to Dx-g-PAMPS4), the graft copolymer Dx-g-PAMPS3, having optimum AMPS concentration at a fixed CAN concentration, has higher intrinsic viscosity indicating that it has longer PAMPS chains. The intrinsic viscosity values of all graft copolymers are reported in Table 1.

4.4. FTIR Spectroscopy. FTIR spectra of the dextran and Dx-g-PAMPS3 are shown in Figure 1. The IR spectra of Dx, AMPS, and Dx-g-PAMPS3 are shown in Figure 2. In the IR spectrum of Dx, a broad peak present at 3423 cm^{-1} and is due to the stretching of –OH. Due to hydrophilic nature of Dx, a peak shows at 1647 cm^{-1} is due to bending of water molecule. The peaks present at 2929 cm^{-1} and 1458 cm^{-1} represent the stretching and bending frequency of $-CH_2$ group. A band at 1157 cm^{-1} confirms the presence of glycosidic linkage (C–O–C) in dextran. The band in the region of 1016 cm^{-1} is due to the C–O stretching. In case of AMPS spectrum, a broad band at 3448 cm^{-1} is due to the collapse of –NH and –OH ($-SO_3H$) stretching frequency. The peaks present at 3039 cm^{-1} and 2992 cm^{-1} represent the C–H stretching frequency of $-CH_3$ and $-CH_2$, respectively. Two bands around 1662 and 1613 cm^{-1} are due to C=O stretching (amide I band) and N–H bending (amide II band), respectively. Two peaks at 1379 and 1300 cm^{-1} are due to asymmetric and symmetrical stretching of (S=O) group. The strong stretching absorption in the region of 1089 to 950 cm^{-1} represent S–O–C group. Dx-g-PAMPS3 shows absorption peaks at 3440 cm^{-1} (–NH and –OH), 2962 cm^{-1} (C–H stretching), 1649 cm^{-1} and 1555 cm^{-1} (amide I and amide II bands), 1461 cm^{-1} (C–H bending), 1395 and 1215 cm^{-1} (asymmetric and symmetric stretching of S=O group), 1160 cm^{-1} (glycosidic linkage), and 1038 to 770 cm^{-1} (S–O–C group). The presence of all the peaks is due to that Dx and PAMPS are present in the graft copolymer (Dx-g-PAMPS3). This proved the grafting of AMPS on Dx backbone.

4.5. ^1H FTNMR Spectroscopy. Figure 2 shows the ^{1}H NMR of Dx, PAPMS, and Dx-g-PAMPS3 in DMSO solvent. In ^{1}H NMR of Dx signals in the region of 4 ppm to 5 ppm are due to the 3 and 6-linked-α-D-glc*p* residues. The complex spectrum in the region of 3-4 ppm resulted from protons in C_2,C_3,C_4, C_5, and C_6. In the case of PAMPS, the resonance peaks are at 7.45 ppm (–NH–), 4.4 ppm ($-CH_2-SO_3H$), 3.06 ppm ($-SO_3H$), 2.70 ppm (–CH–CO), 2.0 ppm ($-CH_2-CH-$), and 1.42 ppm ($-CH_3$). In the proton NMR of graft copolymer (Dx-g-PAMPS3), peaks were observed at 1.43 ppm ($-CH_3$), 1.85 ppm ($-CH_2-$), 2.71 ppm (–CH–CO), 3.2 to 3.7 ppm (proton in C_2,C_3,C_4, C_5, and C_6), 4.4 ppm ($-CH_2-SO_3H$), 4.6 ppm (anomeric proton), and 7.8 ppm (–NH–). The resonance peaks of Dx segment and those of PAMPS repeating unit were also observed in case of graft copolymer. This proves the grafting of PAMPS on Dx backbone.

4.6. Rheological Studies. The rheological studies of aqueous solution of Dx-g-PAMPS1 to Dx-g-PAMPS4 were carried out at concentration of 0.5 wt%. The shear viscosity versus shear rate on two axes in logarithmic scales was plotted. Figure 3 shows that the shear viscosity versus shear rate curves of graft copolymers of all the four graft copolymers in aqueous solution. In these cases, shear viscosity decreased with increase in shear rate and the samples showed shear thinning non-Newtonian behaviour (pseudoplastic). Among the four grades of graft copolymers, the Dx-g-PAMPS3 shows higher shear viscosity in the entire shear rate. This was due to the longer PAMPS chains grafted on to the dextran backbone.

4.7. Scanning Electron Microscopy. Figures 4(a), 4(b), and 4(c) show the scanning electron micrographs of Dx, APMS, and Dx-g-PAMPS3, respectively. The surface morphology of Dx and PAMPS before grafting indicates spherical and small needle-like structures, respectively, which have been changed to fibrillar form after grafting (Dx-g-PAMPS3). The comparison of these figures reveals that grafting has taken place.

4.8. Thermal Analysis. The thermal stabilities of graft copolymers were investigated by thermogravimetric analysis (TGA) in a nitrogen stream at a heating rate of 20°C/min. The TGA curves of Dx, PAMPS, and Dx-g-PAMPS3 were studied in the range of 50°C–600°C (Figure 5). All the polymer samples exhibit three degradation stages. In the first stage, weight loss of Dx, PAMPS, and Dx-g-PAMPS3 was due to water evolving in the range of 50°C–150°C, 50°C–230°C, and 50°C–250°C, respectively. In case of Dx, the second decomposition temperature occurred at 300°C due to degradation of main chain. In case of AMPS and graft copolymer, the weight loss at 250°C was assigned to the decomposition of sulphonic acid groups. In the third stage, combustion of the degraded products began at 350°C. In the last step of degradation of all the polymers, the graft copolymer has less weight loss than the other polymers. The thermal stability of Dx-g-PAMPS3 was higher than dextran and PAMPS beyond 350°C. Before 350°C, Dx-g-PAMPS3 was less stable than PAMPS and dextran.

4.9. X-Ray Diffractometry (XRD). XRD analysis of Dx, PAMPS, and Dx-g-PAMPS3 is shown in Figure 6. Dx and graft copolymer do not show crystallinity in contrast to PAMPS which shows small crystallinity. This is indicative of absence of AMPS monomer in the graft copolymer.

5. Flocculation Study

The plots of supernatant turbidity versus graft polymer dosage for kaolin suspension are given in Table 2. It has been found that graft copolymer Dx-g-PAMPS3 shows better performance (lesser turbidity over lowest polymer dose) than other graft copolymers. This could be explained due to the fact that in graft copolymer, the dangling longer poly-PAMPS chains have better approachability to the contaminant.

6. Conclusion

From the results, it can be concluded that PAMPS is grafted on Dx. Variation in the synthetic parameters results in a series of graft copolymers with varying length of PAMPS chains. The FTIR and FT^{1}H NMR spectra provide a strong proof of grafting. Rheological investigation shows that the graft copolymer having longer PAMPS chains has more shear viscosity. TGA results show that the graft copolymer has more thermal stability than corresponding Dx and PAMPS. Shear viscosity of the aqueous solution of the graft copolymer Dx-g-PAMPS3 is larger than the other graft copolymers. Morphology variation of Dx, PAMPS, and graft copolymers supports the above conclusion. X-ray diffraction pattern shows the amorphous nature of graft copolymer and absence of AMPS in the grafted product. Estimation of flocculation performance in kaolin suspension shows the graft copolymer Dx-g-PAMPS3 as the most effective flocculant. Thus, the graft copolymer in every way is more stable than the base polysaccharide. Thus, by grafting PAMPS onto dextran, shear stable efficient flocculants have been obtained and they can be used for the treatment of wastewater.

Acknowledgment

The authors are grateful to UGC (Grant number: 37-57/2009), New Delhi, for the financial support.

References

[1] C. Nouvel, P. Dubois, E. Dellacherie, and J.-L. Six, "Controlled synthesis of amphiphilic biodegradable polylactide-grafted dextran copolymers," *Journal of Polymer Science A*, vol. 42, no. 11, pp. 2577–2588, 2004.

[2] T. Ouchi, T. Kontani, and Y. Ohya, "Mechanical property and biodegradability of solution-cast films prepared from amphiphilic polylactide-grafted dextran," *Journal of Polymer Science A*, vol. 41, no. 16, pp. 2462–2468, 2003.

[3] T. Ouchi, T. Kontani, R. Aoki, T. Saito, and Y. Ohya, "Characteristic properties of film prepared from poly(L-lactide)-grafted dextran of a relatively high sugar unit content as a degradable biomaterial," *Journal of Polymer Science A*, vol. 44, no. 21, pp. 6402–6409, 2006.

[4] I. Uraz and A. Güner, "Comparison of molecular association of dextran and periodate-oxidized dextran in aqueous solutions," *Carbohydrate Polymers*, vol. 34, no. 3, pp. 127–130, 1997.

[5] H. Wondraczek, T. Elschner, and T. Heinze, "Synthesis of highly functionalized dextran alkyl carbonates showing nanosphere formation," *Carbohydrate Polymers*, vol. 83, no. 3, pp. 1112–1118, 2011.

[6] M. P. Bajgai, D. C. Parajuli, J. A. Ko, H. K. Kang, M.-S. Khil, and H. Y. Kim, "Synthesis, characterization and aqueous dispersion of dextran-g-poly(1,4-dioxan-2-one) copolymers," *Carbohydrate Polymers*, vol. 78, no. 4, pp. 833–840, 2009.

[7] S. Krishnamoorthi, D. Mal, and R. P. Singh, "Characterization of graft copolymer based on polyacrylamide and dextran," *Carbohydrate Polymers*, vol. 69, no. 2, pp. 371–377, 2007.

[8] B. G. De Geest, W. Van Camp, F. E. Du Prez, S. C. De Smedt, J. Demeester, and W. E. Hennink, "Biodegradable microcapsules designed via 'click' chemistry," *Chemical Communications*, no. 2, pp. 190–192, 2008.

[9] N. Pahimanolis, A.-H. Vesterinen, J. Rich, and J. Seppala, "Modification of dextran using click-chemistry approach in aqueous media," *Carbohydrate Polymers*, vol. 82, no. 1, pp. 78–82, 2010.

[10] M. Tang, H. Dou, and K. Sun, "One-step synthesis of dextran-based stable nanoparticles assisted by self-assembly," *Polymer*, vol. 47, no. 2, pp. 728–734, 2006.

[11] R. Y. Cheung, Y. Ying, A. M. Rauth, N. Marcon, and X. Y. Wu, "Biodegradable dextran-based microspheres for delivery of anticancer drug mitomycin C," *Biomaterials*, vol. 26, no. 26, pp. 5375–5385, 2005.

[12] A. P. Rokhade, S. A. Patil, and T. M. Aminabhavi, "Synthesis and characterization of semi-interpenetrating polymer network microspheres of acrylamide grafted dextran and chitosan for controlled release of acyclovir," *Carbohydrate Polymers*, vol. 67, no. 4, pp. 605–613, 2007.

[13] J. Raynaud, B. Choquenet, E. Marie et al., "Emulsifying properties of biodegradable polylactide-grafted dextran copolymers," *Biomacromolecules*, vol. 9, no. 3, pp. 1014–1021, 2008.

[14] S. Krishnamoorthi, D. Mai, and R. P. Singh, "Characterization and solution properties of a partially hydrolyzed graft copolymer of polyacrylamide and dextran," *Journal of Applied Polymer Science*, vol. 110, no. 3, pp. 1297–1303, 2008.

[15] S. Krishnamoorthi, P. Adhikary, D. Mal, and R. P. Singh, "Novel polymeric flocculants based on polyacrylamide grafted dextran in kaolin suspension," *Journal of Applied Polymer Science*, vol. 118, no. 6, pp. 3539–3544, 2010.

[16] P. Adhikary, K. N. Tiwari, and R. P. Singh, "Synthesis, characterization, and flocculation characteristics of polyacrylamide-grafted glycogen," *Journal of Applied Polymer Science*, vol. 103, no. 2, pp. 773–778, 2007.

[17] Y. H. Gad, "Preparation and characterization of poly(2-acrylamido-2-methylpropane- sulfonic acid)/Chitosan hydrogel using gamma irradiation and its application in wastewater treatment," *Radiation Physics and Chemistry*, vol. 77, no. 9, pp. 1101–1107, 2008.

[18] C. Zhang and A. J. Easteal, "Study of free-radical copolymerization of N-isopropylacrylamide with 2-acrylamido-2-methyl-1-propanesulphonic acid," *Journal of Applied Polymer Science*, vol. 88, no. 11, pp. 2563–2569, 2003.

[19] S. J. De Jong, S. C. De Smedt, M. W. C. Wahls, J. Demeester, J. J. Kettenes-van den Bosch, and W. E. Hennink, "Novel self-assembled hydrogels by stereocomplex formation in aqueous solution of enantiomeric lactic acid oligomers grafted to dextran," *Macromolecules*, vol. 33, no. 10, pp. 3680–3686, 2000.

[20] C. Perrino, S. Lee, S. W. Choi, A. Maruyama, and N. D. Spencer, "A biomimetic alternative to poly(ethylene glycol) as an antifouling coating: resistance to nonspecific protein adsorption of poly(L-lysine)-graft- dextran," *Langmuir*, vol. 24, no. 16, pp. 8850–8856, 2008.

[21] Y. Bao, J. Ma, and N. Li, "Synthesis and swelling behaviors of sodium carboxymethyl cellulose-g-poly(AA-co-AM-co-AMPS)/MMT superabsorbent hydrogel," *Carbohydrate Polymers*, vol. 84, no. 1, pp. 76–82, 2011.

[22] A. P. Goodwin, S. M. Tabakman, K. Welsher, S. P. Sherlock, G. Prencipe, and H. Dai, "Phospholipid-dextran with a single coupling point: a useful amphiphile for functionalization of nanomaterials," *Journal of the American Chemical Society*, vol. 131, no. 1, pp. 289–296, 2009.

[23] A. Richard, A. Barras, A. B. Younes, N. Monfilliette-Dupont, and P. Melnyk, "Minimal chemical modification of reductive end of dextran to produce an amphiphilic polysaccharide able to incorporate onto lipid nanocapsules," *Bioconjugate Chemistry*, vol. 19, no. 7, pp. 1491–1495, 2008.

[24] J. Maia, R. A. Carvalho, J. F. J. Coelho, P. N. Simões, and M. H. Gil, "Insight on the periodate oxidation of dextran and its structural vicissitudes," *Polymer*, vol. 52, no. 2, pp. 258–265, 2011.

[25] E. Fathi, N. Atyabi, M. Imani, and Z. Alinejad, "Physically crosslinked polyvinyl alcohol-dextran blend xerogels: morphology and thermal behavior," *Carbohydrate Polymers*, vol. 84, no. 1, pp. 145–152, 2011.

6

Development and Characterization of Semi-IPN Silver Nanocomposite Hydrogels for Antibacterial Applications

A. Chandra Babu,[1] M. N. Prabhakar,[2] A. Suresh Babu,[3] B. Mallikarjuna,[1] M. C. S. Subha,[1] and K. Chowdoji Rao[2]

[1] *Department of Chemistry, Sri Krishnadevaraya University, Anantapur, Andhra Pardesh 515 003, India*
[2] *Department of Polymer Science and Technology, Sri Krishnadevaraya University, Anantapur, Andhra Pradesh 515 003, India*
[3] *Department of Physics, Rayalaseema University, Kurnool, Andhra Pardesh 518 002, India*

Correspondence should be addressed to K. Chowdoji Rao; chowdojirao@gmail.com

Academic Editor: R. J. Linhardt

Sodium carboxymethyl cellulose/poly(acrylamide-co-2-acrylamido-2-methylpropane sulfonic acid) semi-interpenetrating polymer network (semi-IPN) hydrogels were prepared by using free radical polymerization technique. Silver nanoparticles were formed by reduction of silver nitrate in semi-IPN hydrogels with sodium borohydrate at room temperature. UV-visible spectroscopy, thermogravimetrical analysis, X-ray diffractometry, scanning electron microscopy, and transmission electron microscopy techniques were used to characterize the formation of silver nanoparticles in hydrogels. SEM images indicated clearly the formation of group of silver nanoparticles with size range of 10–20 nm. The sizes of silver nanoparticles were also supported by transmission electron microscopy results. The semi-IPN silver nanocomposite hydrogels reported here might be a potentially smart material in the range of applications of antibacterial activity.

1. Introduction

Nanocomposite polymer hydrogels may be defined as cross-linked three-dimensional polymer networks swollen with water or biological fluids in the presence of nanoparticles. The design and development of such materials containing metallic nanoparticles have scientific and technological research interests in recent years due to their unique and versatile properties [1–5]. These properties lead to potential applications in the field of numerous physical, biological, biomedical, and pharmaceutical sectors [6–13] as well as optical, electrical, chemical, and data storage [14–18]. These properties are known for silver in the form of ions, colloidal particles, nanoparticles, metallic silver, and silver compounds, and many works study their use to inhibit the proliferation of microorganisms for medical [19–21], food packaging [22, 23], and water treatment [24, 25] applications. Generally silver ions, as heavy metals, lead to the inactivation of proteins reacting with thiol groups (–SH) on the membrane of bacteria causing the microbial cell death [26–28].

The biological activity of silver, especially the antibacterial property, is size dependent [29]. Thus silver nanoparticles should be small enough to pass through the cell membrane. Silver ions kill microorganisms instantly by blocking their respiratory enzyme systems, while having no negative effect on human cells. Although nanocomposites containing metal NPs have elegant features, the homogeneous dispersion of metal NPs is not easy using a simple process because of their high surface free energy, which may cause agglomeration. Therefore, preparation of such nanocomposites with desired properties (i.e., Ag NPs with convenient size) is an important issue. Thus various strategies have been employed to prepare nanocomposites containing silver nanoparticles with controlled size. One such effective approach reported by Wang et al. [30–32] is the *in situ* reduction of metal ions in the hydrogel networks. In this method the functional groups of the hydrogel network can act as an anchoring agent for the metal ions before *in situ* reduction.

Usually, carboxylic acid groups can be used as a transient anchoring agent for attaching silver cations to polymer hydrogels, and the silver ion is reduced with sodium borohydride to form nanoparticles [33]. This method of preparation depends on the number of functional groups present to stabilize the silver ions, but the nanoparticles are dispersed within the hydrogel matrix. One successful approach that has been employed is the reductive process in homogeneous matrices. In this approach the previously formed crosslinked polymer's swollen with a solution of silver salts and reducing agent, and then reduction takes place within the polymer matrix [34].

This paper focuses on the current design and use of cellulose-based hydrogels (NaCMC/poly(AAm-co-AMPS), which usually couple their biodegradability with a smart stimuli-sensitive behavior. Sodium carboxymethyl cellulose (NaCMC) is an ionic ether of the cellulose with too many major commercial applications. It is a highly water soluble anionic polysaccharide that is widely used in pharmaceuticals, cosmetics, and food industry [35–37]. Then 2-acrylamido-2-methyl-propane sulfonic acid is basically a strong acid [38], which can act as a good hydrophilic monomer, and therefore its copolymers have superswelling properties [39]. In this work we report the results obtained for sodium carboxymethyl cellulose/poly(acrylamide-co-2-acrylam-ido-2-methylpropane sulphonic acid) (NaCMC/poly(AAm-co-AMPS) semi-IPN hydrogels designed as a new antimicrobial biomaterial. Silver containing NaCMC/poly(AAm-co-AMPS) samples were characterized by their structural characteristics, swelling, UV-Vis, TGA, X-RD, and SEM as well as by their antimicrobial performances.

2. Experimental

2.1. Materials. Acrylamido methyl propane sulphonic acid (AMPS) was purchased from Aldrich Chemicals (St. Louis). Sodium carboxymethyl cellulose (NaCMC) was purchased from HiMedia Chemicals, India. Acrylamide (AAm), N,N-methylenebisacrylamide (MBA), ammonium persulfate (APS), and N,N,N,N-tetra methyl ethylenediamine (TEMED) were purchased from S.D. Fine Chemicals, India. Silver nitrate ($AgNO_3$), Sodium borohydrate ($NaBH_4$) were purchased from Merck (Germany). All chemicals were used without further purification, and double distilled water was used for the preparation of all solutions for this study.

2.2. Preparation of Semi-IPN Silver Nanocomposite Hydrogels. NaCMC/poly(AAm-co-AMPS) semi-IPN hydrogels were synthesized by employing free radical polymerization using *N,N*-methylenebisacrylamide as crosslinker and ammonium persulfate/N,N,N′,N′-tetramethylethylenediamine as redox-initiating pair. First, 0.90 gm of AAm and 0.10 g of NaCMC were dissolved in 3 mL of distilled water in a 250 mL beaker under constant stirring. To this reaction mixture, 0.99 gm of AMPS was added as a comonomer. Then 0.09 gm of MBA as a hydrophilic crosslinker and 0.49 gm of APS and 0.86 mM of TMEDA as an initiating pair system were added to this solution. After the addition of the reactants, the temperature of the system was raised to 50°C for 30 min. After the reaction has been completed, the formed hydrogel was immersed in distilled water at room temperature for 24 h to remove the unreacted materials present in the hydrogel network. Finally, the hydrogel was dried at room temperature for 2 days. Similarly, other hydrogels were prepared by the above procedure. Different types of hydrogels prepared along with their feed composition are presented in Table 1.

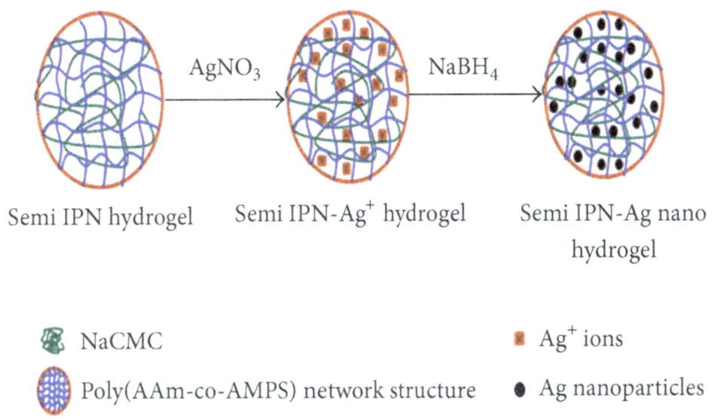

Scheme 1: Schematic representation of formation of Ag nanoparticles in the polymeric network.

The loading of silver nanoparticles into the hydrogel networks has been performed using the following procedure. Accurately weighed dried copolymeric hydrogels were swelled in double distilled water for 3 days. These swollen copolymeric hydrogels were then transferred into another beaker containing 50 mL of aqueous 5 mM $AgNO_3$ solution and then allowed to equilibrating the hydrogel for 24 hours to absorb the silver salt. During this process, most of the silver ions were loaded into free-network spaces of hydrogel. These hydrogels with absorbed silver ions were finally transferred into a beaker containing 50 mL of 10 mM $NaBH_4$ aqueous solution and kept for 2 hours to reduce the silver ions into silver nanoparticles. After reduction of silver ions, the hydrogel was turned into brown colour, which confirms the formation of silver nanoparticles in copolymeric hydrogel matrix as shown in Scheme 1.

2.3. Swelling Studies. Fully dried semiIPN hydrogels, silver nanocomposite hydrogels were accurately weighed and equilibrated in distilled water at 37°C for 3 days. The equilibrium swelling capacity or swelling ratio (Q) of the hydrogels was calculated by employing

$$Q = \frac{W_e}{W_d}, \quad (1)$$

where W_e is the weight of swollen hydrogel and W_d is the dry weight of the semiIPN hydrogel.

2.4. Characterization. UV-visible spectra of semiIPN hydrogel silver nanocomposites (10 mg in 1 mL of distilled H_2O) were characterized using (Lab India UV 3000$^+$) UV-Visible spectrophotometer; X-ray diffraction measurements were also carried out for these nanocomposites using A Siemens D 5000 (Germany) Powder X-ray diffractometer to know the crystallinity of the hydrogel. Thermal properties of semiIPN

Table 1: Various formulation parameters used in the preparation of the hydrogels.

Sample code	AAm (gm)	NaCMC (gm)	AMPS (gm)	MBA (gm)	APS (gm)	TEMED (mM)
SIPN-1	0.9	**0.05**	0.99	0.09	0.49	0.860
SIPN-2	0.9	**0.10**	0.99	0.09	0.49	0.860
SIPN-3	0.9	**0.15**	0.99	0.09	0.49	0.860
SIPN-4	0.9	**0.20**	0.99	0.09	0.49	0.860
SIPN-5	0.9	0.10	**0.19**	0.09	0.49	0.860
SIPN-6	0.9	0.10	**0.29**	0.09	0.49	0.860
SIPN-7	0.9	0.10	**0.49**	0.09	0.49	0.860
SIPN-8	0.9	0.10	**1.22**	0.09	0.49	0.860
SIPN-9	0.9	0.10	0.99	**0.02**	0.49	0.860
SIPN-10	0.9	0.10	0.99	**0.04**	0.49	0.860
SIPN-11	0.9	0.10	0.99	**0.07**	0.49	0.860
SIPN-12	0.9	0.10	0.99	**0.12**	0.49	0.860

hydrogels were evaluated by using SDT-Q600 thermal system at heating rage of 10°C/min under N_2 atmosphere (flow rate 100 mL/min). Morphological variations of dry hydrogel, silver nanocomposite hydrogel were studied using JSM 6400 Scanning Electron Microscope (SEM) (Japan), operating at an acceleration voltage of 15 K volts coating with a thin layer of palladium gold alloy. Particle size of silver nanoparticles was performed out by using a Tecnai F 12 Transmission Electron Microscope (TEM) operating at an acceleration voltage of 15 kV.

3. Results and Discussion

3.1. Swelling Studies. In this study, acryl amide and AMPS were copolymerized in the presence of NaCMC to obtain NaCMC/poly (AAm-co-AMPS) semi-IPN hydrogels, where the NaCMC chains are randomly distributed throughout the gel networks. The special interest in choosing NaCMC as an interpenetrating polymer in hydrogels is due to its viscosity increasing, binding, protecting, reducing, stabilizing, and disintegration properties [40], that is widely used in pharmaceutical applications. The main criterion in selecting AMPS as a comonomer in the preparation of hydrogel is due to the presence of sulphonic acid groups, which are highly water attractive in nature and thereby increase the water absorbing capacity. Hence the hydrogels swell more as the AMPS concentration increases.

Figure 1 shows the swelling studies of various formulations of semi-IPN pure hydrogels and their silver loaded hydrogels. For all these formulations the order of swelling capacity for semi-IPN hydrogels was found to be greater than the silver nanocomposite hydrogels. From Figures 1(a) and 1(b) it is observed that as the amount of NaCMC and AMPS weight ratio increased, the swelling ratio also increased. This can be attributed due to the presence of polar groups (–OH, $-CONH_2$, $-COO^-$, Na^+, and SO_3H) in polymeric chains. The presence of the ionic groups in polymer chains results in increasing of the swelling ratio because the ionic groups are more strongly solvated than the nonionic groups in the aqueous medium.

The effect of crosslinker concentration on the swelling behavior of semi-IPN hydrogels was studied by varying the amount of crosslinker content 0.02–0.12 gm. In Figure 1(c) as the crosslinker concentration is increased the swelling ratio decreased this is because the higher the concentration of crosslinker, the higher is the extent of crosslinking in the polymer network and decrease diameter of pores of resultant hydrogels leading to solvent mobility, and then less will be the swelling ratio. Similar observations have been reported in the literature [41–44].

3.2. UV-Visible Spectroscopy. Generally in silver nanoparticles the conduction band and valence band lie very close to each other in which electrons move freely. These free electrons gives rise to a surface plasmon resonance (SPR) absorption band [45–47] (380–500 nm range) occurring due to the collective oscillation of electrons of silver nanoparticles in resonance with the light wave of UV-Vis spectroscopy. The formation of silver nanoparticles in semi-IPN hydrogel networks can be expected in our spectral analysis. Figure 2 illustrates the absorption spectra for hydrogel loaded with Ag^+ ions in the range of 300 nm and silver nanocomposites in the range of 380–500 nm. A significant improvement in the absorption peak (λ_{max} = 420 nm) was observed due to the surface plasmon resonance (SPR) effect. However, there is no peak observed in pure semi-IPN hydrogel. This spectral analysis confirmed the formation of silver nanoparticles in the semi-IPN hydrogel networks.

3.3. Thermogravimetric Analysis. The semi-IPN silver nanocomposite hydrogels are characterized by thermogravimetrical analysis to determine the percentage of weight loss of pure hydrogel as well as silver nanoparticles in the hydrogel matrix. Figure 3 shows the percentage decomposition of pure hydrogel and silver nanocomposite hydrogel. The pure hydrogel has two decomposition curves at below 500°C and 98% degradation of the hydrogel chains occurs below 800°C. However, in case of silver nanocomposite hydrogel two degradation curves at below 500°C, and only 88% weight

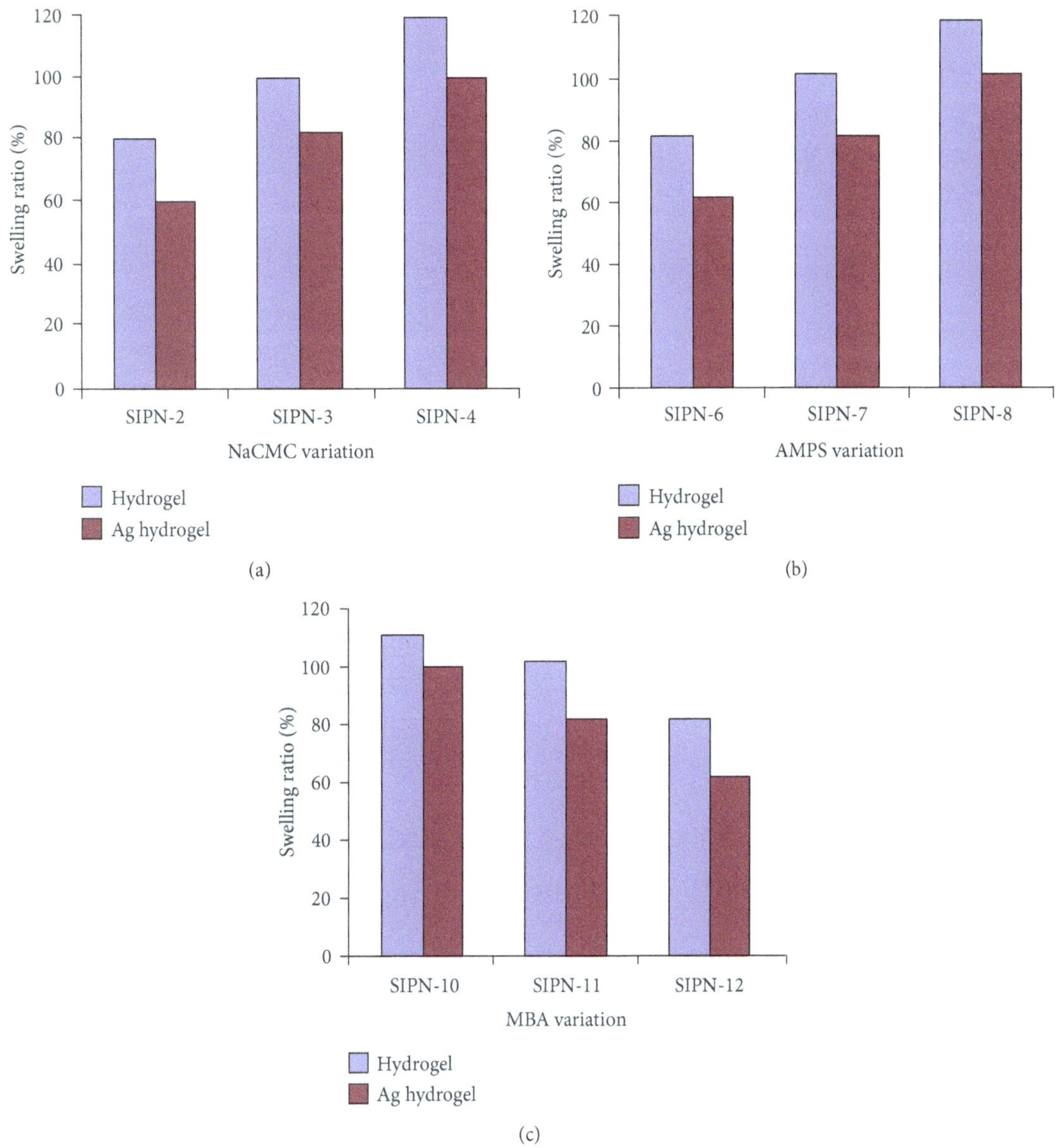

FIGURE 1: Swelling behavior of semiIPN pure hydrogels and silver nanocomposite hydrogels of various formulations.

loss occurred even at 800°C. It is expected that the remaining weight of silver particles is left in Ag hydrogel at 800°C. The difference in decomposition between the hydrogel and silver Nanocomposite hydrogel is found to be 10%, and it confirms the presence of silver nanoparticles in the hydrogel. In the present invention silver nanoparticles play a role in degradation of hydrogel, which restricts the percentage of weight loss.

3.4. X-RD Analysis. The crystallographic nature of the silver nanoparticles in hydrogels was investigated by X-ray diffraction. The X-ray diffraction of pure and silver nanocomposite semi-IPN hydrogels was illustrated in Figure 4. The diffractogram of semi-IPN hydrogel silver nanocomposites is assigned to diffraction at 2θ value about 38°, 44°, 65°, 77°, and 82° to the brags reflections of (111), (200), (220), (311), and (222) lattice planes [48], are observed. These peaks are matched with the face centered cubic (FCC) structure of silver (JCPDS card file no. 04-0783). The intense peak represents to highly crystalline silver nanostructures formed in semi-IPN silver nanocomposite hydrogels. In pure hydrogel there is no peak observed; this is due to the absence of silver nanoparticles in pure semi-IPN hydrogel.

3.5. Scanning Electron Microscopy. The scanning electron micrographs of pure hydrogel and silver nanocomposite hydrogels were shown in Figure 5. The pure hydrogel showed branches and rough surface (Figures 5(a) and 5(b)), whereas silver nanocomposite hydrogels showed a shrunken surface (Figures 5(c) and 5(d)) having uniform distribution of silver nanoparticles in size around 10–20 nm. However there is a pin point variation in the case of silver nanoparticles formed in

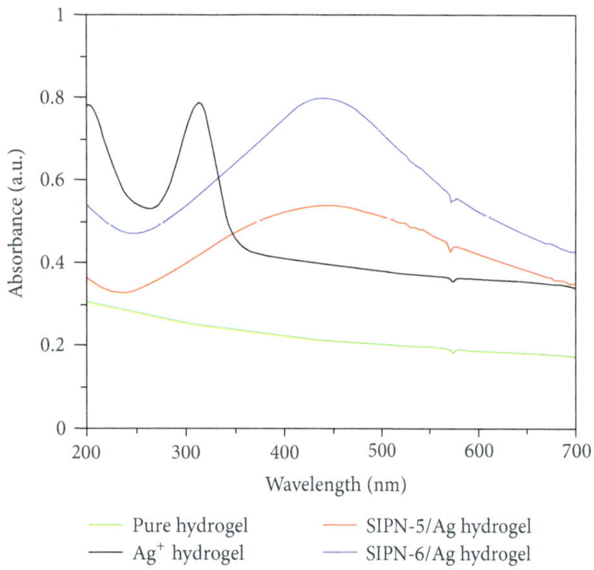

FIGURE 2: UV-visible spectral analysis of pure hydrogel, hydrogel loaded with $AgNO_3$ before reduction, and silver nanocomposite hydrogels.

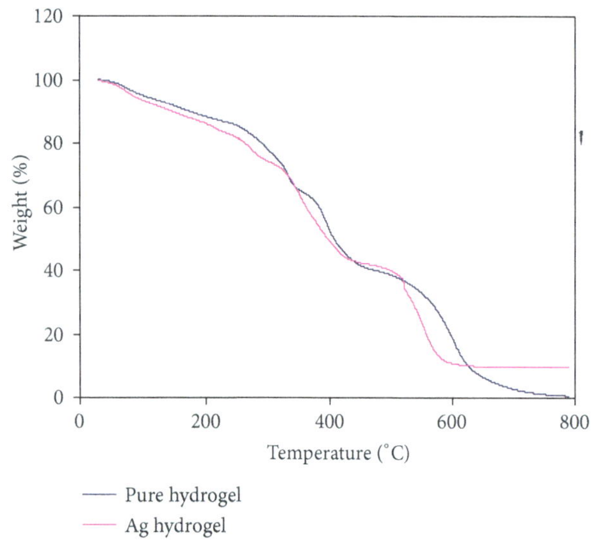

FIGURE 3: Thermogravimetric analysis of pure hydrogel and silver nanocomposite hydrogel.

the gel networks. This clearly indicates the formation of silver nanoparticles along with the polymer chains throughout the gel networks. It reveals that NaCMC polymeric chains control to lower particle size due to effective reduction and stabilization properties. It is quite common that the control or alignment of silver nanoparticles can be achieved by modifying the hydrogel network architectures.

3.6. Transmission Electron Microscopy. TEM image demonstrates a highly uniform distribution of silver nanoparticles (20 nm) as shown in Figure 6. It is clear that the silver

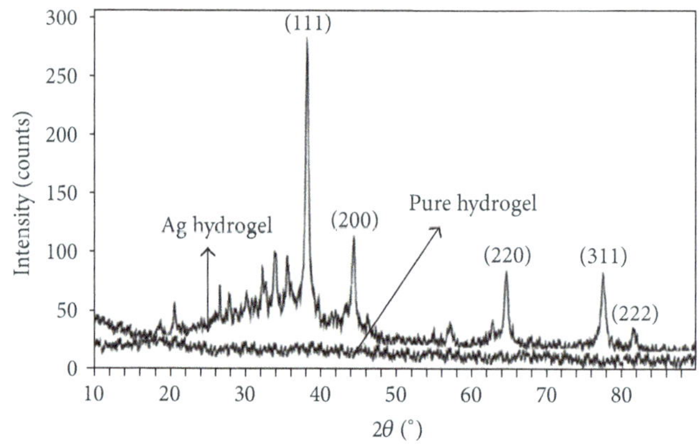

FIGURE 4: X-ray diffraction pattern of pure hydrogel and silver nanocomposite hydrogel.

nanoparticles exhibit obviously no difference in size and morphology because silver (Ag^+) ions were loaded at equilibrium swelling of hydrogel before reduction. The Ag particles were small, regular, when Ag^+ ions were loaded at the equilibrium swelling ratio of hydrogel. It is confirmed that the silver nanoparticles formed in the crosslinked networks are spherical, highly dispersed, and low nanometer in size.

3.7. Antibacterial Studies. In this study the bacterial effect of silver nanocomposite hydrogels mostly depends on the size of particles and the swelling ratio, which have a direct interaction with the bacteria. Antibacterial studies of pure semi-IPN hydrogel and silver nanocomposite hydrogels were tested by paper disc method using *Bacillus subtilis* as shown in Figure 7. The results suggest that SIPN-4, SIPN-3, SIPN-2, SIPN-1 have an antibacterial activity against *Bacillus subtilis* (Gram +ve bacteria), while pure hydrogels were generally inefficient.

The inhibition area follows the order SIPN-4 > SIPN-3 > SIPN-2 > SIPN-1. This behavior is expected for silver nanoparticles of small sizes (1–20 nm) and the same order is expected for swelling ratio. So that these silver nanoparticles could come out easily and could interact with lipid layer of cell membrane, thereby attaching to microbial DNA to prevent bacterial replication, which are responsible for inhibition of bacterial growth.

4. Conclusion

NaCMC/poly (AAm-co-AMPS) semi-IPN silver nanocomposite hydrogels were successfully prepared via a free radical polymerization along with varying doses of polymer, comonomer, and crosslinking agent thereby reduction of silver ions into silver nanoparticles. The developed silver nanoparticles are well characterized by using different techniques to confirm the formation of silver nanoparticles and antibacterial activity on *Bacillus subtilis* bactericides. The swelling capacity of the resulting hydrogels was found to be directly related to NaCMC, AMPS, and MBA ratio

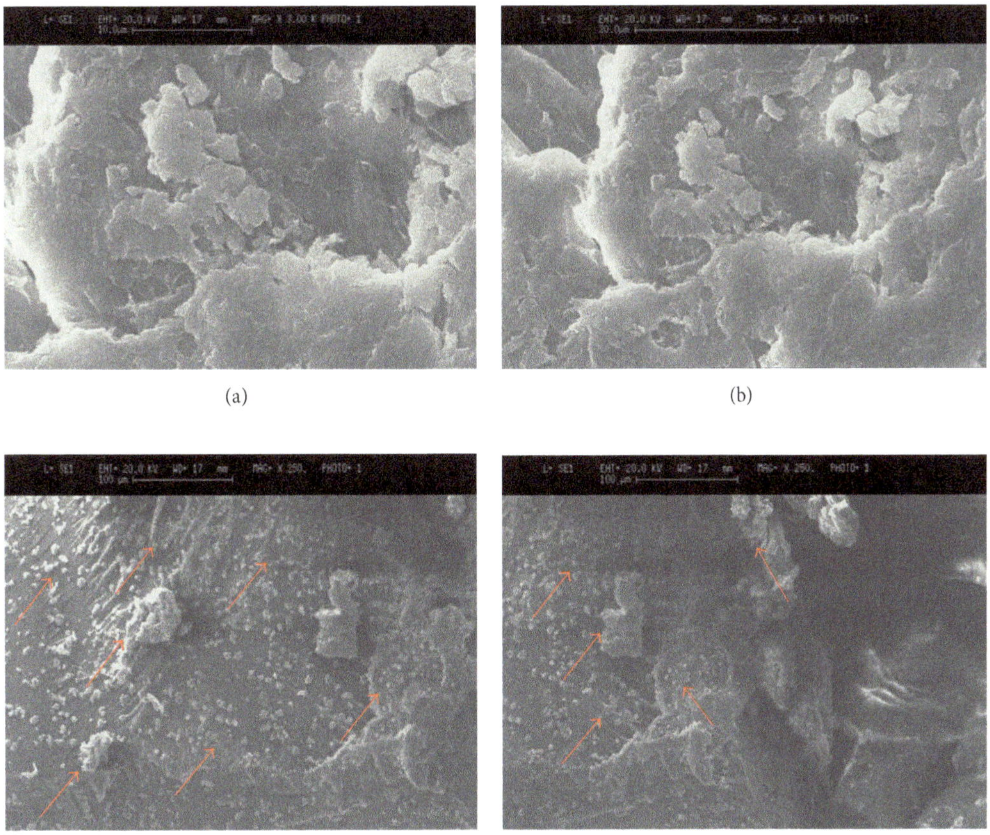

FIGURE 5: SEM images of pure hydrogels (a) and (b), silver nanocomposite hydrogels (c) and (d).

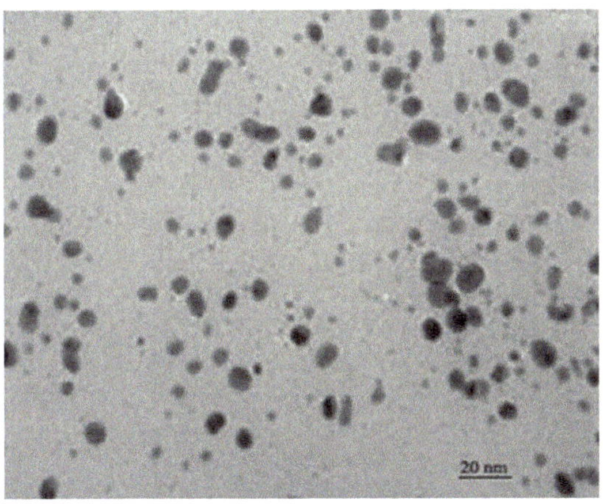

FIGURE 6: TEM image of silver nanocomposite hydrogel.

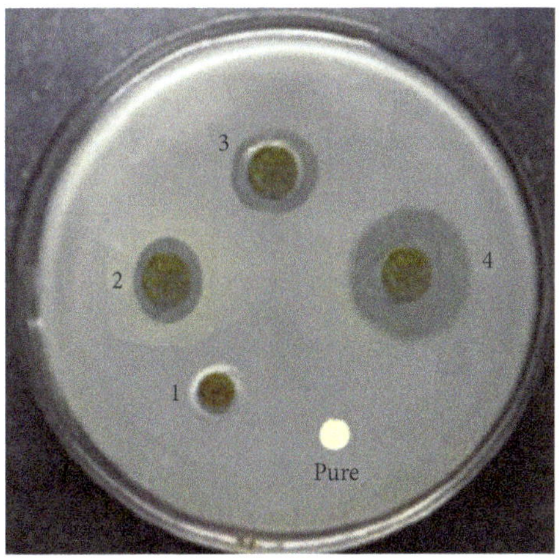

FIGURE 7: Antibacterial activity of silver nanocomposite hydrogels.

used in the polymerization process. UV-visible spectra, thermal analysis, and X-RD data revealed the formation of silver nanoparticles through examination of micrographs obtained using SEM and TEM images. The developed silver nanocomposite hydrogels confirmed that small sizes and abundance of silver nanoparticles determine the antibacterial activity.

Acknowledgments

The authors (A. C. Babu and K. C. Rao) gratefully acknowledge the support by the Defence Research and Development Organization (DRDO) (Sanction letter no. ERIP/ER/1003839 M/1341, dated June 28, 2011) and Ministry of Defense, Government of India, New Delhi, for the financial support.

References

[1] M. M. Demir, M. A. Gulgun, Y. Z. Menceloglu et al., "Palladium nanoparticles by electrospinning from poly(acrylonitrile-co-acrylic acid)-$PdCl_2$ solutions. Relations between preparation conditions, particle size, and catalytic activity," *Macromolecules*, vol. 37, no. 5, pp. 1787–1792, 2004.

[2] S. Porel, S. Singh, S. S. Harsha, D. N. Rao, and T. P. Radhakrishnan, "Nanoparticle-embedded polymer: in situ synthesis, free-standing films with highly monodisperse silver nanoparticles and optical limiting," *Chemistry of Materials*, vol. 17, no. 1, pp. 9–12, 2005.

[3] E. Delamarche, M. Geissler, J. Vichiconti et al., "Electroless deposition of NiB on 15 inch glass substrates for the fabrication of transistor gates for liquid crystal displays," *Langmuir*, vol. 19, no. 14, pp. 5923–5935, 2003.

[4] S. G. Boyes, B. Akgun, W. J. Brittain, and M. D. Foster, "Synthesis, characterization, and properties of polyelectrolyte block copolymer brushes prepared by atom transfer radical polymerization and their use in the synthesis of metal nanoparticles," *Macromolecules*, vol. 36, no. 25, pp. 9539–9548, 2003.

[5] S. Yoda, A. Hasegawa, H. Suda et al., "Preparation of a platinum and palladium/polyimide nanocomposite film as a precursor of metal-doped carbon molecular sieve membrane via supercritical impregnation," *Chemistry of Materials*, vol. 16, no. 12, pp. 2363–2368, 2004.

[6] W. C. W. Chan and S. Nie, "Quantum dot bioconjugates for ultrasensitive nonisotopic detection," *Science*, vol. 281, no. 5385, pp. 2016–2018, 1998.

[7] A. P. Alivisatos, "Semiconductor clusters, nanocrystals, and quantum dots," *Science*, vol. 271, no. 5251, pp. 933–937, 1996.

[8] W. C. W. Chan, D. J. Maxwell, X. Gao, R. E. Bailey, M. Han, and S. Nie, "Luminescent quantum dots for multiplexed biological detection and imaging," *Current Opinion in Biotechnology*, vol. 13, pp. 40–46, 2002.

[9] X. M. Wu, H. Liu, J. Liu et al., "Immunofluorescent labeling of cancer marker Her2 and other cellular targets with semiconductor quantum dots," *Nature Biotechnology*, vol. 21, no. 4, pp. 41–46, 2003.

[10] I. Brigger, C. Dubernet, and P. Couvreur, "Nanoparticles in cancer therapy and diagnosis," *Advanced Drug Delivery Reviews*, vol. 54, no. 5, pp. 631–651, 2002.

[11] F. Forestier, P. Gerrier, C. Chaumard, A. M. Quero, P. Couvreur, and C. Labarre, "Effect of nanoparticle-bound ampicillin on the survival of Listeria monocytogenes in mouse peritoneal macrophages," *Journal of Antimicrobial Chemotherapy*, vol. 30, no. 2, pp. 173–179, 1992.

[12] I. Sondi, O. Siiman, and E. Matijević, "Preparation of aminodextran-CdS nanoparticle complexes and biologically active antibody–aminodextran–CdS nanoparticle conjugates," *Langmuir*, vol. 16, no. 7, pp. 3107–3118, 2000.

[13] O. Siiman, E. Matijevic, and I. Sondi, "Semi conductor nanoparticles for analysis of blood cell populations and methods of making same," U.S. Patent 6, 235, 540 B1.

[14] A. Biswas, O. C. Aktas, U. Schürmann et al., "Tunable multiple plasmon resonance wavelengths response from multicomponent polymer-metal nanocomposite systems," *Applied Physics Letters*, vol. 84, no. 14, pp. 2655–2657, 2004.

[15] P. M. Ajayan, L. S. Schadler, and P. V. Braun, *Nanocomposite Science and Technology*, Wiley-VCH, Weinheim, Germany, 2003.

[16] Y. M. Mohan, K. Lee, T. Premkumar, and K. E. Geckeler, "Hydrogel networks as nanoreactors: a novel approach to silver nanoparticles for antibacterial applications," *Polymer*, vol. 48, no. 1, pp. 158–164, 2007.

[17] Y. Lu, P. Spyra, Y. Mei, M. Ballauff, and A. Pich, "Composite hydrogels: robust carriers for catalytic nanoparticles," *Macromolecular Chemistry and Physics*, vol. 208, no. 3, pp. 254–261, 2007.

[18] J. Y. Ouyang, C. W. Chu, C. R. Szmanda, L. P. Ma, and Y. Yang, "Programmable polymer thin film and non-volatile memory device," *Nature Materials*, vol. 3, no. 12, pp. 918–922, 2004.

[19] K. Yoshida, M. Tanagawa, S. Matsumoto, T. Yamada, and M. Atsuta, "Antibacterial activity of resin composites with silver-containing materials," *European Journal of Oral Sciences*, vol. 107, no. 4, pp. 290–296, 1999.

[20] F. Furno, K. S. Morley, B. Wong et al., "Silver nanoparticles and polymeric medical devices: a new approach to prevention of infection?" *Journal of Antimicrobial Chemotherapy*, vol. 54, no. 6, pp. 1019–1024, 2004.

[21] W. F. Lee and K. T. Tsao, "Preparation and properties of nanocomposite hydrogels containing silver nanoparticles by EX situ polymerization," *Journal of Applied Polymer Science*, vol. 100, no. 5, pp. 3653–3661, 2006.

[22] S. Quintavalla and L. Vicini, "Antimicrobial food packaging in meat industry," *Meat Science*, vol. 62, no. 3, pp. 373–380, 2002.

[23] R. Tankhiwale and S. K. Bajpai, "Graft copolymerization onto cellulose-based filter paper and its further development as silver nanoparticles loaded antibacterial food-packaging material," *Colloids and Surfaces B*, vol. 69, no. 2, pp. 164–168, 2009.

[24] R. Bandyopadhyaya, M. V. Sivaiah, and P. A. Shankar, "Silver-embedded granular activated carbon as an antibacterial medium for water purification," *Journal of Chemical Technology and Biotechnology*, vol. 83, no. 8, pp. 1177–1180, 2008.

[25] C. Maioli, A. Bestetti, A. Mauri, C. Pozzato, and R. Paroni, "Removal of radioisotopes in solution and bactericidal/bacteriostatic sterilising power in activated carbon and metal silver filters," *Environmental Toxicology and Pharmacology*, vol. 27, no. 1, pp. 49–53, 2009.

[26] A. L. Lehninger, D. L. Nelson, and M. M. Cox, *Principles of Biochemistry*, Worth Publishing, New York, NY, USA, 2nd ed edition, 1993.

[27] S. Y. Liau, D. C. Read, W. J. Pugh, J. R. Furr, and A. D. Russell, "Interaction of silver nitrate with readily identifiable groups: relationship to the antibacterial action of silver ions," *Letters in Applied Microbiology*, vol. 25, no. 4, pp. 279–283, 1997.

[28] G. McDonnell and A. D. Russell, "Antiseptics and disinfectants: activity, action, and resistance," *Clinical Microbiology Reviews*, vol. 12, no. 1, pp. 147–179, 2001.

[29] J. R. Morones, J. L. Elechiguerra, A. Camacho et al., "The bactericidal effect of silver nanoparticles," *Nanotechnology*, vol. 16, no. 10, pp. 2346–2353, 2005.

[30] C. Wang, N. T. Flynn, and R. Langer, "Controlled structure and properties of thermoresponsive nanoparticle-hydrogel composites," *Advanced Materials*, vol. 16, no. 13, pp. 1074–1079, 2004.

[31] J. Zhang, S. Xu, and E. Kumacheva, "Photogeneration of fluorescent silver nanoclusters in polymer microgels," *Advanced Materials*, vol. 17, no. 19, pp. 2336–2340, 2005.

[32] A. Biffis, N. Orlandi, and B. Corian, "Microgel-stabilized metal nanoclusters: size control by microgel nanomorphology," *Advanced Materials*, vol. 15, no. 18, pp. 1551–1555, 2003.

[33] Y. Xiang and D. Chen, "Preparation of a novel pH-responsive silver nanoparticle/poly(HEMA-PEGMA-MAA) composite hydrogel," *European Polymer Journal*, vol. 43, no. 10, pp. 4178–4187, 2007.

[34] Y. M. Mohan, K. Lee, T. Premkumar, and K. E. Geckeler, "Hydrogel networks as nanoreactors: a novel approach to silver nanoparticles for antibacterial applications," *Polymer*, vol. 48, no. 1, pp. 158–164, 2007.

[35] T. Heinze and T. Liebert, "Unconventional methods in cellulose functionalization," *Progress in Polymer Science*, vol. 26, no. 9, pp. 1689–1762, 2001.

[36] J. H. Guo, G. W. Skinner, W. W. Harcum, and P. E. Barnum, "Pharmaceutical applications of naturally occurring water-soluble polymers," *Pharmaceutical Science and Technology Today*, vol. 1, no. 6, pp. 254–261, 1998.

[37] K. Pal, A. K. Banthia, and D. K. Majumdar, "Preparation of novel pH-sensitive hydrogels of carboxymethyl cellulose acrylates: a comparative study," *Materials and Manufacturing Processes*, vol. 21, pp. 877–882, 2006.

[38] C. Saykan, R. Coskun, and S. Kirbag, "Poly(crotonic acid-co-2-acrylamido-2-methyl-1-propanesulfonic acid)-metal complexes with copper(II), cobalt(II), and nickel(II): synthesis, characterization and antimicrobial activity," *European Polymer Journal*, vol. 43, pp. 4028–4036, 2007.

[39] M. V. Dinu, M. M. Ozmen, E. S. Dragan, and O. Okay, "Freezing as a path to build macroporous structures: superfast responsive polyacrylamide hydrogels," *Polymer*, vol. 48, no. 1, pp. 195–204, 2007.

[40] S. Conti, L. Maggi, L. Segale et al., "Matrices containing NaCMC and HPMC. 1. Dissolution performance characterization," *International Journal of Pharmaceutics*, vol. 333, no. 1-2, pp. 136–142, 2007.

[41] K. Varaprasad, K. Vimala, S. Ravindra, N. N. Reddy, and K. Mohana Raju, "Development of sodium carboxymethyl cellulose-based poly(acrylamide-co-2acrylamido-2-methyl-1-propane sulfonic acid) hydrogels for in vitro drug release studies of ranitidine hydrochloride an anti-ulcer drug," *Polymer-Plastics Technology and Engineering*, vol. 50, no. 1199, 1207 pages, 2011.

[42] A. Pourjavadi, H. Ghasemzadeh, and F. Mojahedi, "Swelling properties of CMC-g-Poly (AAm-co-AMPS) superabsorbent hydrogel," *Journal of Applied Polymer Science*, vol. 113, no. 6, pp. 3442–3449, 2009.

[43] J. Wu, J. Lin, M. Zhou, and C. Wei, "Synthesis and properties of starch-graft-polyacrylamide/clay superabsorbent composite," *Macromolecular Rapid Communications*, vol. 21, no. 15, pp. 1032–1034, 2000.

[44] A. S. Hoffmann, "Intelligent polymers (in Medicine and Biotechnology)," in *Polymeric Materials Encyclopedia*, J. C. Salamone, Ed., vol. 5, p. 3282, CRC Press, Boca Raton, Fla, USA, 1996.

[45] A. Taleb, C. Petit, and M. P. Pileni, "Optical properties of self-assembled 2D and 3D superlattices of silver nanoparticles," *Journal of Physical Chemistry B*, vol. 102, no. 12, pp. 2214–2220, 1998.

[46] M. A. Noginov, G. Zhu, M. Bahoura et al., "The effect of gain and absorption on surface plasmons in metal nanoparticles," *Applied Physics B*, vol. 86, no. 3, pp. 455–460, 2007.

[47] U. Kreibig and M. Vollmer, *Optical Properties of Metal Clusters*, vol. 25, Springer, Berlin, Germany, 1995.

[48] S. L. Amruth, J. S. Sathish, B. P. Ramachandra, and B. P. J, "Synthesis, characterization and dyeing assessment of some novel disperse azodyes based on 1-(4-amino 2-methyl phenyl)-2-(n- phenyl amino) ethanone onnylon and polyester fabrics," *Journal of Chemical and Pharmaceutical Researc*, vol. 2, pp. 478–482, 2010.

Thermal and Photoresponsive Studies of Starch Modified with 2-(5-(4-Dimethylamino-benzylidine)-4-oxo-2-thioxo-thiazolidin-3-yl)acetic Acid

Ambily Chandran,[1] Sunny Kuriakose,[1] and Tessymol Mathew[2]

[1] *Research and Post Graduate Department of Chemistry, St.Thomas College, Mahatma Gandhi University, Kerala, Pala 686574, India*
[2] *Research and Post Graduate Department of Chemistry, St.George College, Mahatma Gandhi University, Kerala, Aruvithura 686122, India*

Correspondence should be addressed to Sunny Kuriakose, skresearchgroup@rediffmail.com

Academic Editor: Jianjun Li

The present study describes the synthesis of a chromophoric system 2-(5-(4-dimethylamino-benzylidin)-4-oxo-2-thioxo-thiazolidin-3-yl)acetic acid and its incorporation into starch through esterification of the hydroxyl group by the free carboxyl function of the chromophoric system by DCC coupling. The products were characterized by UV-visible, fluorescence, FT-IR, and NMR spectroscopic methods. The newly developed system was subjected to photoresponsive studies such as light absorption, light stabilization and fluorescence emission. The free chromophoric system and the coupled product were also subjected to thermal analysis. The results show that modification enhances the light absorption and light fastening properties of the chromophoric system. Thermal stability of the polymeric system greatly enhances on attaching the chromophoric system. In view of these results the newly developed system is proposed as a nature friendly, green, and photoactive product which could find application in dyes, inks, paints, and so forth.

1. Introduction

Starch is an inexpensive commodity that has been used for food and many nonfood purposes for centuries. Starch, one of the main polysaccharides in the world, has been paid much attention because of its biodegradability and abundance as a renewable resource. It has been widely used as a raw material for biodegradable plastics. The chemical modification of starch is of interest because the modification would not change the fundamental skeleton of starch, would keep the original physicochemical and biochemical properties, and finally would bring new or improved properties. Chemically, starch is composed of two different molecules, amylose and amylopectin. In amylose, the glucose molecules are linked in a “linear” fashion; however, the tetrahedral chemistry of carbon (and the bond angles that result from this chemistry) gives amylose an overall spiral shape. Amylopectin, on the other hand, has a linear arrangement of glucose molecules which includes, at regular intervals, a different kind of linkage between two adjacent glucoses. This different linkage results in the formation of a branched structure and an overall treelike shape for this molecule. Starch consists of amylose and amylopectin, which are polymers of glucose, linear and branched, respectively. Each glucose unit potentially has three reactive hydroxyl groups that are the basis of all derivatizations. Native starches in its derivative forms open a wide scope in pharmaceutical-, food-, and nonfood based industries. Starch modification is generally achieved through derivatization such as esterification, etherification, cross-linking, acid hydrolysis, enzymatic hydrolysis heat treatment, and grafting of starch [1–7]. The recent report on modified starches and its future scope predicted that total consumption will grow to almost 75 million tons by 2012 and the demand for starch by food and nonfood industries in Asia is likely to grow by 4–6 percent per year in low and middle income countries in this region. Despite numerous

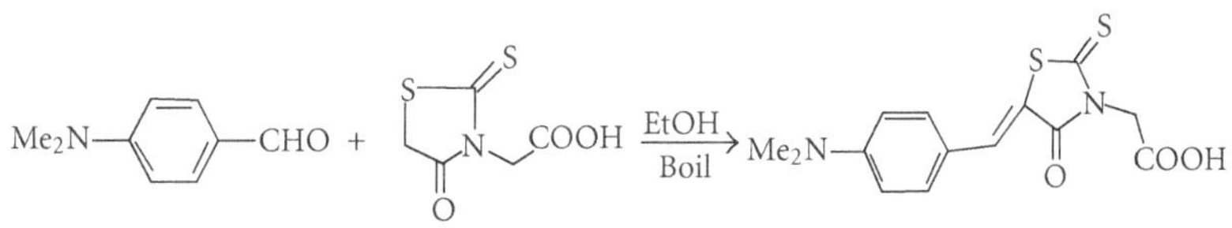

SCHEME 1: Synthesis of 2-(5-(4-dimethylamino-benzylidin)-4-oxo-2-thioxo-thiazolidin-3-yl) acetic acid.

studies that have been undertaken over many years to elucidate the role of modified starch in pharmaceutical-, food-, and nonfood-industry-based application, starch remains an outstanding inscrutable molecule to be explored. Modified starches are utilized in hundreds or even thousands of food, industrial, biofuels, and bioplastic applications [8–13].

In the present study, functional modification of starch has been achieved by DCC coupling. The free hydroxyl group of the biopolymeric core material was esterified with free carboxylic group of the chromophoric system. Thus we could develop a thermo stable and photo responsive system on starch. Modified starches have a very promising industrial prospects on application basis, due to its more biocompatible and environmental friendly nature as compared to synthetic polymers.

2. Experimental

2.1. Materials and Methods: General. Starch was purchased from Merck (Germany). Dimethylaminopyridine (DMAP) and dicyclohexylcarbodiimide (DCC) are commercial samples and are used as purchased from E. Merck India Ltd. P-Dimethylaminobenzaldehyde and rhodanine-N-aceticacid were purchased from Merck (Germany). The solvents used for the study such as dimethylformamide (DMF) were purified by literature procedure. NMR spectra were recorded on a Bruker 500 MHz NMR instrument. IR spectra were recorded on a Shimadzu FT-IR instrument operating in the range 4000–400 cm^{-1}. UV-visible spectra were recorded on a Shimadzu UV-visible-NIR spectrophotometer operating in the range 200–1100 nm.

2.1.1. Synthesis of 2-(5-(4-Dimethylamino-benzylidin)-4-oxo-2-thioxo-thiazolidin-3-yl)acetic Acid. *p*-Dimethylaminobenzaldehyde (1 g) and rhodanine-*N*-acetic acid (1.25 g) were dissolved in ethanol. The mixture was stirred thoroughly for a few minutes. The temperature was raised and the mixture was refluxed for 4 hours with magnetic stirring under an inert atmosphere of argon. The product was filtered, purified by recrystallisation from absolute ethanol, and further purified by column chromatography using silica column and 10 : 3 chloroform-methanol and solvent system, solvent was removed on a vacuum rotary evaporator and characterised by spectroscopic analysis. The yield is noted as 81% [14].

2.1.2. Synthesis of Starch Functionalized with 2-(5-(4-Dimethylamino-benzylidin)-4-oxo-2-thioxo-thiazolidin-3-yl)acetic Acid. 2-(5-(4-dimethylamino-benzylidin)-4-oxo-2-thioxo-thiazolidin-3-yl)acetic acid (2.25 g), starch (1 g), DMAP (200 mg), and DCC (1 g) were separately dissolved in DMF and mixed together. The mixture was stirred at room temperature for 2 hours and at 80°C for 6 hours. The byproduct dicyclohexyl urea (DCU) was removed by warming-cooling-filtration process and the solvents were removed on a vacuum rotary evaporator and dried. It was purified by column chromatography using silica column and 10 : 2 chloroform-methanol solvent system and dried in vacuum [14]. The product was characterised by spectroscopic analysis.

3. Results and Discussion

Spectroscopic analyses such as UV-visible, FT-IR, and ^{1}HNMR have been done in order to visualize the chemical changes in the structure of starch. The green, nature friendly starch functionalized with 2-(5-(4-dimethylamino-benzylidin)-4-oxo-2-thioxo-thiazolidin-3-yl)-acetic acid is a novel photoresponsive system and reported in the present paper for the first time.

3.1. Synthesis and Characterisation of 2-(5-(4-Dimethylamino-benzylidin)-4-oxo-2-thioxo-thiazolidin-3-yl)acetic Acid. *p*-Dimethylaminobenzaldehyde and rhodanin-*N*-acetic acid were condensed together at 80°C yielding 2-(5-(4-dimethyl-amino-benzylidin)-4-oxo-2-thioxo-thiazolidin-3-yl)acetic acid (Scheme 1).

IR spectrum of 2-(5-(4-dimethylamino-benzylidin)-4-oxo-2-thioxo-thiazolidin-3-yl)acetic acid was recorded in the solid state as KBr discs in the operating frequency range 4000–400 cm^{-1}. 3440 cm^{-1}(broad): $\nu_{O\text{-}H}$ (str), 2920 cm^{-1}: $\nu_{C\text{-}H}$ of CH_2, 1716 cm^{-1}: $\nu_{C=O}$ (str), 1573 cm^{-1}: $\nu_{C=C}$ (str), 1379 cm^{-1}: $\nu_{C\text{-}N}$ (str), 1309 cm^{-1}: $\nu_{C=S}$ (str), 1184 cm^{-1}: $\nu_{C\text{-}O}$ (str), 650 cm^{-1}: $\nu_{C\text{-}S}$ (str) (Figure 1).

The proton NMR spectrum of the product was recorded in chloroform using a 500 MHz ^{1}H NMR spectrophotometer. ^{1}H NMR: 10.32 ppm (1H,s: –COOH), 7.42 ppm (2H,d:aromatic proton a), 6.74 ppm (2H,d:aromatic proton b), 4.81 ppm (1H,s:H_c), 3.65 ppm (2H,s:H_d), 3.09 ppm (6H,s:$N(CH_3)_2$) (Figure 2).

3.2. Synthesis and Characterisation of Starch Functionalised with 2-(5-(4-Dimethylamino-benzylidin)-4-oxo-2-thioxo-thiazolidin-3-yl)acetic Acid. The end hydroxyl functionalities of starch were esterified with the free carboxyl group of 2-(5-(4-dimethylamino-benzylidin)-4-oxo-2-thioxo-thiazolidin-3-yl)acetic acid through DCC coupling using DMAP as the catalyst (Scheme 2).

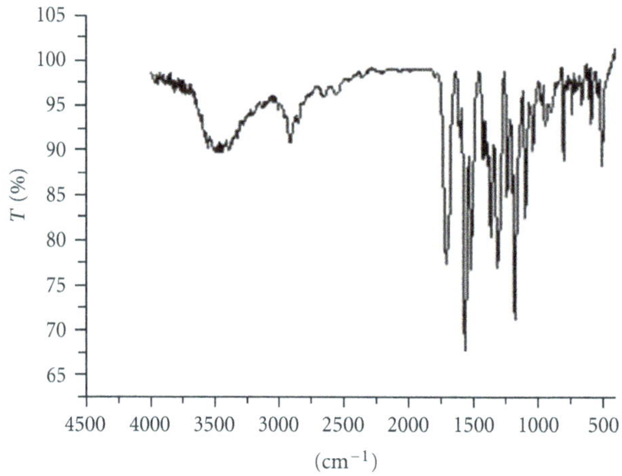

Figure 1: IR spectrum of 2-(5-(4-dimethylamino-benzylidin)-4-oxo-2-thioxo-thiazolidin-3-yl)acetic acid.

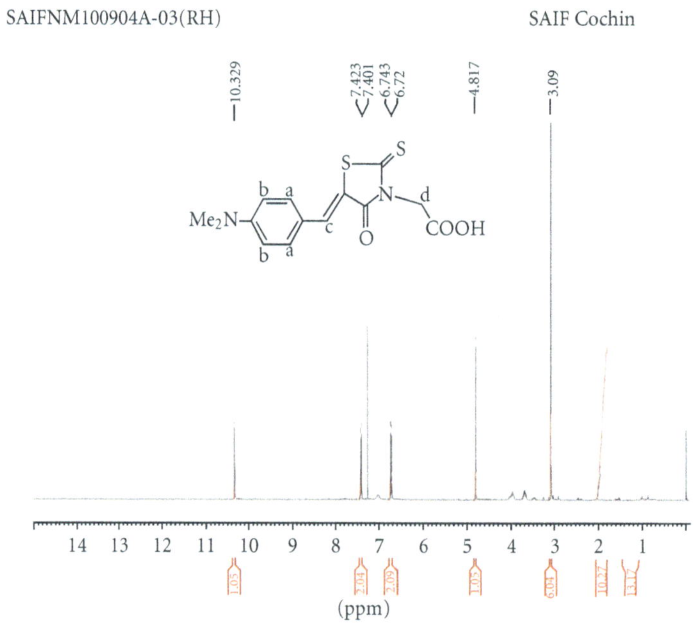

Figure 2: H[1] NMR spectrum of 2-(5-(4-dimethylamino-benzylidin)-4-oxo-2-thioxo-thiazolidin-3-yl)acetic acid.

IR spectrum was recorded in the operating frequency range 4000–400 cm^{-1}. IR(KBr): 3326 cm^{-1}(broad): $\nu_{O\text{–}H}$ (str), 2927 cm^{-1}: $\nu_{C\text{–}H}$ of CH_2, 1706 cm^{-1}: $\nu_{C=O}$ (str), 1627 cm^{-1}(tightly bound H–O–H present in starch), 1579 cm^{-1}:$\nu_{C=C}$ (str), 1334 cm^{-1}: $\nu_{C\text{–}N}$ (str), 1375 cm^{-1}: $\nu_{C=S}$ (str), 811 cm^{-1}: $\nu_{C\text{–}S}$ (str) (Figure 3).

The proton NMR spectrum of the product was recorded in chloroform using a 500 MHz ^{1}H NMR spectrophotometer. ^{1}H NMR: 7. 75 ppm (2H,d:Ha), 6.75 ppm (2H,d:Hb), 5.0 ppm (1H,S: OH group unreacted), 4.90 ppm (1H,s:Hc), 3.55 ppm (2H,s:Hd), 3.09 ppm (6H,S:N$(CH_3)_2$), 1.37–2.17 ppm (m: aliphatic protons of starch). The signal corresponding to COOH proton observed in the proton NMR spectrum of the dye was absent here (Figure 4).

3.3. Light Absorption Properties of Starch Modified with 2-(5-(4-Dimethylamino-benzylidin)-4-oxo-2-thioxo-thia zolidin-3-yl)acetic Acid. The UV-visible spectra of the chromophoric system 2-[5-4-dimethylamino-benzylidine-4-oxo-2-thioxo-thiozolidine-3-yl]acetic acid and starch modified with2-[5-4-dimethylamino-benzylidine-4-oxo-2-thioxo-thiozolidine-3-yl]acetic acid were recorded in chloroform. UV-visible spectra showed an intense band at 474 nm for the chromophoric system and a band centred at 503 nm for the coupled product. The λ_{max} showed a notable red shift of 29 nm on attaching the dye to the starch core (from 474 nm to 503 nm). The HOMO-LUMO energy gap of the conjugated π system is notably reduced on attaching the chromophore onto the core material, and this was

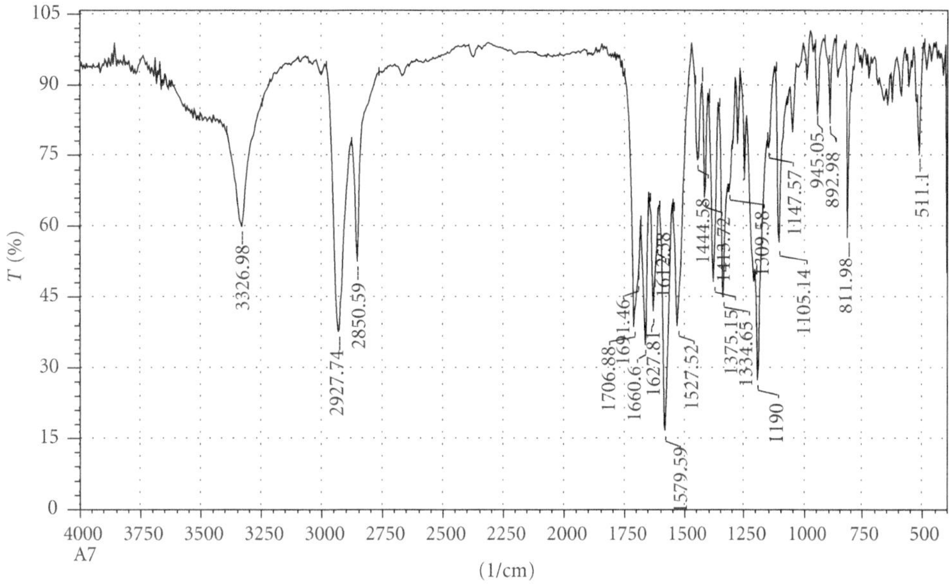

Figure 3: IR spectrum of starch modified with 2-(5-(4-dimethylamino-benzylidin)-4-oxo-2-thioxo-thiazolidin-3-yl)acetic acid.

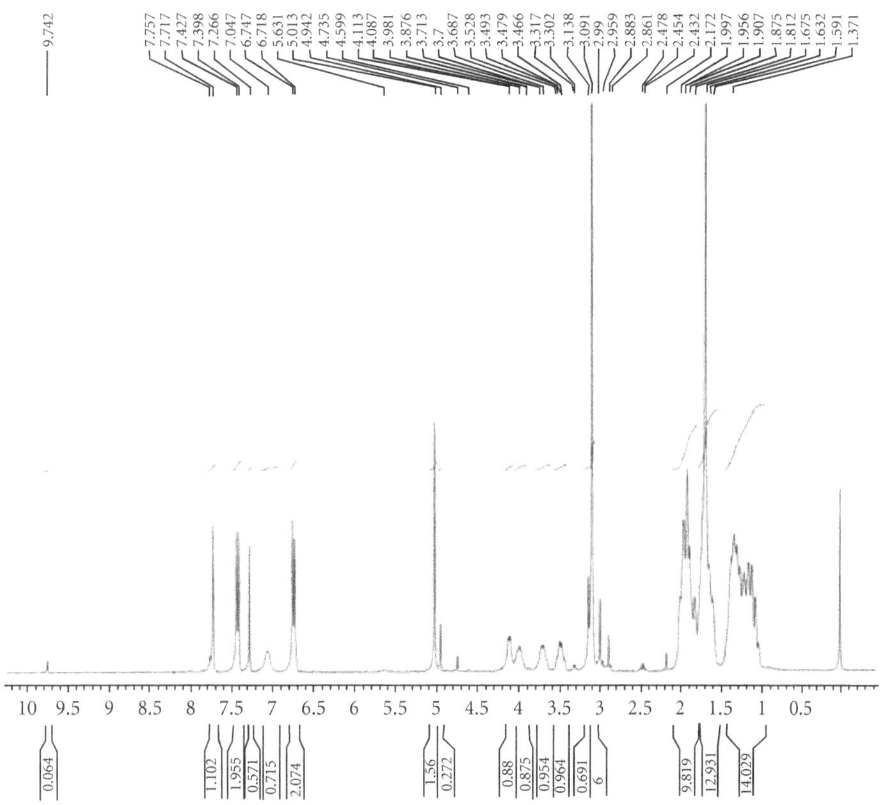

Figure 4: ^{1}H NMR spectrum of starch modified with 2-(5-(4-dimethylamino-benzylidin)-4-oxo-2-thioxo-thiazolidin-3-yl)acetic acid.

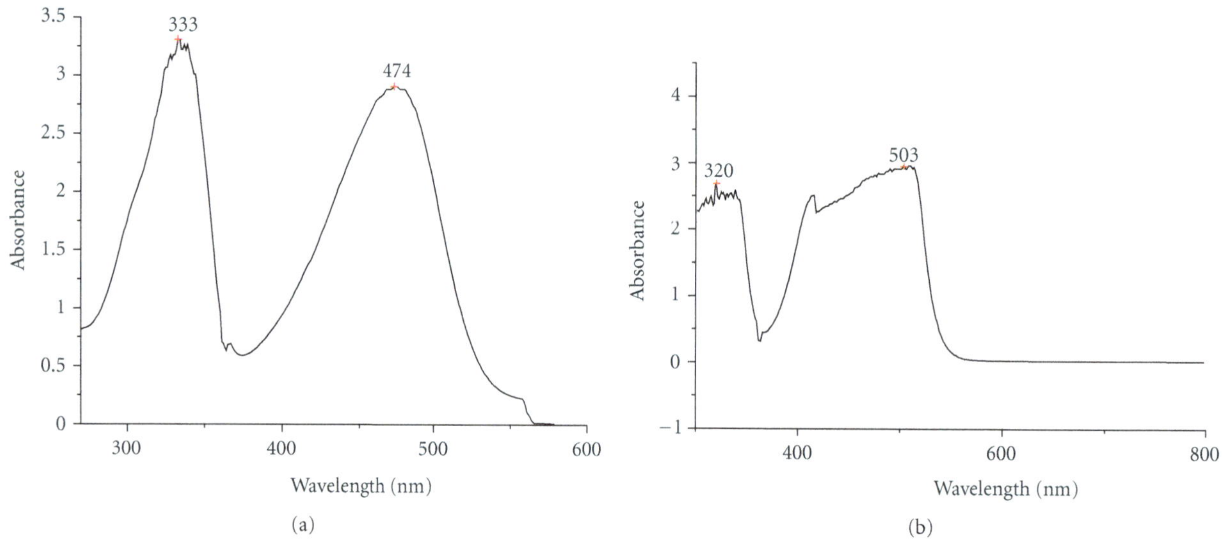

FIGURE 5: UV-visible spectra of (a) 2-[5-4-dimethylamino-benzylidine-4-oxo-2-thioxo-thiozolidine-3-yl]acetic acid and (b)starch modified with 2-[5-4-dimethylamino-benzylidine-4-oxo-2-thioxo-thiozolidine-3-yl]acetic acid.

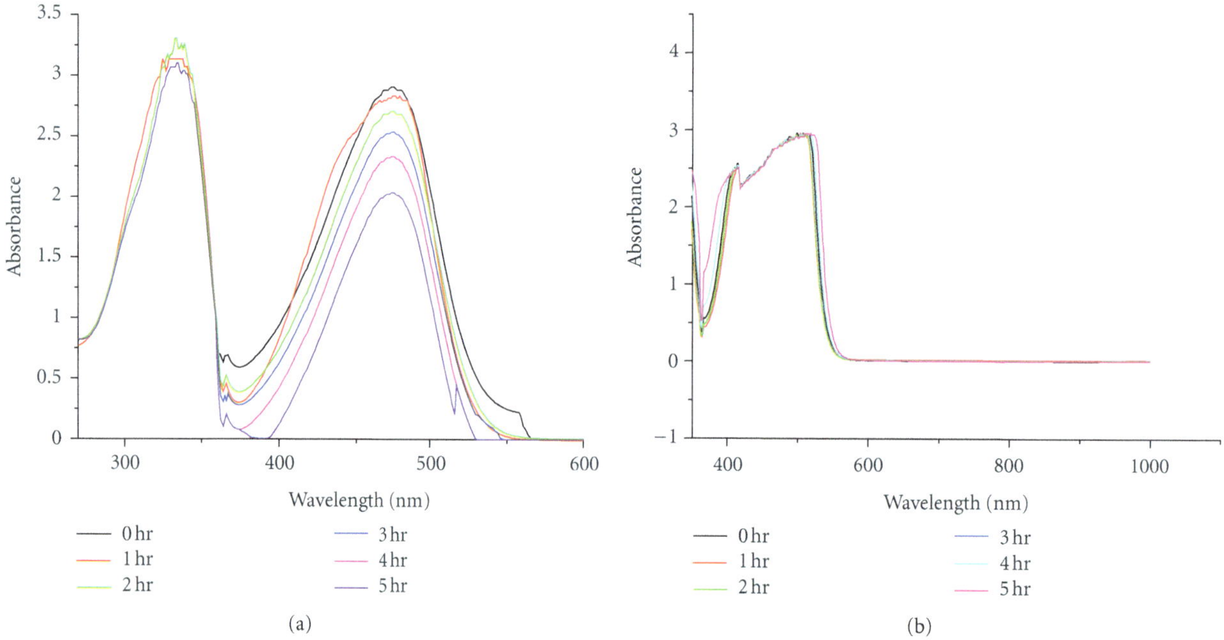

FIGURE 6: Irradiation studies of (a) 2-[5-4-dimethylamino-benzylidine-4-oxo-2-thioxo-thiozolidine-3-yl]acetic acid and (b) starch modified with 2-[5-4-dimethylamino-benzylidine-4-oxo-2-thioxo-thiozolidine-3-yl]acetic acid.

evident from the remarkable red shift noted on attaching the dye to the core material (Figure 5).

4. Light Fastening Studies

The low molecular chromophoric system 2-[5-4-dimethylamino-benzylidine-4-oxo-2-thioxo-thiozolidine-3-yl]acetic acid and its polymer-bound analogue were subjected to light fastening studies. Solutions of the two systems were prepared in chloroform with same molar concentrations. These solutions were subjected to irradiation under visible radiant energy. The changes in the UV-visible absorption spectra as a function of time of irradiation were noted. A sudden, but gradual, decrease in intensity was noted on irradiating the low molecular chromophoric system under visible radiant energy. The intensity of absorption was 2.75 at zero time and this was decreased to 1.8 on prolonged irradiation for 5 hours. The starch modified with 2-[5-4-dimethyl-amino-benzylidine-4-oxo-2-thioxo-thiozolidine-3-yl]acetic

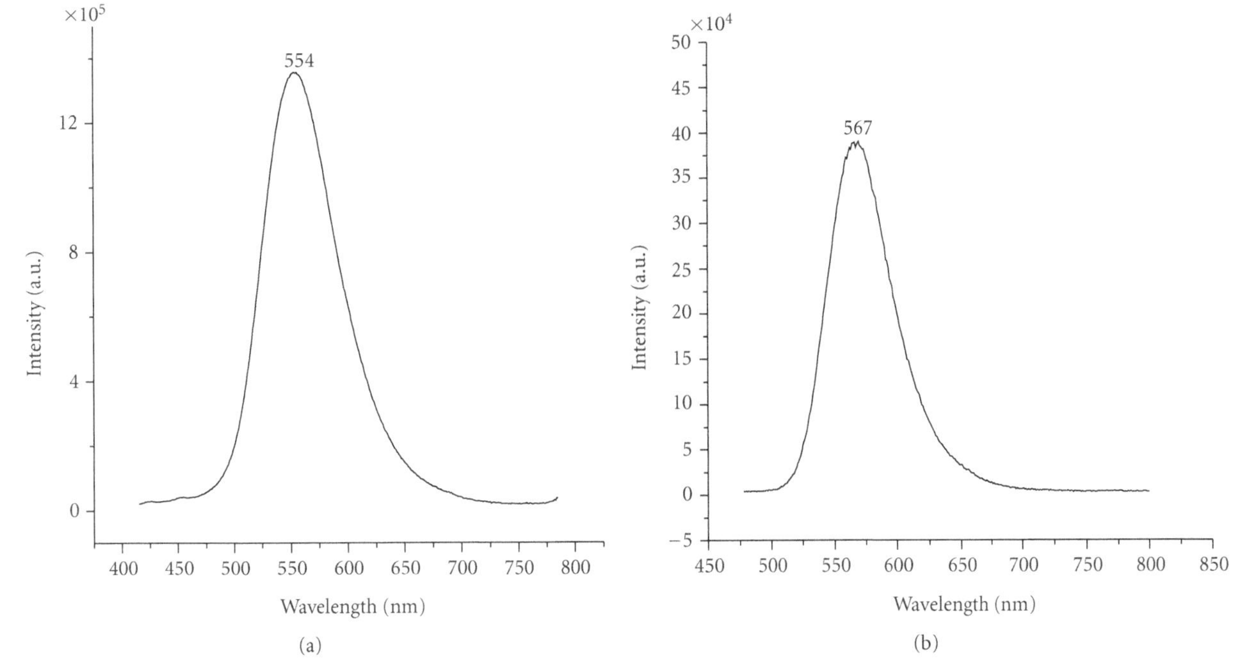

FIGURE 7: Fluorescence spectra of (a) 2-[5-4-dimethylamino-benzylidine-4-oxo-2-thioxo-thiozolidine-3-yl]acetic acid and (b) starch modified with 2-[5-4-dimethylamino-benz ylidine-4-oxo-2-thioxo-thiozolidine-3-yl]acetic acid.

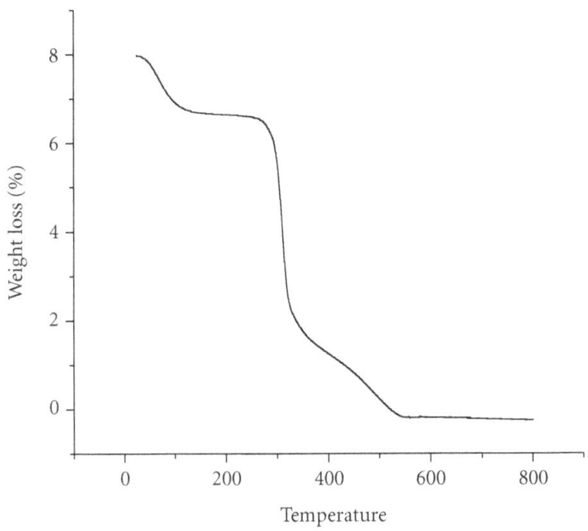

FIGURE 8: TG-DTA curve of starch.

acid showed high stability even after prolonged irradiation. The intensity of absorption was recorded as 3.2 at zero time. This remained nearly the same on irradiation for 5 hours. This shows the efficient light fastening property of the polymer anchored chromophoric system. The irradiation results are shown in Figure 6.

The fluorescence emission spectra of 2-[5-4-dimethylamino-benzylidine-4-oxo-2-thioxo-thiozolidine-3-yl]acetic acid and starch coupled with 2-[5-4-dimethylamino-benzylidine-4-oxo-2-thioxo-thiozolidine-3-yl]acetic acid were recorded in chloroform. The excitation wavelength was

TABLE 1: Thermal analysis data (TG) of starch.

Mass %	Start temp.	End temp.	Weight loss %
80–100	104.53	261.75	10
60–20	277.74	350.70	80

400 nm. The emission maximum for the free chromophoric system was observed at 554 nm and the coupled product was observed at 567 nm. Compared to the pure chromophoric system, the coupled product shows appreciable increase in emission maximum. The intensity of fluorescence emission was also greatly enhanced on attaching to the polymer. The fluorescence spectra of modified as well as pure chromophoric system are shown in Figure 7.

5. Thermal Studies

In the case of the TG curve of starch, a sudden degradation with 90% weight loss occurs at 350°C. The first thermal event occurs at a weight loss of 15%; this is due to the dehydroxylation taking place in the core system (see Table 1). The sudden change at 350°C is due to the degradation of side chains present in the polymer core (Figure 8).

TG of chromophoric system shows four mass loss steps. In the first thermal event between 137–203°C, 36 mass % loss was observed (see Table 2). The higher mass loss above 30 mass % is a consequence of degradation of aromatic chain present in the chromophoric system. In the second mass loss event between 206–245°C, the loss is 17%. This is due to the escape of sulphur group present in the chromophoric system (Figure 9).

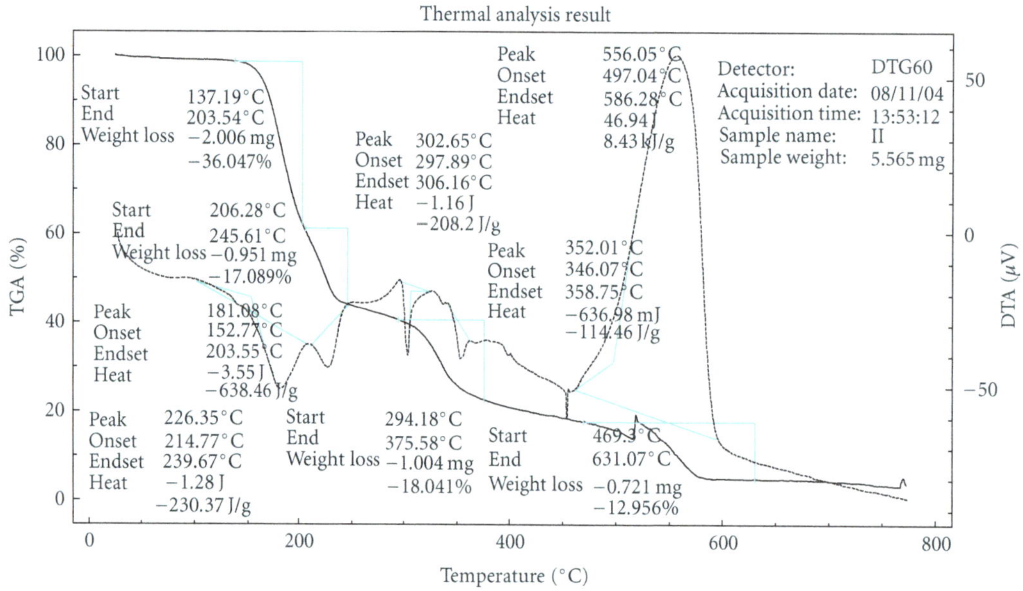

Figure 9: TG-DTA curve of 2-(5-(4-dimethylamino-benzylidin)-4-oxo-2-thioxo thiazolidin-3-yl)acetic acid.

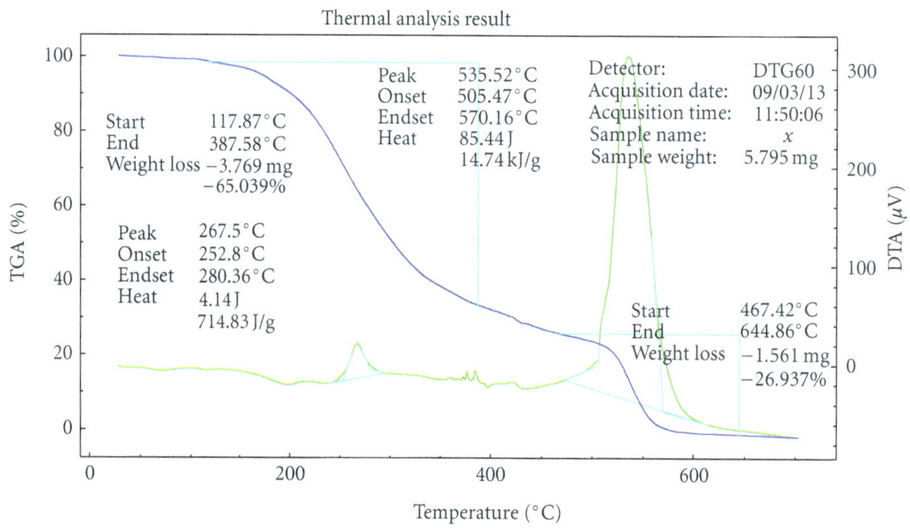

Figure 10: TG-DTA curve of starch modified with 2-(5-(4-dimethylamino-benzylidin)-4-oxo-2-thioxo-thiazolidin-3-yl)acetic acid.

Table 2: Thermal analysis data (TG) of 2-(5-(4-dimethylamino-benzylidin)-4-oxo-2-thioxo-thiazolidin-3-yl)acetic acid.

Mass %	Start temp.	End temp.	Weight loss %
60–100	137.19	203.54	36.04
40–60	206.28	245.61	17.086
20–40	294.18	375.58	18.04

Table 3: Thermal analysis data (TG) of starch functionalised 2-(5-(4-dimethylamino-benzylidin)-4-oxo-2-thioxo-thiazolidin-3-yl)acetic acid.

Mass %	Start temp	End temp	Weight loss %
100–40	117.87	387.58	65.03
40–20	467.42	644.86	26.37

It was noted that TG of modified starch shows two mass loss steps as shown in the curve. In the first thermal event 117–387°C, 65.03% mass loss is observed (see Table 3). The higher mass loss is due to the dehydration of neighbouring OH groups in the glucose ring [15]. In the second mass loss event between 467–644°C, a 26.37 mass % loss is observed (Figure 10).

The thermogravimetric analysis of modified polymer, chromophoric system, and core material were compared. The analysis shows that the degradation temperature was

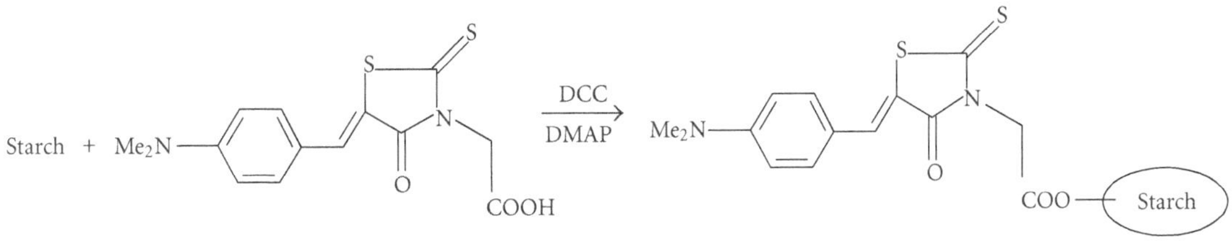

SCHEME 2: Functional modification of starch with 2-(5-(4-dimethylamino-benzylidin)-4-oxo-2-thioxo-thiazolidin-3-yl) acetic acid.

increased in the case of starch modified with 2-(5-(4-dimethylamino-benzylidin)-4-oxo-2-thioxo-thiazolidin-3-yl)acetic acid compared to the core or the low molecular chromophoric system. The overall thermal stability of polymer was also enhanced upon functional modification with the chromophoric system.

6. Conclusion

In the present study the environment friendly natural polymeric system such as starch was used as the core material for developing nature friendly, "green" photoactive system based on polymeric systems. Modified starches have very promising industrial prospects on application basis, due to its more biocompatible and environment friendly nature as compared to synthetic polymers. The paper reports the development of photoactive, thermostable starch moiety. Starch was modified with chromophoric system, namely, 2-(5-(4-dimethylamino-benzylidin)-4-oxo-2-thioxo-thiazolidin-3-yl)acetic acid by DCC coupling between the carboxyl function of chromophoric system and hydroxyl functionalities of starch. The modified products were characterised and compared by spectroscopic techniques. The modified starch shows remarkable red shift and increase in intensity of absorption and emission. Light irradiation studies of the 2-(5-(4-dimethylamino-benzylidin)-4-oxo-2-thioxo-thiazolidin-3-yl)acetic acid and starch modified with 2-(5-(4-dimethylamino-benzylidin)-4-oxo-2-thioxo-thiazolidin-3-yl)acetic acid were compared. The modified system shows excellent light fastening properties. The fluorescence emission maximum of starch coupled with 2-[5-4-dimethylamino-benzylidine-4-oxo-2-thioxo-thiozolidine-3-yl]acetic acid was observed at 567 nm. TGA studies were conducted to determine the effect of esterification on the thermal stability of starch. The results indicated that modified starch has increased thermostability compared to the unmodified starches.

Acknowledgments

The authors thank the Department of Science and Technology (Ministry of Science and Technology), the Government. of India, New Delhi, and University Grant Commission, New Delhi, for financial support by awarding major and minor research projects.

References

[1] R. N. Tharanthan, "Starch—value addition by modification," *Critical Reviews in Food Science and Nutrition*, vol. 45, pp. 371–384, 2005.

[2] A. C. Eliasson, *Starch in Food: Structure, Function and Applications*, CRC Press, Fla, USA, 2004.

[3] J. Bemiller and R. Whistler, Eds., *Starch Chemistry and Technology*, Academic Press, New York, NY, USA, 3rd edition, 2009.

[4] "Global modified starch products derivatives & markets a strategic review," http://www.marketresearch.com/map/prod/2093916.html.

[5] P. Forssell, A. Hamunen, K. Autio, and K. Poutanen, "Hypochlorite oxidation of barley and potato starch," *Starch*, vol. 47, p. 371, 1995.

[6] G. Hellwig, D. Bischoff, and A. Rubo, "Production of cationic starch ethers using an improved dry process," *Starch*, vol. 44, pp. 69–72, 1992.

[7] M. M. Tessler and R. L. Billmers, "Degradable aliphatic polyesters," *Journal of Environmental Polymer Degradation*, vol. 4, p. 85, 1996.

[8] G. Ezra and A. Zilkha, "Grafting of preformed polyethylene oxide on starch," *Journal of Applied Polymer Science*, vol. 13, no. 7, pp. 1493–1496, 1969.

[9] M. Tahan and A. Zilkha, "Anionic graft polymerization of ethylene oxide on starch. I," *Journal of Polymer Science Part A-1*, vol. 7, no. 7, pp. 1815–1819, 1969.

[10] A. Van Warners, E. J. Stamhuis, and A. A. C. M. Beenackers, "Kinetics of the hydroxyethylation of starch in alkaline salt-containing aqueous slurries," *Industrial & Engineering Chemistry Research*, vol. 33, no. 4, pp. 981–992, 1994.

[11] J. L. Jane, L. Shen, and C. Maningat, "Preparation and properties of small-particle cornstarch," *Cereal Chemistry*, vol. 69, pp. 280–284, 1992.

[12] R. N. Waduge, R. Hoover, T. Vasanthan, J. Gao, and J. Li, "Effect of annealing on the structure and physicochemical properties of barley starches of varying amylose content," *Food Research International*, vol. 39, no. 1, pp. 59–77, 2006.

[13] J. F. Matte and J. Doucet, "Recent developments in lignin utilization as wood adhesive: a review," *Cellulose Chemistry and Technology*, vol. 22, no. 1, pp. 71–85, 1988.

[14] C. Ambily, K. Sunny, and M. Tessymol, "Synthesis, characterisation and photoresponsive studies of lignin functionalised with 2-(5-(4-dimethylamino-benzylidin)-4-oxo-2-thioxo- thiazolidin-3-yl)-acetic acid," *Open Journal of Organic Polymer Materials*, vol. 2, no. 4, pp. 63–68, 2012.

[15] J. F. Mano, D. Koniarova, and R. L. Reis, "Thermal properties of thermoplastic starch/synthetic polymer blends with potential biomedical applicability," *Journal of Materials Science: Materials in Medicine*, vol. 14, no. 2, pp. 127–135, 2003.

8

Scanning Electron Microscopy and Kinetic Studies of Ketene-Acetylated Wood/Cellulose High-Density Polyethylene Blends

Yakubu Azeh,[1] Gabriel Ademola Olatunji,[2] and Paul Andrew Mamza[1]

[1] *Department of Chemistry, Ibrahim Badamasi Babangida University, Lapai, Nigeria*
[2] *Department of Chemistry, University of Ilorin, Ilorin, Nigeria*

Correspondence should be addressed to Yakubu Azeh, azehyakubu@yahoo.com

Academic Editor: R. J. Linhardt

Acetylated cellulose and wood cellulose as well as untreated cellulose polyethylene blends were subjected to kinetic studies using water, 0.5 M NaOH, and 0.5 M HCl solutions in order to investigate their absorbent properties at 0.5/1.0 cellulose/wood cellulose/polyethylene matrix. The results of the absorption studies showed that the untreated cellulose and wood cellulose blends absorbed water and the acid and alkali solutions higher than the treated samples, which showed a reduction in acid, alkali, and water uptake. In this work, the effects of acetylation on the morphological studies of the polyethylene blends were obvious. The presence of acetyl groups improved the interfacial bonding between the polymer matrix and cellulose as well as the wood cellulose fibers, as evidenced by scanning electron microscopy (SEM).

1. Introduction

A better understanding of the chemical composition and surface properties/chemistry of natural fibers is necessary for ease of developing natural fiber-reinforced composites/blends. Cellulose is the most abundant biopolymer in nature and its biosynthesis, chemistry, and ultrastructure remain an active field of study. Over the past decades, the interest in sustainability and green chemistry has led to a renewed interest in novel cellulosic materials and composites derived from a variety of cellulosic materials [1]. Cellulose is a homopolysaccharide which is a highly stable chain composed of 12,000 of β-D-glucopyranose monomeric units linked together by covalently bonded $\beta 1 \rightarrow 4$-glycosidic linkages in long chains, resulting in bundles known as microfibrils that can be easily cleaved by mineral acids [2, 3]. Each β-D-glucopyranose unit is oriented at an angle of 180° from the succeeding glucose unit [2]. Native cellulose is composed of amorphous and crystalline domains. The crystalline domains are highly ordered and much more stable to acid hydrolysis compared to their amorphous counterparts [3]. Native cellulose based materials (wood, hemp, cotton, linen, etc.) have been used as construction materials [4, 5]. Cellulose is the basic structural component of all plant fibers, the most important organic biopolymer produced by plants and the most abundant renewable polymer resource available today worldwide [6]. Cellulose is a polymer raw material which has served mankind as a construction material, mainly in the form of solid wood and textile fibers, such as cotton or flax, or in the form of paper and board for many decades. On the other hand, cellulose is a versatile starting material for chemical conversions aiming at the production of value-added products such as cellulose-based surgical threads and films as well as a variety of stable cellulose derivatives used in many areas of industry and domestic life [3, 7]. Cellulose belongs to the class of organic compounds known as carbohydrates and cotton is known to be the purest form of cellulose [8]. Cellulose is a chain polymer and forms microfibrils, which are embedded in a matrix, composed of amorphous hemicelluloses bonded together in relatively short, branching chains and are very hydrophilic [3, 9]. It has been shown that the crystalline domain of cellulose may contain several types of lattices (Iβ, II, III, and IV), depending on the origin of cellulose as well as the chemical treatments used during the processing of the source materials [10]. Cellulose has no taste, is odourless, hydrophilic, is insoluble in water and most organic solvents, is stereogenic, and is highly biodegradable. It can be broken down into its monomer glucose units when treated with certain enzymes or concentrated acids at high temperature. Cellulose requires a temperature of 320°C and pressure of 25 MPa to become amorphous in water [11]. The strong hydrogen bonding network within the cellulose molecules greatly affects the properties of cellulose. For instance, properties such as swelling, absorption, and optical behavior are influenced by the hydrogen bonding. Swelling of cellulose is because the natural fiber exhibits a high hydrophilicity due to attractions and interaction between the hydroxyl groups of fiber components and water molecules which originate from the noncrystalline region and extende to the crystalline region [11, 12]. The swelling of cellulose is to a certain extent reversible when the surrounding humidity level is increased or decreased. Water exchange processes like desorption/absorption with the environment, the reaction with enzymes of bacteria and fungi, and the influence of ultraviolet light truly indicate their importance in the physical, mechanical, and chemical properties of cellulose [9]. Cellulose is a material with unique tensile properties. In its pure form, fibers stronger than Kevlar can be tailored (cellulose is 70–137 GPa while Kevlar is 100 GPa) [3]. Cellulose, wood, and lignocellulosic materials have high reactive surface of hydroxyl groups that facilitate chemical modification to achieve different surface properties which allows for proper interaction with the hydrophobic surface of the most commonly used synthetic polymers [3, 13]. The presence of large hydroxyl groups in the cellulose polymer allows for the preparation of cellulose derivatives by applying the chemistry of alcohols [11]. Derivatization of cellulose using reactions like etherification and esterification for value-added products like ether and ester is a subject of continuing interest because of the wide applications of these products as fibers, fillers, dialysis membranes, adhesives, explosives, thickening agents in foods, moisture-proof coatings, films, toys, cellulosic dyes, and packaging materials [3, 8, 11, 14].

New technologies have been used for the chemical and mechanical modification of cellulosic materials. Some of these techniques have been applied for the production of modified cellulosic materials for our daily and industrial needs. Cellulosic materials are generally strong, hydrophilic, biodegradable, recyclable, and insoluble in water. Typically, the purpose of the modifications is either to reinforce the original properties of cellulose or to add new functionalities in order to create a cellulosic product with specific properties. By virtue of the hydroxyl groups of cellulose and lignocellulosic polymers, they can partially or fully react with various reagents to yield derivatives with industrial and commercial potentials.

Wood is a biopolymer polymeric material considered as a very friendly and useful raw material used extensively for various construction purposes. Wood is a biopolymer with a complex structure and composed of elongated cells whose walls have complex and multilayered structures and constituting cellulose, lignin, and hemicelluloses [9, 13]. These biopolymer makes up the bulk of the cell wall and determines the basic chemistry in wood which explains most of the physical and chemical changes observed in wood, wood-based products, and lignocellulosic materials. Wider applications of wood are restricted due to its susceptibility to certain environmental degradative agents. To improve the performance of wood/lignocellulosic materials for wider applications, the early concept introduced by H. Staudinger is applied to transform the basic functionalities (predominantly –OH groups in cellulose, lignin, and hemicelluloses) in order to have wood/wood-based as well as lignocellulosic products with improved properties [13]. The high moisture sensitivity of the lignocellulosic fiber causes the dimensional instability and limits the use of natural fibers as reinforcement in composite/blend materials [12]. Low interfacial properties between fiber and polymer matrix often reduce their potential as reinforcing agents due to the hydrophilic nature of fibers, which is their major drawback for being used in biocomposite, which reduces natural fiber compatibility with hydrophobic polymer matrix throughout composite fabrication [12, 15]. The inability of wood fibers/cellulosic materials to interact well with polymer components in composite/blends systems causes poor performance of the material and necessitates the use of chemical modification to enhance the surface adhesion between components. Chemical modification is one of the methods that alter basically the structure of wood by optimizing the interface of fibers by introducing new moieties which can efficiently improve surface interlocking within the plastic matrix [9, 12].

Cellulose/lignocellulosic materials are made up of cellulose nanofibers that are extremely strong and have the potentials to replace nonsustainable materials such as plastics and metals. Isolation of nanodimensional cellulose from wood sources and its use in nanocomposites, biocomposite and biodegradable blends will allow the production of much lighter weight materials to replace metals and plastics in various applications [3]. Natural fibers are in general lighter than man-made fibers and tend to have mechanical properties comparable to those of glass fiber reinforced plastics when they are used as fiber composite reinforcements [6].

Wood fibers can be obtained from sawmill chips, saw-dust, wood flour/powder, cutter shavings, and wood pulp/wood residues. Wood fiber acts as more reinforcing in the composite than the wood flour which acts more as a filler [6]. There is a growing interest in natural fiber composite for industrial applications in developed countries, particularly for the production of novel materials with nanoscale dimensions that are also ecofriendly for industrial applications. This is partly due to environmental concern about man-made composite, but mostly due to their potential in cost reduction and biodegradability [6]. The modified wood fiber as reinforcing biopolymer composite has advantages such as renewability, biodegradability, durability, good mechanical, flexural, thermal, and tensile properties and a broad capacity to allow tailoring or grafting of chemical species to improve barrier properties [3, 12, 13, 15].

Because of this ability, nanotechnology represents a major opportunity for wood, wood-based, and lignocellulosic materials to improve their performance and functionality to develop a new generation of products, such as composite/blends and paper products, to new generations of functional nanoscale lignocellulosic materials [16]. For instance, polymer nanocomposites are produced by incorporating only a few percentages of these nanomaterials normally (1–5% cellulose) into the polymer and the improvement is vast due to their large degree of surface area. These nanomaterials are commonly referred to as, for example, nanofillers, nanoparticles, nanoscale building blocks, or nanoreinforcements. Nanocomposites have improved stiffness, strength, toughness, thermal stability, barrier properties, and flame retardancy compared to the pure polymer matrix [16]. Because of the nanometric size effect, these composites show some unique outstanding properties with respect to their conventional microcomposite counterparts.

The properties of nanocomposites materials depend not only on the properties of their individual parents, but also on their morphology and interfacial characteristics. In the particular case of a polymer reinforced with rigid nanofillers, various parameters seem to be of importance in characterizing the fillers: geometrical factors such as the shape, the size, and the aspect ratio; intrinsic mechanical characteristics such as the modulus or the flexibility; surface properties such as specific surface area and surface treatment [16]. Ketene has been used for chemical modification of wood, cellulose, and lignocellulosic materials (natural fibers) through acetylation of accessible hydroxyl groups of the natural fibers [13, 17, 18]. This work is an extension of our previous research work carried out to investigate the effects of ketene-acetylated cellulose and wood cellulose for industrial applications and enhancement of the biodegradation of polyethylene [3, 13].

The procedures for the ketene-modified blends whose SEM images were being studied were obtained based on the method previously described by Azeh et al. [3, 13]. The blends were characterized using scanning electron microscopy (SEM). Consequently, kinetic studies of blends in acid, base, and distilled water were also carried out to investigate their effect on blends.

2. Experimental Procedures

2.1. Materials and Methods. Sodium hydroxide (NaOH) solution (0.5 M) and hydrochloric acid (HCl) solution (0.5 M) were supplied by May and Baker. Distilled water was obtained from our laboratory. All reagents used were of analytical grade. Cellulose/wood cellulose used were obtained as a gift from another research group and used without further purification. Blends used in this work were obtained from our previous research work.

2.2. Acetylation. The process that generated acetylated cellulose/wood cellulose is described elsewhere [3, 13]. One gram of wood cellulose/cellulose was pretreated with 20 mL of 5% NaOH and stirred for 15 h on a magnetic stirrer (mercerization). A viscous suspension was obtained; this was transferred into the reaction flask. The iron pipe adapted as the furnace was preheated for 5 min (to increase heating efficiency of the apparatus), after which 40 mL of dry acetone was distilled. This process generated the ketene that was used to acetylate the samples. The mixture was cooled to room temperature while stirring with a magnetic stirrer. It was filtered and the residues washed until neutral with distilled water and then air dried. The filtrates had two layers and were separated using a separating funnel and stored at room temperature for further treatment. The dried residues were milled in a porcelain mortar and sieved through a 2 mm size nylon sieve. The fine particles were collected and stored at room temperature prior to IR and NMR analyses. Other wood cellulose acetylation products were obtained from metals (magnesium turnings, zinc filings, copper metal, and sphalerite dust) using the process previously described.

2.3. Dissolution of Polyethylene (Packaged Water Sachet) and Blending Procedures. The dissolution of high-density polyethylene (packaged water sachets) follows the method described by [19]. In brief, 1 g pieces of nonprinted portion was placed in a reaction flask and 30 mL of toluene was added. The contents were placed on a hot plate at 100°C, and a thermometer was inserted. The sachets swelled at 60°C and then rapidly dissolved at 80°C to give a clear solution. Blending of high-density polyethylene and acetylated cellulose/wood cellulose is thus described. Into the clear solution of high-density polyethylene sachets was introduced 0.5 g of each ketene-treated and untreated cellulose and wood cellulose separately while stirring for 5 min and then spread on clean metal plates to form a film. The thin films formed were peeled off the metal plates to afford acetylated cellulose and wood cellulose-polyethylene blends [3].

2.4. Characterization of Blends Using Scanning Electron Microscopy (SEM). Surface morphology of the blends was characterized using a microscope, Carl Zeiss Axioskop 50. connected to a Canon PC 1200 Digital Camera operating at EHT voltage of 20 kV.

2.5. Water Absorption Test. The kinetics of water absorption was studied using a water absorption test. The sample dimensions were 3 × 20 × 200 mm in diameter. Prior to the test, the samples were dried in an oven at 45°C for 20 min in order to remove any water moisture content. Then, the samples were totally immersed in distilled water, 0.5 M HCl, and 0.5 M NaOH at various hour intervals. After these periods of immersion, samples were removed at fixed time intervals, wiped with filter paper to remove surface adhered water, and weighed using an analytical balance. The sorption of water, acid, and alkali by the blends at time t was calculated using

$$\frac{M_t - M_o}{M_o} \times 100, \tag{1}$$

where M_t is the weight of blends after 24 h immersion in solution while M_o is the initial weight of blends.

3. Results and Discussion

3.1. A Morphological Study of Ketene-Acetylated Wood Cellulose and Cellulose Blends before and after Acetylation. Acetylation of cellulose and wood cellulose as described by

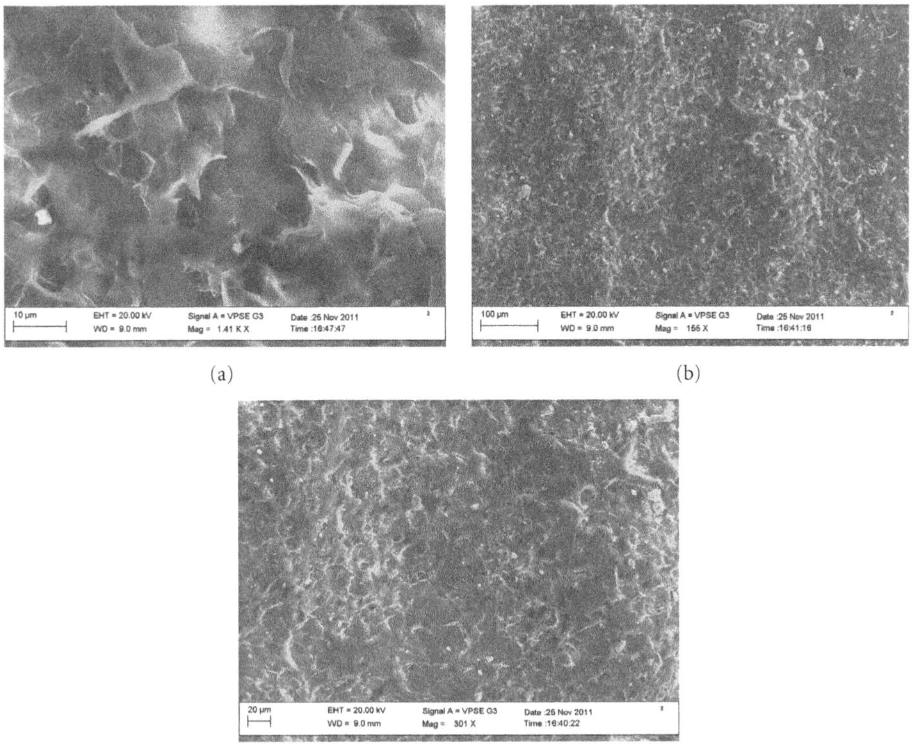

(a) (b) (c)

FIGURE 1: (a) SEM image of ketene-treated lignocellulosic fiber (wood cellulose). (b) SEM image of ketene-treated lignocellulosic fiber (wood cellulose). (c) SEM image of ketene-treated lignocellulosic fiber (wood cellulose).

Azeh et al. [3, 13] results in removal of hemicelluloses in wood and transformation of cellulose Iα to cellulose IIβ. This treatment has resulted in the following morphological changes observed in the SEM micrographs of both treated and untreated samples. Acetylation has led to a decrease of the fiber cross-section, thereby smoothing its outer surface. That is, the transformation cellulose Iα to cellulose IIβ by acetylation is irreversible at both the crystallographic and morphological level as shown in Figures 1(a) and 1(c).

The scanning electron microscopy (SEM) images showed the crystalline pattern of cellulose and the interfacial interactions of acetylated samples with polyethylene, which explains the dispersibility of treated cellulose attributed to acetyl moieties on cellulose surface as they interact well with polyethylene. The absence of cavities and pulled-out fibers confirmed that the interfacial bonding between acetylated cellulose, wood cellulose, and the polymer (polyethylene) matrix was good and strong. Because of the transformation of hydrophilic surfaces of wood cellulose and cellulose to hydrophobic surfaces through acetylation, the polymer blends showed better interfacial adhesion between the fibers and the polymer matrix and these surfaces were smooth. As a result, there were no fracture surfaces and pull-out fibers. The crystalline surface of blends formed from treated samples is indicative of the removal of the amorphous regions of cellulose in these samples by mercerization and acetylation [14, 20]. Figures 1(a) and 2(b) illustrate the scanning electron micrograph of polyethylene blends reinforced with ketene-acetylated wood cellulose and cellulose at 0.5 g cellulose/1.0 g polyethylene ratio in toluene. Figures 1(a) and 2(b) showed that the acetylation of fiber surfaces has changed the blend surfaces by reducing disordered regions in the fibers. A comparison of acetylated cellulose fiber blend surface with that of the blend before treatment showed that surface roughness of blends after treatment is lower than that of untreated blends. A high decrease in surface roughness of blends in acetylated samples is a direct evidence of interfacial compatibility of fiber and matrix surfaces [15, 20]. An increase in the degree of crystallinity with respect to the degree of acetylation which is due to the removal of lignin and extractible and small weight percent gains (WPGs) have been reported to increase the crystallinity of cellulose [12]. In this work, ketene which is known to give small weight percent gains was used for the acetylation of wood cellulose and cellulose, and this effect can be seen in the morphology of the blends being studied [21].

Figures 3(a) and 3(c) illustrate the SEM images of blends reinforced with untreated wood cellulose and cellulose samples. The blends showed numerous cavities and pull-out fibers can be seen and they were also not smooth. There were more observable cellulose microbundles, and also the localized bunch of fibers and patches can be seen which are indicative of the poor dispersion of untreated cellulose within the polyethylene matrix. Thus, the surface of the blends reinforced with untreated cellulose appeared to be dominated by pull-out amorphous fiber surfaces. Similar

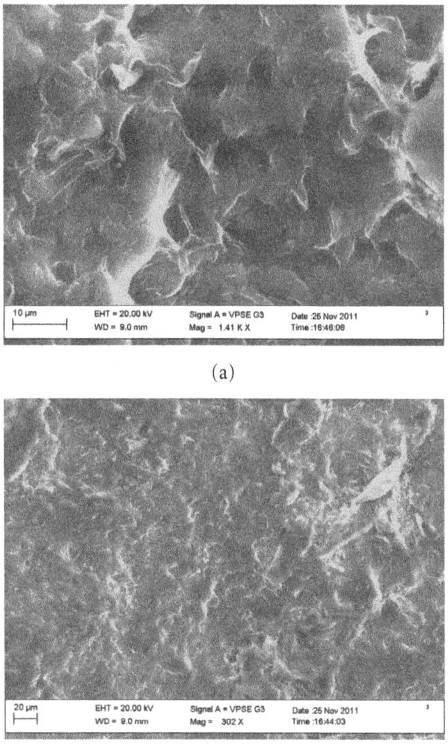

(a)

(b)

FIGURE 2: (a) SEM image of ketene-treated cellulose blend. (b) SEM image of ketene-treated cellulose blend.

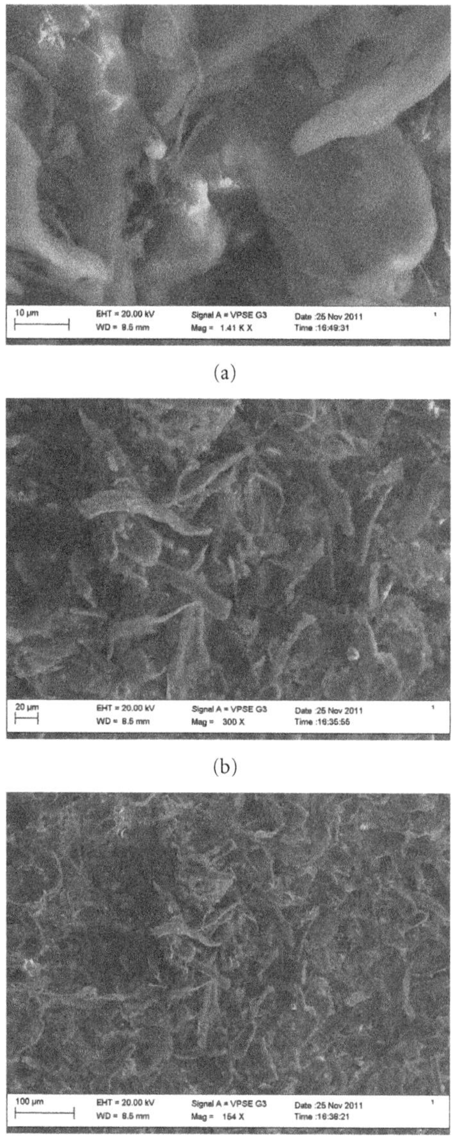

(a)

(b)

(c)

FIGURE 3: (a) Untreated cellulose polyethylene blends. (b) Untreated cellulose polyethylene blend. (c) Untreated cellulose polyethylene blend.

surfaces have been reported [15, 18]. There were also no even distribution of fiber within the polymer matrix and the blend surfaces were rough indicating the presence of disordered cellulose regions. It is also observed that the cellulose particles in the fiber form were oriented randomly, and as the fiber-matrix interfacial adhesion was poor, the blends showed some gaps and signs of a pull-out tendency. These is attributed to the biofiber, hydrophilic nature, which reduces their compatibility with hydrophobic polymer matrix throughout composite/blend fabrication [15]. A comparison of acetylated wood cellulose and cellulose blends surface with that of the blends before treatment showed that surface roughness of blends after treatment is lower than that of untreated blends. The surface of the treated blends displayed the crystallinity of treated cellulose as well as wood cellulose and their dispersibility in the polymer matrix.

3.2. Effect of Acetylation on Water Absorption of Ketene-Acetylated Wood Cellulose and Cellulose Blends. The use of fibers as reinforcing material and fiber loading on synthetic polymers has effect on water absorption of biocomposites/blends [15]. Since biofibers are hydrophilic in nature due to the presence of numerous hydroxyl functionality, acetals, and ether linkages in cellulose structure, the free hydroxyl groups in cellulose, hemicelluloses, and lignin interact with water molecules by hydrogen bonding. Therefore, as expected, the addition of cellulosic material to the synthetic polymer to form biocomposites/blends increased the water absorption of the biocomposites/blends. Because polyethylene (HDPE) is hydrophobic and the wood cellulose and cellulose are hydrophilic, the absorption of water is a function of fiber alone. As the fiber loading increased, the cellulose content increased, which in turn resulted in the absorption of more water. It has been reported that the hydrophilicity of the fibers can be reduced by a suitable chemical treatment [20]. Table 1 illustrates the percent uptake of the solvent by both treated and untreated cellulose and wood cellulose samples obtained after 24 h immersion in distilled water, 0.5 M NaOH, and 0.5 M HCl solutions, respectively.

However, acetylation is known to reduce the hygroscopicity of cellulosic materials, and since wood cellulose and cellulose have been acetylated, it is expected that the blends reinforced with acetylated cellulosic materials will absorb less water molecules as compared to the untreated samples (Table 1). Accessible hydroxyl groups in the fibers

TABLE 1: Percent solvent uptake of treated and untreated blend samples.

	Treated cellulose blend	Treated wood cellulose blend	Untreated cellulose blend
NaOH (0.5 M)	14.29	14.29	71.43
HCl (0.5 M)	42.86	42.86	85.71
H_2O	12.86	12.86	57.14

have been converted to acetyl moieties through acetylation. The highest percent solvent uptake was found with blends reinforced with untreated cellulose, which was attributed to the hydrophilic (presence of polar groups like hydroxyl, acetals, and ether linkages) nature of cellulose fibers in the cellulose structure. According to George et al. [20] the hydrophilicity of cellulose fibers can be reduced by a suitable chemical treatment. The results obtained in this work (Table 1) agree with his findings. The treated cellulose blends showed a lower degree of solvent absorption when compared to the untreated cellulose blends. As a result of the chemical treatment, the ketene reacted with the hydroxyl group of the cellulose by a mechanism well established to form acetylated cellulose and wood cellulose, respectively. Blending of the like surfaces gave rise to covalent bond formation between the hydrophobic acetylated cellulose and the hydrophobic polymer matrix and thus established a good fiber/matrix bonding interaction. Consequently, the chance of having hydroxyl groups coming into contact with water molecules was reduced as can be seen from the absorption test.

4. Conclusion

In this work, we have been able to further prove that acetylation of cellulose and wood cellulose using ketene was successful [13]. Kinetic studies showed that the treated blends showed a reduction in hygroscopicity implying that the blends are fit for use as biodegradable packaging materials. The uptake of water, acid, and alkali solutions was due to the hydrophilicity of cellulose. Morphological studies using SEM showed that fiber chemical treatment using ketene at 0.5 g/1.0 g fiber/matrix loading also improved the fiber-matrix surface interaction, which led to better incorporation of the fiber with the matrix.

References

[1] J. Zhang, T. J. Elder, Y. Pu, and A. J. Ragauskas, "Facile synthesis of spherical cellulose nanoparticles," *Carbohydrate Polymers*, vol. 69, no. 3, pp. 607–611, 2007.

[2] H. B. T. Abdulaziz, *Reactive extraction of sugars from oil palm fruit bunch hydrolysate using naphthalene-2-boronic acid [M.S. thesis]*, Graduate Faculty Universiti of Sains Malaysia, 2007.

[3] A. Yakubu, T. M. Umar, and S. S. D. Mohammed, "Chemical modification of microcrystalline cellulose: improvement of barrier surface properties to enhance surface interactions with some synthetic polymers for biodegradable packaging material processing and applications in textile, food and pharmaceutical industry," *Advances in Applied Science Research*, vol. 2, no. 6, pp. 532–540, 2011.

[4] G. Rodionova, M. Lenes, Ø. Eriksen, and Ø. Gregersen, "Surface chemical modification of microfibrillated cellulose: improvement of barrier properties for packaging applications," *Cellulose*, vol. 18, no. 1, pp. 127–134, 2011.

[5] R. J. Moon, A. Martini, J. Nairn, J. Simonsen, and J. Youngblood, "Cellulose nanomaterials review: structure, properties and nanocomposites," *Chemical Society Reviews*, vol. 40, no. 7, pp. 3941–3994, 2011.

[6] M. Sercer, P. Raos, and M. Rujnic-Sokele, Processing of wood-thermoplastic composites, http://www.jobwerks.com/news/Archives/iwpc.pdf.

[7] A. Biswas, B. C. Saha, J. W. Lawton, R. L. Shogren, and J. L. Willett, "Process for obtaining cellulose acetate from agricultural by-products," *Carbohydrate Polymers*, vol. 64, no. 1, pp. 134–137, 2006.

[8] G. A. Olatunji, "The eighty-eight inaugural lecture," in *Journey to the Promised Land: The Travails of an Organic Chemist*, p. 17, Department of Chemistry, Faculty of Science, University of Ilorin, Ilorin, Nigeria, 2009.

[9] B. Mohebby, "Application of ATR infrared spectroscopy in wood acetylation," *Journal of Agricultural Science and Technology*, vol. 10, no. 3, pp. 253–259, 2008.

[10] L. M. Ilharco, A. R. Garcia, J. Lopes da Silva, and L. F. Vieira Ferreira, "Infrared approach to the study of adsorption on cellulose: influence of cellulose crystallinity on the adsorption of benzophenone," *Langmuir*, vol. 13, no. 15, pp. 4126–4132, 1997.

[11] I. Filpponen, *The synthetic strategies for unique properties in cellulose nanocrystals materials [Ph.D. thesis of Philosophy]*, Graduate Faculty of North Carolina State University, Wood & Paper Science, Raleigh, NC, USA, 2009.

[12] A. K. Bledzki, A. A. Mamun, M. Lucka-Gabor, and V. S. Gutowski, "The effects of acetylation on properties of flax fibre and its polypropylene composites," *Express Polymer Letters*, vol. 2, no. 6, pp. 413–422, 2008.

[13] A. Yakubu, G. A. Olatunji, O. Sunday, and A. Olubunmi, "Ketene acetylated wood cellulose for industrial applications in wood-base and polymer industry," *Journal of Environmental Science and Technology*, vol. 5, no. 3, pp. 168–176, 2012.

[14] G. T. Ciacco, D. L. Morgado, E. Frollini, S. Possidonio, and O. A. El Seoud, "Some aspects of acetylation of untreated and mercerized sisal cellulose," *Journal of the Brazilian Chemical Society*, vol. 21, no. 1, pp. 71–77, 2010.

[15] S. Samira, A. I. Nor, A. Sanaz, W. Y. Wan Md Zin, and A. R. M. Zaki, "Effects of fibre esterification on fundamental properties of oil palm empty fruit bunch fibre/poly (butylenes adipate-co-terephthalate) biocomposites," *International Journal of Molecular Sciences*, vol. 13, no. 2, pp. 1327–1346, 2012.

[16] S. Kamel, "Nanotechnology and its applications in lignocellulosic composites, a mini review," *Express Polymer Letters*, vol. 1, no. 9, pp. 546–575, 2007.

[17] R. M. Rowell, R. H. S. Wang, and J. A. Hyatt, "Flakeboards made from Aspen and Southern Pine Wood flakes Reacted with Ketene," *Journal of Wood Chemistry and Technology*, vol. 6, no. 3, pp. 449–471, 1986.

[18] A. G. Supri and B. Y. Lim, "Effect of treated and untreated filler loading on the mechanical, morphological, and water absorption properties of water hyacinth fibers-low density polyethylene composites," *Journal of Physical Science*, vol. 20, no. 2, pp. 85–96, 2009.

[19] S. A. Abdulkareem and B. Garba, "Novel application of polymer dissolution technique," *Nigerian Journal of Pure and Applied Sciences*, vol. 20, pp. 1799–1485, 2005.

[20] J. George, S. S. Bhagawan, and S. Thomas, "Effects of environment on the properties of low-density polyethylene composites reinforced with pineapple-leaf fibre," *Composites Science and Technology*, vol. 58, no. 9, pp. 1471–1485, 1998.
[21] V. Tserki, N. E. Zafeiropoulos, F. Simon, and C. Panayiotou, "A study of the effect of acetylation and propionylation surface treatments on natural fibres," *Composites A*, vol. 36, no. 8, pp. 1110–1118, 2005.

Partial Chemical Characterization of Immunomodulatory Polysaccharides from *Plantago palmata* Hook. f. s. Leaves

Gabriel Biringanine,[1] Moustapha Ouedraogo,[2] Bernard Vray,[3] Anne Berit Samuelsen,[4] and Pierre Duez[1]

[1] *Laboratoire de Pharmacognosie, de Bromatologie, et de Nutrition Humaine, Faculté de Pharmacie, Université Libre de Bruxelles (ULB), Campus Plaine CP 205/9, Boulevard du Triomphe, 1050 Bruxelles, Belgium*
[2] *Laboratoire de Pharmacologie et de Toxicologie, Unité de Formation et de Recherche en Sciences de la Santé, Université de Ouagadougou, Ouagadougou 03, Burkina Faso*
[3] *Laboratoire d'Immunologie Expérimentale, Faculté de Médecine, Université Libre de Bruxelles (ULB), Campus Erasme, Route de Lennik, 1070 Bruxelles, Belgium*
[4] *Department of Pharmaceutical Chemistry-Pharmacognosy, School of Pharmacy, University of Oslo, P.O. Box 1068, Blindern, 0316 Oslo, Norway*

Correspondence should be addressed to Pierre Duez, pduez@ulb.ac.be

Academic Editor: Jianjun Li

A previous work on *Plantago palmata* polysaccharides (PS) attributed immunomodulatory properties of leaves to a polysaccharide fraction (PS50) that stimulated NO and TNF-α production by interferon gamma- (IFN-γ-) activated macrophages. The present work aims to elucidate the chemical structure of these immunomodulatory polysaccharides. Size exclusion chromatography showed that the active polymers present an active fraction with a very high molecular weight (about 1200 kDa). These polysaccharides are pectic in nature, with a predominantly unbranched galacturonan domain and with a domain bearing side chains that consist of highly branched arabinan, galactan, and/or arabinogalactan. Comparatively to the well-known *Plantago major* biologically active PS, *Plantago palmata* PS50 contained less arabinogalactan-proteins (AGPs) and had a different composition in glucose, galactose, and galacturonic acid. DNA contamination of the polysaccharide was estimated at about 0.04%, a concentration much lower than those reported immunomodulatory in hyaluronic acid preparations (3 to 15%). Therefore, the eventuality of a contaminating DNA-mediated biological activity could be ruled out.

1. Introduction

Plantago palmata grows in the humid mountain regions of intertropical Africa at 1800 to 3000 m [1] and is widely used in traditional medicine in Burundi, Rwanda and South Kivu (Democratic Republic of Congo). Its leaves are often used as herbal remedies in the treatment of a number of diseases such as wounds, hepatitis, parasitic or infectious diseases, and pregnancy, and lactation troubles. They are also used for the improvement of health after disease [2, 3]. A survey of the literature indicates a certain convergence between these medicinal uses (antibacterial, antifungal, antiviral and, immunomodulatory) and those of other *Plantago* species [4–11]. This led us to investigate three groups of *Plantago* metabolites possibly related to these traditional uses and biological activities, that is, iridoids, phenylethanoid glycosides and polysaccharides [3, 12].

We previously showed that *Plantago palmata* leaves (i) contain appreciable amounts of iridoids and phenylethanoid glycosides, with aucubin and acteoside as major compounds [2]; (ii) present immunomodulatory properties, which could be attributed to polysaccharides (PS) [3].The effects of crude extracts and fractions obtained by high performance liquid chromatography/size exclusion chromatography (SEC) were investigated by the measurement of nitric oxide (NO), tumor necrosis factor-alpha (TNF-α), and interleukin-10 (IL-10) production using interferon gamma- (IFN-γ-) activated macrophages and nonactivated macrophages as control. IFN-γ-activation of J774 cell line macrophages synergized with a raw 50°C aqueous extract from leaves (PS50) and one of its fractions obtained by SEC (fraction F2) to produce both NO and TNF-α, but not IL-10. Eventual contamination of PS extracts and fractions by lipopolysaccharide (LPS) was ruled out by showing the same pattern of response on peritoneal macrophages harvested from C3H/Hej mice, a strain known to be a very low responder to LPS.

Due to their potential biological activities, such as antioxidant, antiulcer, hepatoprotective, antitumor, and immunostimulatory effects, polysaccharides from natural sources have attracted more attention in recent years [13–17]. To the best of our knowledge, primary structures, spatial shapes, and functions of PS are intimately related. Therefore as a prerequisite to better understand immunomodulatory activities of *P. palmata* polysaccharides, but also to develop parameters for quality control, the present work aims to investigate their primary chemical structures.

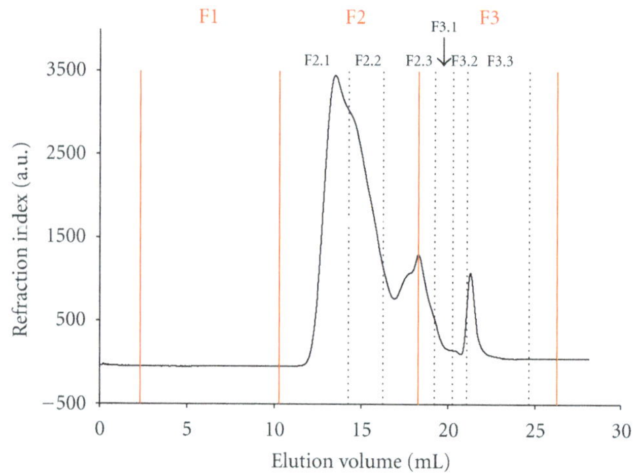

Figure 1: SEC chromatogram of *P. palmata* PS50. Columns, SHODEX OH-pack SB-804-HQ and OH-pack SB-806M-HQ (serially connected; exclusion limits, 1000 and 20000 kDa, resp.); mobile phase, 5 mM NaCl (aq) at 0.8 mL/min. Three fractions (F1 to F3) and 6 subfractions (F2.1 to F2.3) were collected for biological testing.

2. Experimental

2.1. Plant Material and Polysaccharides (PS) Extraction. *Plantago palmata* seeds were harvested in the Democratic Republic of Congo and grown in a greenhouse (Experimental Garden Jean Massart, Université Libre de Bruxelles, Belgium). After three months of culture, fresh leaves were collected, immediately immersed into acetone, and stored in this solvent for a maximum of three months. The material was washed three times with acetone and the extraction of PS was performed as previously described [18] with some modifications. Briefly, dried and crushed leaves (17.6 g) were extracted by water (1 × 1000 mL) at 50°C for three hours and filtered. The extract was concentrated to 20 mL at 30°C under vacuum. The residual crushed leaves were further incubated in water at 100°C (1 × 1000 mL) and similarly treated. Both extracts were dialyzed (Medicell International Ltd., London, MW 12–1400 Da) against distilled water for 48 h at 4°C and lyophilized to yield extracts designated as PS50 and PS100, respectively; filtration on a Millipore cellulose filter (0.22 μm) yielded extracts PS50F and PS100F. These extracts were maintained at −18°C. Biological testing of both extracts, before filtration, showed a synergistic effect between IFN-γ and PS50 only; this effect was abolished by the filtration step. For structure comparison purposes, the same extraction was performed on *Plantago major* leaves.

2.2. Fractionation of the PS50 Crude Extract. Two mg of the PS crude extract were dissolved by swelling in 1 mL of 5 mM NaCl (aq) (24 h), sonicating (2 min) and vortexing (1 min) to obtain an homogeneous solution; 25 μL of this solution were injected in a HPLC Agilent 1100 system (Hewlett Packard, Waldbronn, Germany) equipped with a Agilent 1100 RID detector, and two polyhydroxymethacrylate gel permeation columns (SHODEX OH-pack SB-804-HQ followed by a OH-pack SB-806 M-HQ, 300 × 8 mm i.d.; exclusion limits, 1000 and 20000 kDa; Alltech Associates, Deerfield, USA) connected in series with a PWXL 12 μm, 40 × 6 mm i.d. guard column (Tosoh Bioscience, Montgomeryville, USA). The system was driven by a ChemStation A.09.01 software equipped with the GPC add-on A.02.01 (Agilent). The mobile phase was 5 mM NaCl (aq) at a flow-rate of 0.8 mL/min. Three fractions and six subfractions were collected according to the elution volume (Figure 1). After four injections, the collected fractions were pooled and dialysed (silver-A-Lyzer, Rockford, USA) at 4°C against distilled water for 3 days, renewing the dialysis bath every 24 hours, and dialysates were lyophilized. This size exclusion chromatography yielded a fraction significantly more active than PS50, fraction F2, but the subsequent subfractionations were unsuccessful to concentrate the activity [3].

2.3. Estimation of the Molecular Weights. After a 24 hours equilibration of the size exclusion chromatography (SEC) system with the mobile phase, a series of calibration standards were injected (5 mg/mL solutions of pullulan P5, P-20, P-100, P-200, P-400, P-800 in the mobile phase; 25 μL injection; 2 injections per calibrator), covering a 5.8 to 853 kDa M_w range (Shodex standard P-82 kit). The ChemStation GPC add-on A.02.01 (Agilent) was run on recorded chromatograms to establish a calibration curve "log M_{max} *versus* Elution volume" (fitting with a quadratic function) and the molecular weights of the polymers were estimated as number average molecular weights (M_n), weight average molecular weights (M_w), Z average molecular weights (M_z), and peak molecular weights (M_p).

2.4. Reaction with the Yariv β-Glucosyl Antigen. 3.5 mL of an agarose suspension (1% in 0.15 M NaCl and 0.02% w/v NaN_3 (aq)) were added with 10 μg/mL (aq) of the Yariv reagent (Biosupplies Australia) and cast on glass plates to form a 1-mm gel. After solidification, 1.2-mm i.d. wells were dug into the gel and filled with 0.8 μL of tested PS solutions

(0.1 to 1 μg/well). The plates were incubated overnight in a humidified atmosphere and red halos were compared with those obtained from an Arabic gum suspension (Federa, Brussels, Belgium) applied at the same concentrations.

2.5. Determination of the Monosaccharides Composition by HPLC. HPLC was performed upon prechromatographic derivatization with 3-methyl-1-phenyl-2-pyrazolin-5-one (MPP) according to Honda et al. [19]). Each PS (1 mg) or fraction (200 μg) was hydrolyzed in 1 mL 2 M trifluoroacetic acid (aq) (100°C, 1 hour), dried under vacuum (SpeedVac) and taken up in 3 portions of 500 μL methanol (30 sec sonication). The combined methanolic solutions were added or not with 10 μL of the internal standard (I.S., talose, 1 mg/mL (aq)), dried, taken up in 150 μL water, and filtered.

Individual solutions of monosaccharides (arabinose, fucose, glucose, galactose, mannose, rhamnose, xylose, glucuronic acid, and galacturonic acid) were dissolved in methanol (1 mg/mL), mixed in different proportions to prepare calibrators (2 to 40 μg of each ose), evaporated to dryness and dissolved in 150 μL water.

Hydrolyzed polymers or calibrators were added with 20 μL of a 0.5 M methanolic solution of 3-methyl-1-phenyl-2-pyrazolin-5-one (Acros Organics, New Jersey, USA) and 20 μL of 0.3 M NaOH (aq), incubated 2 h at 70°C, neutralized by 20 μL 0.3 M HCl (aq) and extracted 3 times with 500 μL of butyl ether (10 s vortex, 5 min centrifugation at 2000 rpm). The aqueous phase was then dried under vacuum (SpeedVac) and redissolved in 150 μL water; 10 to 25 μL of this solution were injected in a HPLC Agilent 1100 system equipped with a Agilent 1100 DAD detector set at 254 nm and a 250 × 4.6 mm i.d. Atlantis dC_{18} column (Waters, Ireland). The mobile phases A (triethylamine : acetonitrile : water, 0.4 : 10: 90, v/v) and B (acetonitrile : water, 60 : 40, v/v) were pumped at 1 mL/min with the following gradient: 0 min, 10% B; 22 min, 28% B; 55 min, 28% B; 56 min, 10% B; 65 min, 10% B.

2.6. Determination of the Monosaccharide Composition by TLC. Hydrolyzed PS (5 μg/spot) and standards (0.5 μg/spot) were applied on silicagel 60 F_{254} plates (Merck, Germany). The plates were eluted with acetonitrile: water (80 : 20) (2 developments over 9 cm with mobile phase renewing), dried, sprayed with a mixture 0.2% w/v naphtoresorcinol in ethanol : H_2SO_4 (96 : 4, v : v) and heated at 105°C for 15 min.

2.7. Linkage Analysis. Prior to linkage analysis by methylation and GC-MS [20], the uronic acid groups of the PS were reduced. PS50 (1 mg) or its fractions (200 μg) were treated by sodium borodeuteride ($NaBD_4$) after activation with carbodiimide [21]; the use of borodeuteride allows to differentiate the masses of the reduced uronic acids from the corresponding neutral monosaccharides [22]. The reduced PS were then methylated by methyl iodide and hydrolyzed in 2.5 M TFA (aq) (100°C, 1 hour). After reduction by $NaBD_4$ and acetylation with acetic anhydride [20], the resulting partially methylated alditol acetates were analyzed by GC-MS (I.S.: inositol) on a Hewlett-Packard methyl silicone capillary column (12 m × 0.22 mm i.d., 0.33 μm film) in the E.I. ionization mode, as previously described [22].

Methanolysis was performed according to the method of Chambers and Clamp [23] with modifications as previously described [18], using 4 M HCl in anhydrous methanol for 24 h at 80°C; the methyl glycosides were then silylated and analyzed by GC [18].

2.8. Measurement of DNA Contamination. Two mg of the crude PS extract of *Plantago major* or *Plantago palmata* were dissolved by swelling overnight at 4°C in 8 mL buffer (10 mM Tris, 0.2 M NaCl, 1 mM EDTA (aq), pH 7.4). A series of dilutions of this solution were analyzed with the kit DNA quantitation Hoechst 33258 bisbenzimide (Bio-Rad), following the manufacturer's protocol (reference: calf thymus DNA).

3. Results and Discussion

3.1. Harvesting and Extraction. Preliminary extraction trials on *Plantago palmata* leaves that were dried upon harvesting resulted in a brownish material, heavily contaminated with chlorophyll. Colorless material could not be separated, which may indicate an enzymatic activity during the drying step, probably leading to some covalent bonding. Postharvesting treatment with acetone overcame the problem, removing chlorophyll and other colored low molecular weight lipophilic material and inactivating enzymes, so that white PS fractions could be obtained.

3.2. Fractionation of the PS50 Crude Extract. Figure 1 shows a typical SEC chromatogram obtained for *Plantago palmata* PS50, indicating the collected fractions; biological data [3] indicate that fraction F2 is the most active one.

Referring to a calibration curve obtained with pullulan standards, the molecular weights of PS50 and PS50F were estimated (Table 1). Three peaks could be discerned from the chromatogram of the active PS50, a very high molecular weight (M_w about 1200 kDa), and 2 lower weight (M_w about 270 kDa and 12 kDa, resp.) polymers.

As there is no obvious difference between the active PS50 and inactive PS50F, these data do not allow explaining the detrimental effect of the filtration step on biological activity [3].

As SEC determinations are not dependent on the molecular mass but on the hydrodynamic volume of the solute [24], these mass values however need confirming by an absolute method based on multiangle laser light scattering. Indeed arabinogalactan-proteins (AGPs) are known for globular structures which may strongly influence the retention on SEC columns [25].

3.3. Reaction with the Yariv β-Glucosyl Antigen. As arabinogalactans are important immunomodulatory substances, their presence within the polysaccharides has been investigated [26, 27]. The Yariv β-glucosyl reagent precipitates some AGPs of vegetal cell walls, especially those with a high proportion of 1,3,6-linked galactose residues [28, 29]. As this reagent presents a still unexplained specificity for polymers with bioactive structures, the formation of a red precipitate

Table 1: Molecular weights of polymers obtained from *Plantago palmata* leaves by aqueous extraction (50°C, 3 hours), concentration and extensive dialysis (PS50) and further filtration (PS50F) on a Millipore cellulose filter (0.22 μm). SEC analysis referring to a calibration curve obtained with pullulan standards ($n = 3$).

Polymer	Elution volume (mL)	M_n (g/mol)	M_w (g/mol)	M_z (g/mol)	M_p (g/mol)	Polydispersity (M_w/M_n)
PS50	11.8 to 13.8 mL[a]	1129867 (±39000)	1192500 (±51000)	1255767 (±65000)	1190167 (±37000)	1.05
	13.8 to 16.8 mL[b]	222537 (±16000)	271913 (±12000)	322567 (±8000)	303167 (±12000)	1.22
	16.8 to 19.6 mL[c]	39485 (±650)	11818 (±720)	14856 (±760)	9895 (±550)	1.24
PS50F	11.8 to 13.8 mL	1189650 (±9000)	1278550 (±14000)	1369650 (±20000)	1262000 (±13000)	1.07
	13.8 to 16.8 mL	230515 (±1000)	277390 (±620)	325485 (±605)	304600 (±6000)	1.20
	16.8 to 19.6 mL	10338 (±250)	12860 (±720)	16068 (±960)	10597 (±3300)	1.24

M_n = Number average molecular weight; M_w = Weight average molecular weight; M_z = Z average molecular weight; M_p = Peak molecular weight.
[a]Corresponds to fraction F2-1 (Figure 1).
[b]Corresponds to fraction F2-2 plus part of F2-3 enclosing the tail of the major peak.
[c]Comprehends the peak skimmed from the major peak (corresponds to the end of fraction F2-3 plus part of F3-1).

Table 2: Monosaccharides constitutive of the PS extracted from *Plantago palmata* and *Plantago major* (% of total carbohydrates).

	Sample		
	PS50 from *Plantago major*[a]	PS50 from *Plantago palmata*[a]	Fraction F2 from *Plantago palmata*[b]
Man	5.7% (5.8%–5.5%)	5.4% (4.9%–6.0%)	27.7%
Ara/Xyl[c]	27.8% (28.8%–26.8%)	26.8% (25.5%–28.0%)	17.7%
Fuc	Traces	Traces	n.d.
Glc	20.7% (21.0%–20.3%)	4.8% (4.7%–4.9%)	16.9%
Gal	30.0% (30.8%–28.7%)	39.2% (37.4 %–41.0%)	13.3%
Rha	7.6% (5.6%–9.6%)	9.6% (10.9 %–8.3%)	12.5%
GlcA	Traces	n.d.	n.d.
GalA	8.5% (8.1%–8.8%)	14.2% (16.5% –11.9%)	11.8%

[a]Data from 2 experiments in duplicate; in brackets, the values from the 2 experiments.
[b]Data from 1 experiment in duplicate.
[c]TLC shows that the ratio Ara/Xyl is approximately 9 : 1 in both *Plantago* (Figure 4).

is a recognized indication for the presence of bioactive AGPs [30].

As AGPs with immunomodulatory (anticomplementary) activity have been shown in *Plantago major* [18, 22], we compared its AGPs content with *P. palmata* (Figure 2). Although the method is only semiquantitative (rapid saturation of the halo with increasing concentrations), it is clear that both *Plantago* contain AGPs, but with a significantly lower amount in *P. palmata*.

3.4. Determination of the Monosaccharides Composition. The composition in monomers was determined by HPLC with 3-methyl-1-phenyl-2-pyrazolin-5-one (MPP) prechromatographic derivatization (Figure 3, Table 2). The method was found linear in the investigated range (2 to 40 μg monomers derivatized, typical $R^2 > 0.990$ with or without I.S.), allowing the determination of neutral monosaccharides and uronic acids in a single step, without the need for acid groups reduction. The resolution was however lower than in GC and we could separate neither rhamnose from the intended internal standard, talose, nor arabinose from xylose. As data computed with or without I.S. were very similar, talose was

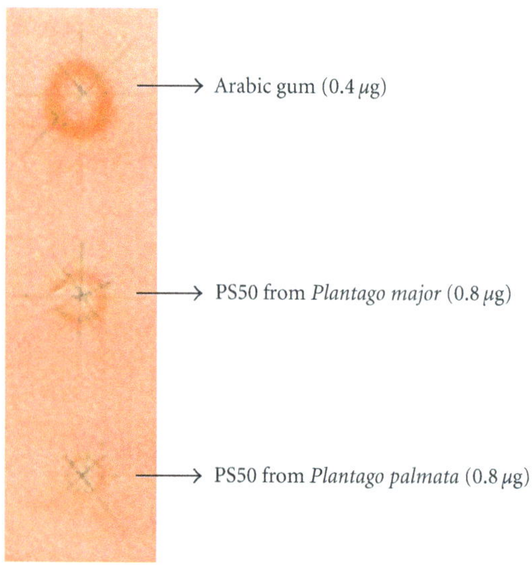

Figure 2: Precipitation of the Yariv β-glucosyl antigen by PS50 extracted from *Plantago major* or *palmata* (1% agarose gel added with 10 μg/mL of the Yariv reagent (aq); 0.8 μL/well of tested PS solutions; overnight incubation).

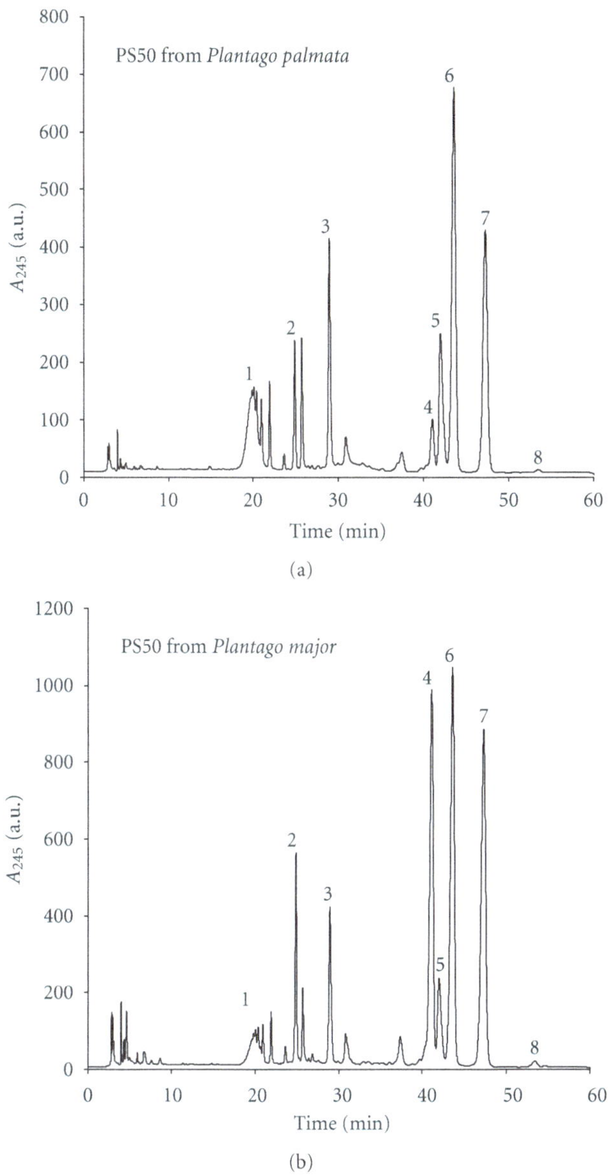

Figure 3: HPLC chromatogram of monosaccharides constitutive of the PS50 extracted from *Plantago palmata* and *Plantago major* (Atlantis dC_{18} column; gradient of mobile phases A (water: acetonitrile : triethylamine, 90 : 10: 0.4, v/v) and B (water : acetonitrile, 40 : 60, v/v)). *Peaks:* 1, excess MPP reagent; 2, mannose; 3, rhamnose; 4, glucose; 5, galacturonic acid; 6, galactose; 7, xylose + arabinose; 8, Fucose.

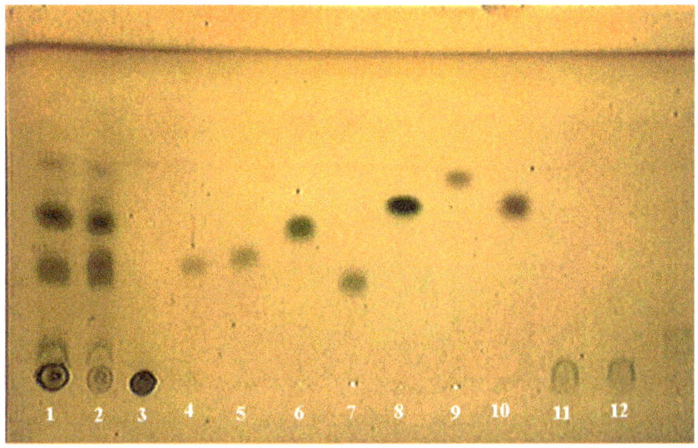

Figure 4: TLC chromatogram of monosaccharides constitutive of the PS50 extracted from *Plantago palmata* and *major* (silicagel 60 F_{254}, elution with acetonitrile: water (80 : 20), revelation with naphtoresorcinol-H_2SO_4). *Lanes* : 1, hydrolyzed PS50 from *P. palmata*; 2, hydrolyzed PS50 from *P. major;* 3, nonhydrolyzed PS50 from *P. palmata*; 4, glucose; 5, mannose; 6, arabinose; 7, galactose; 8, xylose; 9, rhamnose; 10, fucose; 11, galacturonic acid; 12, glucuronic acid.

omitted from the analysis. Both pentoses could be readily resolved by TLC (Figure 4) and GC after methanolysis which both showed that the ratio ara/xyl was approximately 9 : 1 in the PS50 from *Plantago major* and *Plantago palmata*. Both PS50 had a roughly similar composition, compatible with pectins; there were however some differences and *Plantago palmata* PS50 showed a different content in glucose and, to a lesser extent, in galactose and galacturonic acid.

3.5. Linkage Analysis. The major linkage is 1,4-GalA followed by 1,3,6-Gal. Three major pectic polysaccharides are known to exist in all cell walls, (i) an unbranched homogalacturonan domain (compatible with the observed high ratio of 1,4-GalA in PS50); (ii) a highly diversified branched rhamnogalacturonan-I domain with arabinan/galactan/arabinogalactan lateral chains; and (iii) a highly conserved rhamnogalacturonan-II domain [30]. The linkages found (Table 3) are typical for homogalacturonan regions (1,4-GalA) and hairy regions of pectins [31]. The rhamnose residues are part of the pectin backbone whereas neutral side chains composed of arabinogalactans, galactans and arabinans are linked to position 4 of the 1,2-linked rhamnose residues. The galactose residues are 1,3-, 1,6-, and 1,3,6-linked while the majority of the arabinose residues are terminally linked indicating the presence of arabinogalactan. The presence of such structures was confirmed by the positive reaction with Yariv antigen. Most of the linkages found in PS50 were also detected in PS50 fraction 2 but could not be quantified due to the small sample size available for analysis.

From previous studies, immunomodulatory activities of pectin polysaccharides seem to depend on the presence of arabinogalactan side chains [32–34]. This might also be the case in the present study, but it remains to confirm this by isolation of pectin hairy regions from *Plantago palmata* PS50.

The fragmentation by size exclusion chromatography allows to partly resolve the complex and high molecular weight PS fraction but yields only small amounts of polymers, insufficient for complete characterization. The elucidation of structure probably requires additional methods, based on anion exchange chromatography, enzymatic degradation of the polymer and mass spectroscopy of fragments. Such a glycomic approach however still requires major developments, comparatively to genomic and proteomic. Indeed, the many linkages/branching possibilities and the similar *M/Z* of many of the monomers limit analytical possibilities; moreover the available glycosidases still either lack specificity or are significantly contaminated with undesired activities.

Table 3: Products of the PS extracted from *Plantago palmata* (% mole) after reduction, methylation, hydrolysis, and acetylation, followed by GC-MS.

Monosaccharide	Linkage	PS50 from *Plantago palmata* (%)	Fraction 2 from *Plantago palmata* PS50[a]
	Terminal	7.3	x
	1, 2	1.3	x
Ara *f*	1, 3	1.3	x
	1, 5	0.6	x
	Total	10.5	
Fuc *p*	Terminal	3.4	
	Terminal	3.8	x
	1, 3	2.1	x
Gal	1, 6	5.0	x
	1, 3, 6	13.8	x
	Total	24.7	
	Terminal	1.4	x
Glc	1, 4	6.3	x
	Total	7.7	
	Terminal	1.5	
	1, 2	1.0	
Rha *p*	1, 3	0.4	x
	1, 2, 4	1.1	
	Total	4.0	
	Terminal	0.5	x
Xyl *p*	1, 4	1.4	
	Total	1.9	
	Terminal	0.8	x
	1, 4	38.7	xx
GalA	1, 3, 4	0.5	
	1, 2, 4	0.4	
	Total	40.4	
GlcA	Terminal	1.9	x

[a]% linkage not quantifiable due to the minute amount of PS analyzed; the x indicates that the linkage was effectively detected.

3.6. Polysaccharides Contamination by DNA. DNA contaminating hyaluronic acid preparations has been shown to be responsible for the induction of proinflammatory cytokines IL-12 and TNF-α in human monocytes [35]. The DNA contents of *Plantago palmata* and *Plantago major* PS50 were 0.04 ± 0.01% and 0.08 ± 0.01%, respectively. These concentrations are much lower than those reported active in hyaluronic acid preparations (3 to 15%) [35]. This most probably rules out the eventuality of a contaminating DNA-mediated biological activity on J774 cells.

4. Conclusion

The new polysaccharide with immunomodulatory activities isolated from *Plantago palmata* has a very high molecular weight (1200 kDa) and is pectic in nature, with a predominantly unbranched galacturonan domain and with a domain bearing side chains that consist of highly branched arabinan, galactan, and/or arabinogalactan. Comparatively to the well-known *Plantago major* biologically active polysaccharide (PS), *Plantago palmata* PS50 contain less arabinogalactan-proteins and has different contents of glucose, galactose, and galacturonic acid. It would be interesting to compare the biological activities of both *Plantago*.

Acknowledgments

The authors are grateful to Mrs. Helson-Cambier and Mrs. M. Faes for their precious assistance for the chromatographic developments and to Mr. O. Vaillant for maintaining excellent conditions of work in the laboratory.

References

[1] S. Lisowski, F. Malaisse, and J. J. Symoens, *Flore d'Afrique Centrale, Zaire-Rwanda-Burundi, Spermatophytes, Plantaginaceae*, Université Nationale du Zaire, Lubumbashi, Jardin Botanique National de Belgique, Bruxelles, Belgium, 1972.

[2] G. Biringanine, M. T. Chiarelli, M. Faes, and P. Duez, "A validation protocol for the HPTLC standardization of herbal products: application to the determination of acteoside in leaves of *Plantago palmata* Hook. f.s.," *Talanta*, vol. 69, no. 2, pp. 418–424, 2006.

[3] G. Biringanine, B. Vray, V. Vercruysse, R. Vanhaelen-Fastré, M. Vanhaelen, and P. Duez, "Polysaccharides extracted from the leaves of *Plantago palmata* Hook.f. Induce nitric oxide and tumor necrosis factor-α production by interferon-γ-activated macrophages," *Nitric Oxide*, vol. 12, no. 1, pp. 1–8, 2005.

[4] C. Long, C. Moulis, E. Stanislas, and I. Fouraste, "Aucuboside and catapol in *Plantago lanceolata* L., *Plantago major* L., *Plantago media* L. leaves," *Journal de Pharmacie de Belgique*, vol. 50, no. 6, pp. 484–488, 1995.

[5] M. D. Bowers, "Variation in iridoid glycosides in a population of *Plantago patagonica* Jacq. (Plantaginaceae) in Colorado," *Biochemical Systematics and Ecology*, vol. 24, no. 3, pp. 207–210, 1996.

[6] K. Darrow and M. D. Bowers, "Phenological and population variation in iridoid glycosides of *Plantago lanceolata* (Plantaginaceae)," *Biochemical Systematics and Ecology*, vol. 25, no. 1, pp. 1–11, 1997.

[7] F. Fons, A. Gargadennec, A. Gueiffier, J. L. Roussel, and C. Andary, "Effects of cinnamic acid on polyphenol production in *Plantago lanceolata*," *Phytochemistry*, vol. 49, no. 3, pp. 697–702, 1998.

[8] J. Bruneton, *Pharmacognosie: Phytochimie. Plantes Médicinales*, Tec & Doc, Paris, France, 1999.

[9] D. H. Paper and M. Marchesan, "Plantain (*Plantago lanceolata* L.): its introduction, analysis, constituents, pharmacology and standardisation," *Zeitschrift fur Phytotherapie*, vol. 20, no. 4, pp. 231–238, 1999.

[10] N. Rønsted, E. Göbel, H. Franzyk, S. R. Jensen, and C. E. Olsen, "Chemotaxonomy of *Plantago*. Iridoid glucosides and caffeoyl phenylethanoid glycosides," *Phytochemistry*, vol. 55, no. 4, pp. 337–348, 2000.

[11] A. B. Samuelsen, "The traditional uses, chemical constituents and biological activities of *Plantago major* L. A review," *Journal of Ethnopharmacology*, vol. 71, no. 1-2, pp. 1–21, 2000.

[12] R. Slaveska-Raicki, V. Rafajlovska, V. Rizova, and I. Spirevska, "HPTLC determination of gallic acid and tannin in extracts of

bearberry leaves," *Journal of Planar Chromatography*, vol. 16, no. 5, pp. 396–401, 2003.

[13] R. Li, W. C. Chen, W. P. Wang, W. Y. Tian, and X. G. Zhang, "Extraction, characterization of *Astragalus* polysaccharides and its immune modulating activities in rats with gastric cancer," *Carbohydrate Polymers*, vol. 78, no. 4, pp. 738–742, 2009.

[14] M. Zhang, S. W. Cui, P. C. K. Cheung, and Q. Wang, "Antitumor polysaccharides from mushrooms: a review on their isolation process, structural characteristics and antitumor activity," *Trends in Food Science and Technology*, vol. 18, no. 1, pp. 4–19, 2007.

[15] L. Xia, X. Liu, H. Guo et al., "Partial characterization and immunomodulatory activity of polysaccharides from the stem of *Dendrobium officinale (Tiepishihu) in vitro*," *Journal of Functional Foods*, vol. 4, no. 1, pp. 294–301, 2012.

[16] Y. Sun, H. Liang, X. Zhang, H. Tong, and J. Liu, "Structural elucidation and immunological activity of a polysaccharide from the fruiting body of *Armillaria mellea*," *Bioresource Technology*, vol. 100, no. 5, pp. 1860–1863, 2009.

[17] Y.-X. Sun and J. F. Kennedy, "Antioxidant activities of different polysaccharide conjugates (CRPs) isolated from the fruiting bodies of *Chroogomphis rutilus* (Schaeff.: Fr.) O. K. Miller," *Carbohydrate Polymers*, vol. 82, no. 2, pp. 510–514, 2010.

[18] A. B. Samuelsen, B. S. Paulsen, J. K. Wold, H. Otsuka, H. Yamada, and T. Espevik, "Isolation and partial characterization of biologically active polysaccharides from *Plantago major* L.," *Phytotherapy Research*, vol. 9, no. 3, pp. 211–218, 1995.

[19] S. Honda, E. Akao, S. Suzuki, M. Okuda, K. Kakehi, and J. Nakamura, "High-performance liquid chromatography of reducing carbohydrates as strongly ultraviolet-absorbing and electrochemically sensitive 1-phenyl-3-methyl-5-pyrazolone derivatives," *Analytical Biochemistry*, vol. 180, no. 2, pp. 351–357, 1989.

[20] M. J. McConville, S. W. Homans, J. E. Thomas-Oates, A. Dell, and A. Bacic, "Structures of the glycoinositolphospholipids from *Leishmania major*. A family of novel galactofuranose-containing glycolipids," *Journal of Biological Chemistry*, vol. 265, no. 13, pp. 7385–7394, 1990.

[21] J. B. Kim and N. C. Carpita, "Changes in esterification of the uronic acid groups of cell wall polysaccharides during elongation of maize coleoptiles," *Plant Physiology*, vol. 98, no. 2, pp. 646–653, 1992.

[22] A. B. Samuelsen, B. S. Paulsen, J. K. Wold, S. H. Knutsen, and H. Yamada, "Characterization of a biologically active arabinogalactan from the leaves of *Plantago major* L.," *Carbohydrate Polymers*, vol. 35, no. 3-4, pp. 145–153, 1998.

[23] R. E. Chambers and J. R. Clamp, "An assessment of methanolysis and other factors used in the analysis of carbohydrate-containing materials," *Biochemical Journal*, vol. 125, no. 4, pp. 1009–1018, 1971.

[24] K. Štulík, V. Pacáková, and M. Tichá, "Some potentialities and drawbacks of contemporary size-exclusion chromatography," *Journal of Biochemical and Biophysical Methods*, vol. 56, no. 1–3, pp. 1–13, 2003.

[25] L. Picton, I. Bataille, and G. Muller, "Analysis of a complex polysaccharide (gum arabic) by multi-angle laser light scattering coupled on-line to size exclusion chromatography and flow field flow fractionation," *Carbohydrate Polymers*, vol. 42, no. 1, pp. 23–31, 2000.

[26] P. Ghildyal, T. E. Grønhaug, A. Rusten et al., "Chemical composition and immunological activities of polysaccharides isolated from the malian medicinal plant syzygium guineense," *Journal of Pharmacognosy and Phytotherapy*, vol. 2, no. 6, pp. 76–85, 2010.

[27] M. Peters, M. Kauth, O. Scherner et al., "Arabinogalactan isolated from cowshed dust extract protects mice from allergic airway inflammation and sensitization," *Journal of Allergy and Clinical Immunology*, vol. 126, no. 3, pp. 648.e4–656.e4, 2010.

[28] M. A. Jermyn and Y. M. Yeow, "A class of lectins present in the tissues of seed plants," *Australian Journal of Plant Physiology*, vol. 2, pp. 501–531, 1975.

[29] G.-J. van Holst and A. E. Clarke, "Organ-specific arabinogalactan-proteins of *Lycopersicon peruvianum* (Mill) demonstrated by crossed electrophoresis," *Plant Physiology*, vol. 80, pp. 786–789, 1986.

[30] T. Bacic, "Herbal arabinogalactan-(proteins): structure, biosynthesis & biological function in plant. Development and industrial use," in *Proceedings of the 51st Annual Congress of the Society for Medicinal Plant Research ("Fascination, Facts, Fucture")*, Society for Medicinal Plant Research (Gesellschaft fur Arzneipflanzenforschung-GA), 2003.

[31] F. Voragen, G. Beldman, and H. Schols, "Chemistry and enzymology of pectins," in *Advanced Dietary Fibre Technology*, B. V. McCleary and L. Prosky, Eds., pp. 379–398, Blackwell Science, London, UK, 2001.

[32] H. Yamada, "Pectic polysaccharides from Chinese herbs: structure and biological activity," *Carbohydrate Polymers*, vol. 25, no. 4, pp. 269–276, 1994.

[33] X.-S Wang, Y. Zheng, J. Zuo, and J. Fang, "Structural features of an immunoactive acidic arabinogalactan from *Centella asiatica*," *Carbohydrate Polymers*, vol. 59, no. 3, pp. 281–288, 2005.

[34] A. B. Samuelsen, B. S. Paulsen, J. K. Wold et al., "Characterization of a biologically active pectin from *Plantago major* L.," *Carbohydrate Polymers*, vol. 30, no. 1, pp. 37–44, 1996.

[35] M. C. Filion and N. C. Phillips, "Pro-inflammatory activity of contaminating DNA in hyaluronic acid preparations," *Journal of Pharmacy and Pharmacology*, vol. 53, no. 4, pp. 555–561, 2001.

10

The Study of the Reaction of Pectin-Ag(0) Nanocomposites Formation

Nadezhda Viktorovna Ivanova, Natalya Nikolaevna Trofimova, Lydmila Alexandrovna Es'kova, and Vasilyi Anatolievich Babkin

A.E. Favorsky Institute of Chemistry, Russian Academy of Sciences, Siberian Branch, Irkutsk 664033, Russia

Correspondence should be addressed to Natalya Nikolaevna Trofimova, natrof@inbox.ru

Academic Editor: Yuriy A. Knirel

Pectin polysaccharides (PSs) were isolated from a bark of *Larix sibirica* Ledeb. Structure of PS fragments determined by chemical transformations, chromatography, and spectroscopic analyses was found to be a linear galacturonane comprising 1,4-linked α-D-Gal$_p$A residues and a rhamnogalacturonan I (RG-I). The fifth part of galacturonane residues was methyl esterified at at C-2 and/or C-3 and C-6 atoms. Some of RG-I side chains were identified as arabinogalactan subunits with highly branched structure consisting of linear backbone with $\rightarrow$ 3,6)-β-D-Gal$_p$-(1 $\rightarrow$ residues, substituted at C-6 by neutral side chains. This side chains contained $\rightarrow$ 2,5)-α-L-Ara$_f$-(1 $\rightarrow$ and $\rightarrow$ 3,5)-α-L-Ara$_f$-(1 $\rightarrow$ residues and terminal arabinose in the pyranose and furanose form. It was found that "pectin-Ag(0)" nanobiocomposites were formed via the interaction between PS aqueous solutions and silver nitrate, with PS playing both reducing and stabilizing functions. It was shown that the content of Ag(0) particles in "pectin-Ag(0)" depended on the reaction conditions and can range from 0.1 to 72 %, the size of Ag(0) particles being 3–27 nm. Using ^{13}C NMR technique, it was revealed that PS underwent destructive changes and they they were more considerable, more than the lot of Ag(I) that was inputed into the reactionary medium.

1. Introduction

Two species of the genus *Larix* Mill, *Larix sibirica* Ledeb and *Larix gmelinii* (Rupr.) Rupr., are considered to be the most abundant trees in the Russian Federation and the total stock of their wood recourses exceeds 26 billion m^3. Traditionally, the main economic benefit of the larch wood is a manufacture of roundwood (timber), the value of which is determined by a high quality of lumber obtained from this tree. Sometimes, the larch wood can be used in insignificant quantities in pulp-and-paper manufacture to produce pulp. Currently, about 40% of this valuable wood (bark, sawdust) is utilized as wastes. Such inexpedient approach to the utilization of larch wood does not allow using the richest potential of this renewable raw material. Meanwhile, biologically active compounds contained in larch biomass can be successfully applied for the manufacture of medical, food, and agricultural products. The development of complex technologies for chemical processing of larch biomass and waste timber will considerably raise the economic value of this biological resource.

However, the larch bark does not find industrial application. Annual volume of waste produced by wood-processing industry and the pulp-and-paper enterprises is more than 30 million m^3. It represents a serious environmental challenge because the bark is poorly exposed to biodegradation. At the same time, the chemical compounds of the bark can be used as a source of valuable biologically active substances including polysaccharides.

Earlier we have showed that the content of pectin polysaccharides in the larch bark is about 12% from the weight of absolutely dry raw material (a.d.r.m.) and the larch bark can be promising alternative raw material. The method of pectin isolation has been developed, and common physical and chemical characteristics and membranotropic activity of pectin have been investigated.

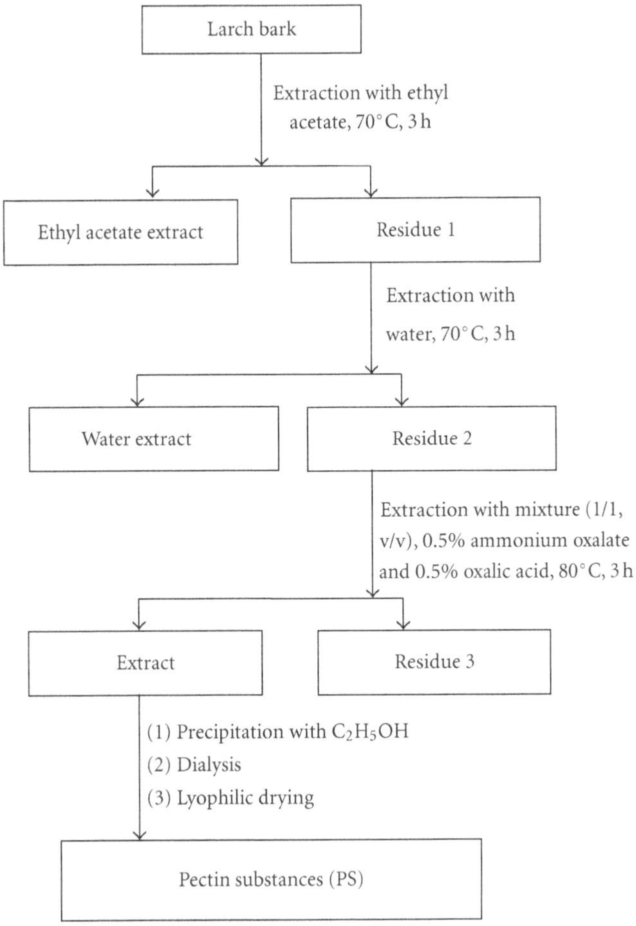

Figure 1: Scheme of pectin substances extraction from larch bark.

It has been established and patented that the larch bark pectin can possess reducing and stabilizing properties in the formation of nanobiocomposites with precious metals ions.

The present work aims at the following.

(1) The study of general characteristics of pectin substances (PS) extracted from bark of *L. sibirica* Ledeb. and structure investigation of its main linear and side chains.

(2) The study of PS interaction with silver nitrate and elucidation of alteration of pectin structural characteristics during the formation of "pectin-Ag(0)" nanobiocomposite.

2. Methods and Materials

Pectin substances were extracted from the bark of *L. sibirica* Ledeb. according to the scheme depicted in Figure 1. Larch air-dried bark (500 g) which was preliminary grinded and treated with ethyl acetate was extracted with distilled water at 70°C for 3 h. The residue of raw material was poured into a mixture (1 : 1, *v/v*) 0.5% aqueous solution of ammonium oxalate and 0.5% aqueous solution of oxalic acid and heated at 80°C for 2 h. The extract was concentrated. Polysaccharides were precipitated with triple volume of ethyl alcohol or acetone and dried with lyophilization. As a result, PS was obtained.

PS (50 mg) was dissolved in 20 mL of water. Pektinaza (2 mg, Sigma, USA) aqueous solution was added. The mixture was temperature-controlled at 37°C for 3 h. Then, a reaction mixture was heated for 5 min in water bath at 100°C. Coagulated protein was separated by centrifugation. Obtained supernatant was concentrated up to 5 mL, 96% ethyl alcohol was added (4 volumes). Deposition was separated by centrifugation. Alcoholic supernatant was concentrated and analyzed by paper chromatography (PC).

Paper chromatography was carried out on "Filtrak FN-13" paper with a descending method in n-butanol-pyridine-water system (volume correlations 6 : 4 : 3, resp.). To define carbohydrates, the paper was poured with aniline phthalate and heated at 105°C.

Glucuronic acids content in PS was defined according to the reaction with 3,5-dimethyl phenol in the presence of concentrated H_2SO_4, protein using Lowry method [1] with the calibrating schedule for a bovine serum albumin 80000 Da.

Gas-liquid chromatography (GLC) was carried out on a Hewlett-Packard 4890A (the USA) chromatograph equipped with flame-ionization detector, RTX-1 (0.25 mm × 30 m) capillary column, argon carrier gas, 1:60 dumping. Temperature rate: 175°C (1 min)–250°C (2 min), Δ3°/min.

Total acid hydrolysis PS (5 mg) was carried out with the implementation of 2 M trifluoroacetic acid (TFA) (2 mL) which contained *mio*-inositol (1 mg/mL). The mixture in soldered ampoule was heated for 5 h at 100°C, and the acid was removed by repeated dry evaporation with methanol addition.

Ion-exchange chromatography PS (100 mg) was carried out on DEAE-cellulose (25 × 2 cm) column. NaCl solutions were used as an eluent with increasing concentration (0.01 M–1 M, 60 mL/h elution's speed, fractions' selection by 12 mL). Pick correspondent fractions at the output bents were combined, dialyzed, and lyophilized. As a result, PS 1–4 fractions were obtained. Monosaccharides of each fraction after hydrolysis of PS 1–4 and acetilation were defined by GLC.

Monosaccharides acetilation: each PS 1–4 fraction was dissolved in 1 M ammonia solution (1 mL) and 5 mg of $NaBH_4$ was added. The mixture was kept at room temperature during one day. Then, $NaBH_4$ abundance was destroyed by adding 2-3 drops of concentrated acetic acid; 0.2 mL of dry pyridine and acetic anhydride were added to a dry residue. The mixture was being acetilized at 100°C during 1 h. The solution was evaporated up to pyridine and acetic anhydride abundance removal, first adding 1 mL of toluene and then 1 mL of methanol. The obtained acetate mixture of PS 1–4 polyol fractions was dissolved in 0.2 mL of dry chloroform and moved quantitatively to Appendorf tubes, concentrated up to 0.1-0.2 mL, and analyzed with GLC method.

PS (5 mg) partial acidic hydrolysis was carried out with 0.05 M TFA (2 mL) which contained *mio*-inositol (1 mg/mL). The mixture in soldered ampoule was heated at 100°C for 3 h. The acid was removed by repeated dry evaporation with methanol addition to give PS-s.

PS (5 mg) partial acidic hydrolysis was carried out with 0.01 M TFA (2 mL) which contained *mio*-inositol (1 mg/mL). The mixture in soldered ampoule was heated at 100°C for 3 h. The acid was removed by repeated dry evaporation with methanol addition to afford PS-g.

IR spectra were registered on a FT-IR (RAM II) Brukker Vertex 70 spectrometer in pellets with KBr.

^{13}C NMR spectra were recorded on a Bruker DRX 500 instrument (Germany) using 3–5% carbohydrate solution in D_2O at 55 and 70°C; the internal standard was DMSO-d6.

Synthesis of PS-g1 and PS-g1 nanobiocomposites "pectin-Ag(0)" was carried out according to the [2]. To 2 mL of the aqueous solution containing 1 g pectin, added 2 mL of 0.12–5.92 mmol silver nitrate solution and kept at room temperature for 30 min, added 10% sodium hydroxide solution up to pH 11, boiled on a water bath of 15 min and filtered through the paper filter. Obtained products were purified from low-molecular impurity by reprecipitations from ethanol, dried in vacuum over $CaCl_2$. The silver content in nanobiocomposites was determined titrimetrically with ammonium thiocyanate after preliminary metal translation in the ionic form by treatment with nitric acid by [3].

UV-Vis spectra of nanobiocomposite aqueous water solutions were recorded on a Perkin Elmer Lambda 35 instrument in ultraviolet and visible areas.

3. Results and Discussions

Pectin substances are abundant in land and water plants, as well as some freshwater algae [6]. Being an important component of cell walls, they are involved in ion exchange, water metabolism, and cell wall structure formation. They stimulate seed germination and germ growth, provide turgor, and so forth.

Unique physicochemical properties of pectin make it indispensable material in medical, food, and cosmetic industries as gelling agent, thickener, stabilizer, and dietary fiber. Since recently, it is extensively used as a matrix carrier for biologically active components in drugs. Pectins exert diverse physiological activities such as immunomodulating, hepatoprotective, anticarcinogenic, and antimetastatic, that allow them to be applied as drugs and biologically active food additives.

The pectins are typically isolated from economically valuable pants, for example, citrus, apple, sugar beet, and sunflower head pith. Other plants such as amaranth, small mallow, duckweed, SILENE, and coffee beans, have been reported as potential sources of pectins [7–10].

The methods for isolation of pectin polysaccharides from plant tissues are numerous. Among the classic ones is hydrolysis extraction of dry raw material particles of certain sizes [11] with hot water, organic and inorganic acid solutions as well as salts, alkali or their mixtures as extracting solutions. Basic parameters of the pectin isolation process such as a raw material preprocessing, hydromodulus, temperature, extraction duration, medium pH and precipitator can be varied depending on characteristics of the raw material [12].

3.1. Isolation and Structural Study of Main and Side Chains of PS.

Pectin substances were extracted from the bark of *L. sibirica* Ledeb. according to the scheme depicted in Figure 1. Pectinase treatment of a PS sample, isolated from larch bark by the above method and further analysis of the hydrolysis products by paper chromatography (PC), has shown that significant destruction of PS occurs to form free D-galacturonic acid. High positive value of the rotation angle (+245.3°, *c* 0.1, H_2O) indicates α-D-configuration of D-galactopyranosyluronic acid residues.

Table 1 shows main maxima of the absorption bands in the IR spectra of PS and their assignment proving PS pectin nature of the samples [13]. Thus, enzyme hydrolysis and IR spectroscopy data prove that the polysaccharide isolated from larch bark refers to pectin group.

To determine the sugar composition of PS the total acid hydrolysis has been carried out. Monosaccharide identification in hydrolysate has been performed by gas-liquid chromatography. It has been shown that the sample consists of galacturonic acid and monosaccharides of arabinose, galactose, rhamnose, glucose, mannose, and (in minor quantity) xylose (Table 2). Dominating monosaccharides are galactose and arabinose in a ratio of 2.7 : 1.

TABLE 1: Absorption band maximums in IR spectra of PS and their assignment.

Frequency (ν, cm^{-1})	Assignment
3460	ν(OH), ν(H_2O)
3260	ν(NH)
2962, 2872	ν(CH_3)
2573	ν(OH),
1730	$\nu(C{=}O)_B$ COOH
1640	$\delta(OH)_\lambda$
1540	δ(NH)
1380–1450	δ(C–CH_3), ν(C–O) pyranose rings
1331	δ(OH) in pyranose rings
1265	ν(C–O) in esters
1150	ν(C–O–C)
1095	ν(C–C)
1027	ν(C–OH)
890	δ(C1–H) in glucopyranose ring
766, 629, 528	Pulse vibrations of pyranose ring

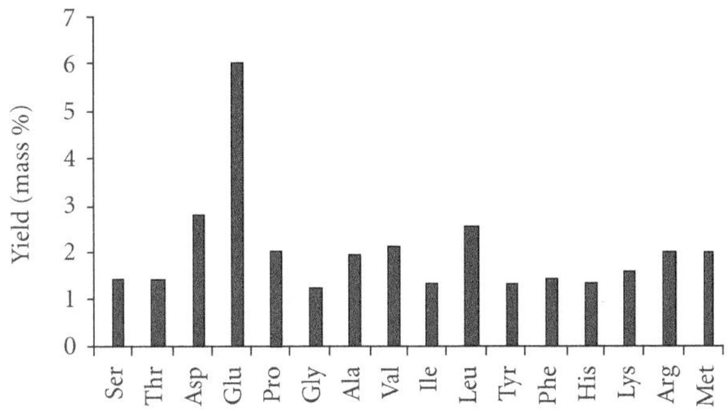

FIGURE 2: Amino acid composition of PS proteins.

Partial acid hydrolysis of PS by 0.05 M TFA gives polysaccharide PS-s representing a linear polysaccharide with insignificant amount of side chain subunits. It is also confirmed by the decrease of the relative content of neutral monosaccharides of PS-s in comparison with their content in initial polysaccharide PS (GLC analysis of polyols peracetates of the corresponding sugars in the studied substances, Table 2) and increase of galacturonic acid content from 38.4% in PS up to 88% in PS-s.

Values of chemical shifts (CSs) of carbon atoms in ^{13}C NMR spectrum of PS-s (Table 3) as compared to data [4, 5] correspond to those of carbon atoms in D-galacturonic acid residues in pyranose constituting linear fragment of the pectin molecule (pectin core).

The signal of anomeric carbon atom at 101.9 ppm indicates both the (1 → 4)-bonding between D-galacturonic acid residues and α-configuration of C-1 anomeric atom. Signal at 176.2 ppm is assigned to C-6 atom and points to the free carboxyl group in D-galacturonic acid residue. Besides, galacturonic acid residues esterified by methoxyl are present the PS-s molecule that follows from signals with CS at 172.2 ppm ($\underline{C}$-6-OCH_3) and 54.4 ppm (–$O\underline{C}H_3$). Ratio of integrated signal intensities of carbon atoms in the methoxyl and carboxyl groups is indicative of high degree of galacturonan methoxylation.

Ion exchange column chromatography of PS on DEAE cellulose by 0.01–0.2 M sodium chloride aqueous solutions has yielded four polysaccharide fractions, PS 1–4, whose chemical characteristics are summarized in Table 4.

In the fractions PS-1 and PS-2, galactose and arabinose are predominant (18.26/52.96% and 11.65/30.83%, resp.), thus, they belong to acidic arabinogalactans. Acidic character of PS is caused bythe presence of D-galacturonic acid residues, the content of which in PS-1 is 5 times less than in PS-2, while in PS-3 and PS-4 it is a major monosaccharide that allows PS to pectins to be assigned. Content of neutral monosaccharides in PS-4 is minimum as compared to other fractions (3 mass %).

All the fractions contain protein substances that are not eliminated by gel filtration (Table 4). Probably, the protein and polysaccharide compounds are strongly associated, or their molecular weights are close to each other.

Amino-acid composition of PS proteins has been studied. Major components of proteins are glutamic acid (6%) and aspartic acid (2.8%), while total content of amino acids with aliphatic side chains (glycine, alanine, valine, isoleucine, leucine) is 9% (Figure 2).

As mentioned above, full acid hydrolysis of PS with 2 M TFA has delivered galacturonan (PS-s). A milder treatment of PS with 0,01 M TFA gives PS-g.

According to the ^{13}C NMR data, PS-g is a pectin polysaccharide. The spectrum contains typical signals of galacturonic acid residues, namely, pronounced signals of anomeric carbon atoms at 100.4 and 100.9 ppm, and signals of the carboxyl carbon atoms at 171.4, 166.5, and 53.7 ppm, the latter two being of carbon atoms in uronic acid residues methoxylated by C-2 and C-3 atoms (Table 5). Intensities and spectral positions of signals at 68.9, 70.8, 78.9, and 72.2 ppm correspond to literature data for α-D-Gal$_p$A residues connected by 1 → 4 bonds. A ratio of integral signal intensities of the carboxyl and methoxyl carbon atoms is 1 : 5 which testifies higher methoxylation degree of PS.

In the ^{13}C NMR spectrum of PS-g sample there were upfield signals at 17.9 and 18.13 ppm belonging to C-6 atoms in terminal rhamnose residues and in polysaccharide chains, respectively. The integral intensities of these signals and those of C-2 and/or C-3 and C-6 carbon atoms for galacturonan residues at 166.42 and 171.33 ppm were found to have a ratio of 1 : 5. The total integral intensity of signals for anomeric C-1 atoms for rhamnose and total integral intensity of signals of anomeric atoms of s were equal to each other, that is, they had the same ratio for rhamnose galacturonan residue in the chain. According to data [14], signals at 99.7, 77.6, 70.8, 82.5, 68.9, and 17.9 ppm are assigned to C-1, C-2, C-3, C-4, C-5, and $\underline{C}H_3$ carbon atoms in → 2,4)-α-L-Rha$_p$-(1 → residues.

Thus, according to the ^{13}C NMR spectral data, linear fragment of pectin polysaccharide PS-g is rhamnogalacturonan (RG-I), where D-galacturonic acid residues in pyranose form with α-configuration of anomeric center are linked by 1 → 4-bonds. One-fifth of galacturonan residues at C-6 atom

TABLE 2: Monosaccharide compositions determined by GLC analysis of PS and PS-s samples.

Sample	Content, %						
	Gal A	Rha	Ara	Xyl	Man	Glc	Gal
PS	38.4	1.6	6.8	traces	2.5	3.7	18.4
PS-s	87.86	0.97	2.75	traces	0.73	0.82	3.02

TABLE 3: Chemical shifts values of carbon atoms in D-galacturonic acid residues in the ^{13}C NMR spectrum of PS-s (δ, ppm, D_2O) and of [4, 5].

Residue	C-1	C-2	C-3	C-4	C-5	C-6	C-6–(OCH_3)	–O$\underline{C}H_3$
→ 4)-α-D-Gal$_p$A-(1 → of PS-s	101.9	68.9	72.1	79.2	73.4	176.2	172.2	54.4
→ 4)-α-D-Gal$_p$A-(1 → of apple pectin [5]	100.7	69.4	69.8	79.4	72.2	175.2	—	—
→ 4)-β-D-Gal$_p$A-(1 → [4]	103.4	72.4	74.3	78.3	76.0	175.2	—	—

TABLE 4: Chemical characterization of PS sample after DEAE-cellulose fractioning.

Sample*	Yield, %	Content, %							
		Gal$_p$A	Protein	Monosaccharides					
				Rha	Ara	Xyl	Man	Glu	Gal
PS-1	12.1	5.67	6.9	traces	18.26	1.54	2.53	5.95	52.92
PS-2	5.9	29.12	7.3	0.53	11.65	1.02	2.71	8.81	30.83
PS-3	17.0	65.93	5.7	1.91	4.45	0.75	1.18	1.12	9.42
PS-4	37.0	79.87	3.6	0.35	0.93	0.18	0.24	0.21	1.06

Note: *PS-1 isolated with use of 0.01 M NaCl solution, PS-2—0.1 M NaCl solution, PS-3 and PS-4—0.2 M NaCl solution.

TABLE 5: Chemical shifts of carbon atom signals in ^{13}C NMR spectrum of PS-g.

Residue	C1	C2	C3	C4	C5	C6	–O$\underline{C}H_3$ ($\underline{C}H_3$–)
→ 4)-α-D-Gal$_p$A-(1 →	100.4	68.9	70.8	78.9	72.2	171.4	—
2-Me-α-D-Gal$_p$A-(1 →	100.9	166.5	69.6	78.9	73.8	171.4	53.7
3-Me-α-D-Gal$_p$A-(1 →	100.9	68.9	166.5	78.9	73.8	171.4	53.7
→ 2,4)-α-L-Rha$_p$-(1 →	99.7	77.6	70.8	82.5	68.9	—	17.9
β-D-Gal$_p$-(1 →	104.64	71.7	74.1	69.6	76.1	62.0	—
→ 6)-β-D-Gal$_p$-(1 →	104.38	71.7	73.8	69.6	74.3	70.8	—
→ 3,6)-β-D-Gal$_p$-(1 →	104.64	71.7	82.5	69.57	74.1	71.4	—
α-L-Ara$_f$-(1 →	108.6	80.7	78.9	84.9	62.0	—	—
β-L-Ara$_p$-(1 →	101.1	69.6	n.d.	n.d.	n.d.	—	
→ 3,5)-α-L-Ara$_f$-(1 →	108.6	80.7	84.9	83.2	67.8	—	—
→ 2,5)-α-L-Ara$_f$-(1 →	108.0	84.9	77.6	83.2	67.8	—	—

Note. n.d.: not determined.

is esterified by methoxy groups. The ratio between 2,4-substituted rhamnopyranosyl and galacturonosyl residues (1 : 5) indicates that the main chain structure of the pectin polysaccharide is highly branched at C-4 atoms of rhamnopyranosyl residues.

Further ^{13}C NMR studies of PS-g sample shows that arabinogalactan subunits are present in rhamnogalacturonan as side chains. In the ^{13}C NMR spectrum of PS-g sample, in the region of anomeric carbon signals, apart from the signals of anomeric carbons in galacturonopyranosyl residues of the galactan core, appear the signals at 101.1, 104.38, 104.64, and 108.6 ppm. According to [15], intensities and values of CS can be assigned to signals of anomeric carbon atoms in β-L-Ara$_p$, α-L-Ara$_f$, and β-D-Gal$_p$ residues. The most upfield signal (δ 101.1 ppm) is attributable to terminal β-L-Ara$_p$ residues. Signals at 104.38 and 104.64 ppm belong to C-1 in β-D-Gal$_p$ residues while CS values of C-2, C-3, C-4, C-5, and C-6 atom signals have been determined by the comparison with literature data for β-D-galactopyranosyl residues. Bonding at C-3 and C-6 positions of β-D-galactopyranose is proved by downfield shifts of these signals by 8.7 and 8.8–9.4 ppm, respectively, due to glycosylation of these atoms as compared to their positions in nonsubstituted 1 → 3,6 bonded β-D-Gal$_p$ residues. Signals at 08.6 ppm, like those at 80.7, 78.9, 84.9, and 62.0 ppm, are assigned to terminal α-L-arabinofuranose. According to signal intensities of anomeric atoms of arabinose and galactose, a ratio of these monosacharides is 1 : 2.

Hence, according to spectral data for PS-g fragment of the pectin polysaccharide from larch bark, highly branched arabinogalactan is detected as side chains consisting of linear chains with → 3,6)-β-D-Gal$_p$-(1 → residues with branching at C-6 atoms. Side chains of arabinogalactan fragment contain terminal arabinose both in the pyranose and in furanose form as well as → 2,5)-α-L-Ara$_f$-(1 → and → 3,5)-α-L-Ara$_f$-(1 → residues as intermediate fragments.

It is known that the reactivity of polysaccharide molecule is due to terminal sugars, mainly localized in side chains [16]. In addition, conformation features of the macromolecule, caused by intramolecular stabilization bonds between functional groups in side chains, are responsible for biological activity of a polysaccharide.

Arabinogalactan is a prevailing polysaccharide of larch wood; hence, its presence in the polysaccharides bark is owing to biogenetic reasons. It has been reported that arabinogalactan is present in the pectin substances of cell walls of plants both as associable bonding and accompanying component [17].

3.2. Study of Ag(I) Redox Reaction and Ag(0) Nanoparticles Formation and Stabilization in Pectin Polysaccharide Matrix. Significant interest to nanosize metals is caused by their high technological potential as important magnetic materials, catalysts, nonlinear-optical medium, and biologically active agents. So, for example, it has been established that silver nanoparticles possess rare combination of valuable features, that is, unique optical properties due to surface plasmon resonance, highly developed surface, catalytic activity [18], and antimicrobial activity, which is even more expressed than in ionic silver [19].

One of the widespread approaches to the synthesis of metal nanoparticles involves the reduction reaction of metal ions in a polymeric solution. As a rule, the high-molecular compound (polysaccharide) employed in this case acts as a protective polymeric screen ensuring both the size of metal nanoparticles and stabilization of the nanobiocomposite formed [20]. Borohydrides, aluminium hydrides, aminoboranes, hypophosphites, hydroquinone, formalin, light, and radiation are used as reducing agents here [21].

Another approach to the nanocomposites synthesis is based on the nanocomposites self-organization, where the polymers play a role of reducing agent and nanostabilizing medium [22]. In this case, synergism of properties of the polymeric matrix (biological activity, hydrodynamical characteristics) and those of the metal core (optical, biological, thermophysical, electric) takes place which provides for promising performance characteristics of the nanocomposites formed. According to this approach, nanobiocomposites have been prepared using natural polysaccharides: arabinogalactan [22], galactomannan, carboxymethylcellulose, geparin [23], sea seaweed polysaccharides [24], pectin [2], and so forth.

Study of the Ag(I) redox reaction and Ag(0) nanoparticles formation and stabilization in pectin polysaccharide matrix was carried out depending on the reaction conditions, in particular, pH value, initial reagents ration (metal salt/pectin), and reaction time.

The absorption spectra of pectin and silver nitrate aqueous solutions at different reaction time are depicted in Figure 3(a). It has been found that at pH 3.5 reduction reaction Ag(I) proceeds very slowly. It is supported by the appearance of a bond in the absorption spectrum at λ 280–470 nm only 24 h after the reaction beginning (Figure 3(a)). Wide maximum of the low intensive bond indicates the formation of primary centers of metal silver. Besides, the reaction rate is so slow that even 96 h is not enough for the generation of totally recovered Ag(0) centers. For the reaction of pectin with Ag(I) at pH 7, a symmetric bond at λ_{max} 420 nm is observed in the electron spectra thus proving the formation of Ag(0) nanoparticles (Figure 3(b), line 2). Total conversion of silver cations, which is experimentally evaluated by the growth of the absorption bond intensity, is reached for about 24 h. The reduction Ag(I) with pectin at pH 11-12 occurs instantly just after the reagents mixing (Figure 3(b), line 4 and Figure 3(c) line 1) and is completed in 30 min. The reduction under these conditions is also accompanied by variations of Ag(0) particle sizes which follows from the shift in Plasmon pick position into shortwave region by 10 nm (Figure 3(c), line 5).

Thus, the increase in rate of Ag(I) reduction with pectin at elevated pH value allows one to assume the direct participation of OH^- in this reaction.

In the alkaline medium, polysaccharides are known to undergo diverse transformations [25–27] such as depolymerization, alkaline hydrolysis, and oxidizing destruction.

^{13}C NMR spectroscopy data (Tables 5 and 6) indicate that these processes take place during the interaction of PS with Ag(I) in the alkaline medium. The data of ^{13}C NMR spectrum of PS-g1, containing 20% of Ag(0), are given in Table 6.

The comparison of ^{13}C NMR spectra of PS-g1 and PS-g has shown that PS-g1 represents a partially destructed acidic polysaccharide. This is evidenced from the presence (^{13}C NMR) of characteristic signals of anomeric carbons at 100.3 ppm in the → 4)-D-Gal$_p$A-(1 → and 6-Me-α-D-Gal$_p$-(1 → residues in rhamnogalacturonan (Table 6). At the same time, the number of C-6 substituted acid residues (8 : 1) as compared to their content in the starting polysaccharide (5 : 1) has been decreased. Also, the signals of carbon atoms in galactopyranosyl residues that have different locations in linear chain, that is, terminal 1 → 6- and 1 → 3,6-substituted galactopyranosyl residues, have been detected in the spectrum. The values of chemical shifts of carbon atoms at 108.6, 80.7, 78.9, 84.9, and 62.0 ppm are assigned to C-1-6 atoms of terminal α-L-Ara$_f$. RG-I region of the PS-g1 sample also undergoes destructive changes. First, its labile side chains containing highly branched arabinozil fragments involve in the alkaline hydrolysis. The fact of the destruction is also supported by the presence (in ^{13}C NMR spectra of PS-g1 and PS-g samples) of signals of carbon atoms assigned to galactopyranosyl residues linked by 1 → 3- and 1 → 3,6-bonds. Arabinose residues in furanoside cycles are present only as terminal residues.

The structure of PS-g2 differs from that of PS-g1 by the content of Ag(0) (72%). According to ^{13}C NMR data (Table 7), carbohydrate counterpart of the former represents

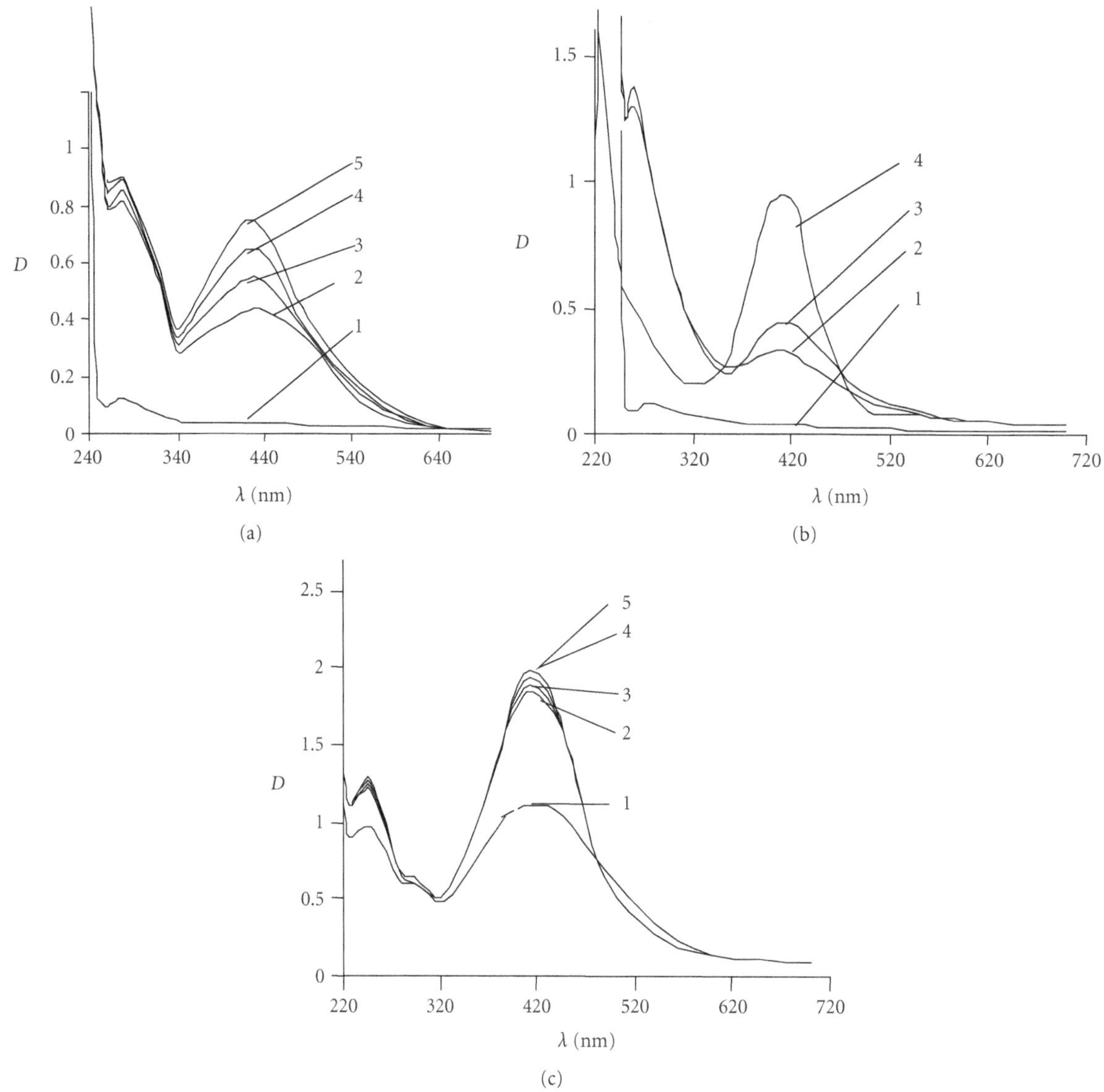

FIGURE 3: Absorption spectra of a mixture of aqueous solutions of pectin (0.5%) and silver nitrate (0.1%) in a ratio of 1 : 1 depending on (a) reaction duration: 1 min (1), 24 h (2), 48 h (3), 72 h (4), 96 h (5); (b) pH value: 3.5 (1), 7 (2), 9.7 (3), 11.5 (4); (c) reaction duration at pH 11.5: 1 min (1), 30 min (2), 60 min (3), 180 min (4), 24 h (5).

a product of deeper destruction of the initial polysaccharide PS-g. First of all, the signals of carbon atoms, characteristic for acid polysaccharides and its methoxylated derivatives [28], are absent in low-field region of the spectrum at 166–177 ppm. Also, the signals of carbon atoms of rhamnose residues are not observed in the spectrum. It has been established that the main structural component of PS-g2 sample is the linear fragment comprising 1 → 3-linked galactopyranosides residues and bearing the side fragments in the form of terminal β-L-Ara$_p$ and β-D-Gal$_p$ residues and/or olygosaccharides with 1 → 3-linked galactopyranosides residues and attached by β-L-Ara$_p$ and β-D-Gal$_p$ terminal residues.

Thus, the comparison of structural characteristics of the initial pectin with those of "pectin-Ag(0)" nanobiocomposites PS-g1 and PS-g allows one to assume that in the reaction studied the pectin polysaccharide acts a reducing agent and also undergoes destructive changes. Pectin carboxy-groups and other groups (–OH, =O) are contained in structure of side chains of RG-I participate in formation of zero-valence silver participates.

The destruction of PS-g1 backbone in rhamnosyl residue is observed upon the addition of larger quantities of Ag(I) to the reaction. As a result, RG-I is abstracted, the main chain of which is less destructed by this moment and stabilizes the Ag(0) nanoparticles.

Efficiency of stabilization functions has been estimated by influence of quantitative ratio of the initial pectin substances taken in reaction: metal salt (mmol)/pectin (1 g) on the content of Ag(0) nanoparticles and their sizes in formed nanobiocomposites "pectin-Ag(0)." It has been shown that in area 0.1–2.5 mmol AgNO3/1 g of pectin the content of zero-valent silver in received nanobiocomposite is indirectly

Table 6: Chemical shifts (δ, ppm, D_2O) of carbon atom signals of the carbohydrate residues in ^{13}C NMR spectrum of PS-g1 sample.

Residue	C-1	C-2	C-3	C-4	C-5	C-6	$-OCH_3$
→ 4)-α-D-Gal$_p$A-(1 →	100.3	69.5	70.2	78.0	72.2	176.8	
6-Me-α-D-Gal$_p$A-(1 →	100.3	69.5	70.0	78.0	74.1	174.7	55.4
→ 2,4)-α-L-Rha$_p$-(1 →	99.7	77.6	71.0	82.5	72.8	18.0	
β-D-Gal$_p$-(1 →	104.7	71.7	74.1	69.5	76.6	62.5	
→ 6)-β-D-Gal$_p$-(1 →	105.1	71.7	74.1	69.5	72.7	70.3	
→ 3,6)-β-D-Gal$_p$-(1 →	105.1	71.7	82.7	70.1	72.7	71.7	
α-L-Ara$_f$-(1 →	108.6	80.7	78.9	84.9	62.0	—	

Table 7: Chemical shifts (δ, ppm, D_2O) of carbon atoms of the carbohydrate residues in ^{13}C NMR spectrum of PS-g2 sample.

Residue	C-1	C-2	C-3	C-4	C-5	C-6
β-D-Gal$_p$-(1 →	104.8	71.7	74.2	69.6	76.6	62.5
→ 6)-β-D-Gal$_p$-(1 →	104.8	71.7	74.2	69.6	72.8	70.1
→ 3,6)-β-D-Gal$_p$-(1 →	104.8	71.7	82.8	70.1	74.9	71.7

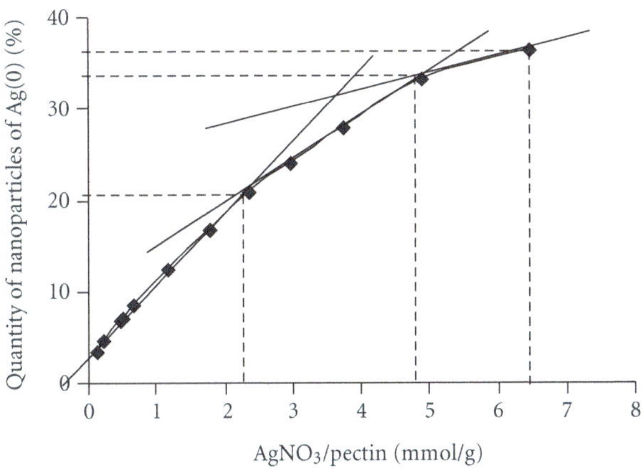

Figure 4: Influence of a ratio of initial metal salt/pectin on the content of Ag(0) in formed nanobiocomposites.

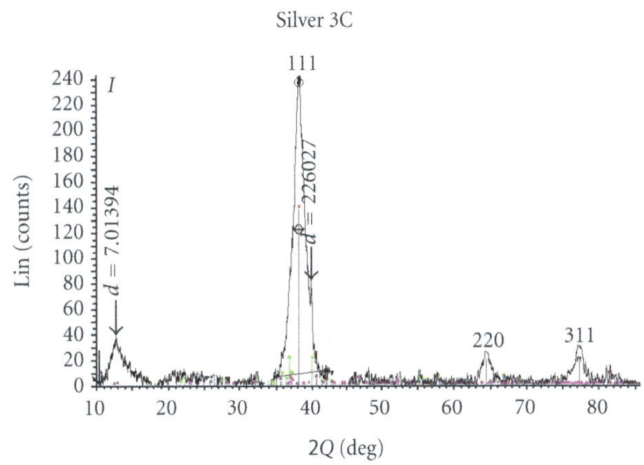

Figure 5: Diffractogram of precipitate formed in the reaction medium at use of initial components in area from 2.5 up to 6.5 mmol of $AgNO_3$/1 g of pectin.

proportional relationship of quantity the salt taken in treatment (Figure 4). In these conditions the size of received silver nanoparticles, according to data electronic microscopy, is 5–7 nm. With further increase of the initial ratio of metal salt/pectin the bends in the curve are observed, which suggests that a part of the resulting Ag(0) is not involved in formation nanobiocomposites. Results of the Roentgen phase analysis of this insoluble phase (Figure 5) showed that it is a metal silver with well-formed crystalline cubic structure. At a ratio of initial components from 5 up to 6.5 mmol AgNO3/1 g of pectin the gain of Ag(0) in nanosize area is not significant, and allows to receive the nanobiocomposite "pectin-Ag(0)" with the content of Ag(0) nanoparticals at 36.4% whose size is in area from 6 up to 15 nanometers.

The content of Ag(0) nanoparticals in nanobiocomposite "pectin-Ag(0)" can be increased up to 72% using the most nanobiocomposite as an initial component. It should be noted that thus the size of nanoparticals in nanobiocomposite increases and is from 4 up to 27 nm.

4. Conclusions

In conclusion, the interaction of aqueous solutions of pectin with silver nitrate affords "pectin-Ag(0)" nanobiocomposites, where pectin plays reducing and stabilizing roles. The reaction rate essentially increases when pH values of the reaction mixture are close to alkaline ones. The ratio of the starting reactants influences the content of Ag(0) in the nanobiocomposites and their sizes: the more is Ag(I) per 1 g of pectin, the lower amount of Ag(0) nanoparticles is formed.

It has been established (^{13}C NMR) that when 0.1–2.5 mol of $AgNO_3$ and 1 g of pectin are employed in the reaction, the PS structure is preserved. The increase of Ag(I) amount up to 6.5 mmol leads to the pectin destruction with abstracting the side fragments and destruction of rhamnogalacturonan core. Under ratio of $AgNO_3$/pectin equaling 6.5 mmol/1 g, the formation of the nanocomposite is stopped due to the total destruction of pectin polysaccharide.

Thus, in the reaction of "pectin-Ag(0)" nanobiocomposites formation, pectin acts as having reducing and stabilizing functions and regulates the sizes of Ag(0) nanoparticles.

References

[1] O. H. Lowry, N. J. Rosebrough, A. L. Farr, and R. J. Randall, "Protein measurement with the Folin phenol reagent," *The Journal of Biological Chemistry*, vol. 193, no. 1, pp. 265–275, 1951.

[2] V. A. Babkin, N. V. Ivanova, N. N. Trofimova et al., "Isolation of pectin, possessing membrane stabilizing activity and ability to reduce silver ions, from larch wood," Silver nanocomposites stabilized by pectin. Pat RU, 2403263, 2010.

[3] A. P. Krechkov, *The Foundations of Analytic Chemistry*, vol. 2, Khimya, 1978.

[4] R. G. Ovodova, O. A. Bushneva, A. S. Shashkov, A. O. Chizhov, and Y. S. Ovodov, "Structural studies on pectin from marsh cinquefoil *Comarum palustre* L," *Biokhimiya*, vol. 70, no. 8, pp. 1051–1062, 2005.

[5] P. Odonomažig, D. Balga, and A. Ebrigerova, "Structures of pectin polysaccharides isolated from the Siberian apricot (*Armaniaca sibirica* Lam)," *Carbohydrate Research*, vol. 226, pp. 353–358, 1992.

[6] Y. S. Ovodov, V. V. Golovchenko, E. A. Gunter, and S. V. Popov, *Pectin Substances in Plants of European North of Russia*, Uro RAN, Ekaterinburg, Russia, 2009.

[7] S. T. Minzanova, V. F. Mironov et al., "Technological aspects of receiving fodder additives from the amarant," *Doklady Chemistry*, vol. 3, pp. 363–366, 2007.

[8] V. N. Bubenchikov, I. L. Drozdova, and V. I. Kochkarov, *Method of Isolation of the Polysaccharides Possessing Anti-Inflammatory Activity*, Pat RU, 2002.

[9] R. G. Ovodova, S. V. Popov, O. A. Bushneva et al., "Branching of the galacturonan backbone of comaruman, a pectin from the marsh cinquefoil *Comarum palustre* L," *Russian Journal of Bioorganic Chemistry*, vol. 71, no. 5, pp. 652–671, 2006.

[10] M. K. Gelgay, L. V. Donchenko, and A. I. Reshetnyak, "Innovative technology of pectin from secondary sources of raw material after processing coffee," *New Technologies*, vol. 6, pp. 15–18, 2008.

[11] G. B. Aimukhomedova and N. P. Shelukhina, *Pectin Substances and Their Determination Methods*, Ilim, Frunze, Russia, 1964.

[12] N. V. Ivanova, O. V. Popova, and V. A. Babkin, "Analysis of influecing of the different factors on an output and some characteristics of pectin materials of a Larch bark," *Khimiya Rastitel'nogo Syr'ya*, no. 4, pp. 43–46, 2003.

[13] K. Nakanisi, *Infrared Spectra and Structure of Organic Compounds*, Moscow, Russia, 1965.

[14] R. G. Ovodova, V. V. Golovchenko, A. S. Shashkov, S. V. Popov, and Y. S. Ovodov, "Structural studies and physiological activity of lemnan, a pectin from *Lemna minor L*," *Russian Journal of Bioorganic Chemistry*, vol. 26, no. 10, pp. 669–676, 2000.

[15] S. Karacsonyi, V. Kovacik, J. Alfoldi et al., "Chemical and 13C-n.m.r. studies of an arabinogalactan from *Larix sibirica* L," *Carbohydrate Research*, vol. 134, no. 2, pp. 265–274, 1984.

[16] S. D. Achubaeva, *Chemical Reactions of Pectinaceous Substances*, Ilim, Frunze, Russia, 1984.

[17] A. E. Clarke, R. L. Anderson, and B. A. Stone, "Form and function of arabinogalactans and arabinogalactan-proteins," *Phytochemistry*, vol. 18, no. 4, pp. 521–540, 1979.

[18] Y. A. Krutyakov, A. A. Kudrinskiy, A. Y. Olenin, and G. V. Lisichkin, "Synthesis and properties of silver nanoparticles: advances and prospects," *Russian Chemical Reviews*, vol. 77, no. 3, pp. 233–257, 2008.

[19] A. A. Kornevskii, V. V. Sorokin, and G. I. Karavaiko, "Co-operating of silver ions with the cages of Canlida utilis," *Microbiologia*, vol. 6, pp. 1085–1092, 1993.

[20] O. E. Litmanovich, "Psevdomatrix sintez of polimer-metal nanocomposite sols: interakion of macromolecules with metal nanoparticles," *Chinese Journal of Polymer Science C*, vol. 58, no. 7, pp. 1370–1396, 2008.

[21] A. D. Pomogailo, "Polymer-immobilised nanoscale and cluster metal particles," *Uspekhi Khimii*, vol. 66, no. 8, pp. 785–791, 1997.

[22] B. A. Trofimov, B. G. Sukhov, G. P. Aleksandrova et al., "Nanocomposites with magnetic, optical, catalytic, and biologically active properties based on arabinogalactan," *Doklady Chemistry*, vol. 393, no. 4-6, pp. 287–288, 2003.

[23] B. A. Trofimov, B. G. Sukhov, G. P. Aleksandrova et al., "Polyfunctional self-organized hybrid nanobiocomposetes on basis of natural polymers," http://edu-cons.net/atlas_last/doc/302/1(2).pdf.

[24] I. N. Yurkova, D. A. Panov, and V. A. Ryabushko, "Study of optical properties of nanobiocomposites on silver and marine alga polysaccharides base," *Uchenye Zapiski Tavricheskogo Universiteta Series Biologiya and Khimiya*, vol. 22, no. 1, pp. 203–207, 2009.

[25] O. P. Golova and N. V. Nosova, "Oxidative-alkaline destruction of cellulose," *Russian Chemical Reviews*, vol. 42, no. 4, pp. 743–767, 1973.

[26] P. A. Gachonidze, "Saccharine acids," *Russian Chemical Reviews*, vol. 49, no. 4, pp. 743–767, 1980.

[27] B. P. Nikolskii, *Novyi Spravochnik Chimika I Technologa*, 2009.

[28] I. V. Norgov, *Progress in Carbohydrate Chemistry*, Nauka, Moscow, Russia, 1985.

Certain Rheological Aspects of Functionalized Guar Gum

Meenu Kapoor,[1] Dhriti Khandal,[1] Ruchi Gupta,[1] Pinklesh Arora,[1] Geetha Seshadri,[1] Saroj Aggarwal,[1] and Rakesh Kumar Khandal[2]

[1] *Shriram Institute of Technology, Gautum Buddh Technical University, I.E.T. Campus, Sitapur Road, Lucknow 226 021, India*
[2] *Gautum Buddh Technical University, I.E.T. Campus, Sitapur Road, Lucknow 226 021, India*

Correspondence should be addressed to Rakesh Kumar Khandal; dr.r.k.khandal@gmail.com

Academic Editor: R. J. Linhardt

Guar gum and its derivatives are highly important industrial hydrocolloids as they find applications in various industrial sectors. Guar is a polymer of high molecular weight and its aqueous solutions exhibit unique rheological properties, which has led to its wide acceptance by the industry. In certain industrial applications low molecular weight guar and its derivatives are needed, and conventionally chemical depolymerisation of guar is carried out for this purpose. Radiation processing is a novel and green technology for carrying out depolymerization and can be an ideal substitute for chemical depolymerisation technique. In order to study the effect of radiation on guar derivatives, three types of derivatives have been taken in the present study: carboxymethyl, hydroxyethyl, and methyl guar. The effect of 1–50 KGy radiation dose on the rheological behavior of these derivatives has been studied, and the results have been described in the present paper. The effect on storage and loss modulus with respect to frequency and effect on viscosity with respect to shear rate have been discussed in detail.

1. Introduction

Guar gum is a polygalactomannan found in the endosperm of the seeds of the plant *Cyamopsis tetragonolobus*. It is a hydrophilic heteropolysaccharide of mannose and galactose monomer units where the mannose forms the main linear chain of the polymer and the galactose forms the pendant branches. The mannose units are linked together by β-1,4 glycosidic bonds and the galactose units are linked to mannose through α-1,6 glycosidic bonds [1]. The mannose to galactose ratio of guar gum has been reported to be approximately 2 : 1. Thus, every second mannose unit bears a branch of galactose unit.

Guar gum is a high molecular weight polysaccharide (generally more than 2 million g/mol) that can form extensive intermolecular H-bonding resulting in high solvation and thereby increased viscosity [2]. These attributes of guar gum find it varied applications in the industry [3–11]. Guar gum functions as a thickener, emulsion stabilizer, gelling agent, film former, or texture modifier.

Despite the unique rheological behavior there are certain limitations of guar gum which need to be overcome to ensure its effective use. The excessive hydrophilicity of guar gum along with the high molecular mass prevents complete or homogeneous solvation or hydration of the polymer. The resulting two-phase aqueous solution has certain regions showing increased viscosity and other regions showing the solid phase of undissolved guar gum. The other drawback of guar gum is the low thermal and shear stability, as observed in other natural polysaccharides, which reduces the durability and imposes temperature and heat conditions for its effective performance.

In order to overcome the disadvantages of guar gum, its modification is carried out mainly by the functionalization of the hydroxyl groups present on guar backbone. Various functionalities can be introduced on guar by carrying out etherification of guar with different reagents as shown in Table 1.

Another aspect of guar gum is its high molecular weight, which may not be always desirable. In several industrial applications, low molecular weight guar is used, which is conventionally produced by degradation of native guar by following chemical route. A novel method for degradation of guar can be radiation processing technology. A comparison

Table 1: Etherification reactions of guar.

S. no.	Reaction	Reagent
1	Carboxymethylation	Monochloroacetic acid
2	Cationization	Quaternary ammonium salts
3	Methylation	Methyl chloride
4	Hydroxyethyl/propylation	Ethylene/propylene oxide

of the methodology followed for the degradation of guar by conventional route and by radiation processing is given in Figure 1.

The most important property of guar gum is its unique rheology, which makes it an invaluable industrial product. The rheological behavior of guar has been a subject for study for scientists world over. Venkataiah and Mahadevan [12] investigated the aqueous solutions of guar gum as well as its nonionic hydroxypropyl and anionic carboxymethyl derivatives for their flow properties in the range of low to moderately high shear rates and observed the transition from Newtonian fluid to pseudoplastic fluid. Wientjes et al. [13] studied the linear viscoelastic behavior of guar gum solutions as a function of frequency, temperature, polymer concentration and molecular weight and revealed the importance of different relaxation mechanisms. Aubry and Moan [14] focused on the rheological effect of hydrophobically modified hydroxypropyl guar gum and demonstrated that the linear and nonlinear rheological behaviors of such associating systems were dependent on the nature, concentration, and distribution along the chain of hydrophobic junctions. Zhang et al. [15] have studied the semidilute solutions of hydroxypropyl guar gum (HPGG) with respect to their viscosity changes under various HPGG concentrations added salts and temperatures could be well described by the Cross viscosity model.

In the present paper, the effect of irradiation on guar and its carboxymethyl, hydroxypropyl, and methyl derivatives has been studied in respect to the change in the rheological behavior.

2. Materials and Methods

The methodology followed for studying the effect of radiation processing on guar and its derivatives involved the following steps.

2.1. Guar Gum and Derivatives. The sample of guar gum was procured from a local supplier. The carboxymethyl, hydroxypropyl, and methyl derivatives of D.S./M.S. 0.1 and 0.2 were synthesized by well-established techniques as published previously [16, 17].

2.2. Irradiation of Samples. Guar gum and its derivatives (hydroxypropyl, carboxymethyl, and methyl of DS 0.1 and 0.2) were sealed in plastic bottles, which were packed in a cardboard box. The box was then irradiated at different doses (1, 2.5, 5, 10, 20, and 50 KGy) at the Co-60 gamma irradiation facility available within the institute. All specimens were finely grounded to pass 200 μm sieves before subjecting them to irradiation. The irradiation was carried out on 200 g batch size in case of all the specimens at each dose.

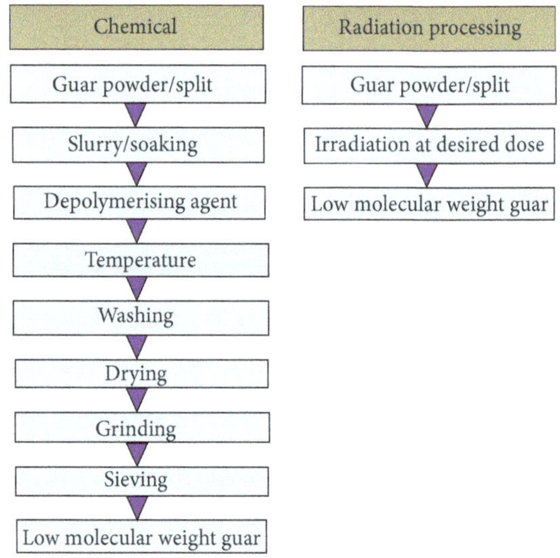

Figure 1: Comparison of the processing steps involved in depolymerisation of guar by Chemical Synthesis Route.

2.3. Preparation of Solutions. The moisture content of every sample was determined by heating the sample in an air circulating oven till constant weight. Aqueous solutions of 1% were prepared after applying the moisture correction factor. The desired amount of sample was weighed and dissolved in required amount of water using either an overhead motorized stirrer or a magnetic stirrer. Three hours of stirring was given to each sample before allowing it to stand overnight at ambient conditions. The following day the solution was considered ready for the rheological evaluation.

2.4. Rheological Evaluation. The rheology study of 1% aqueous solutions of guar gum and its hydroxypropyl (M.S. = 0.1 and 0.2), carboxymethyl (D.S. = 0.1 and 0.2), and methyl derivatives (D.S. = 0.1 and 0.2) was carried out on the Rheometer (MARS, Thermo Fisher) at 25°C temperature. The linear viscoelastic range was determined from the amplitude sweep by varying the shear stress at a constant frequency of 1 Hz. Next, the frequency sweep was carried out at a shear stress (or amplitude value) lying in the linear viscoelastic region. The effect of shear rate on the viscosity of samples was also determined by studying the change in viscosity with varying shear rate from 1 to 30 rpm.

3. Results and Discussion

Guar gum and its derivatives of carboxymethyl, hydroxypropyl, and methyl derivatives with 0.1 and 0.2 DS/MS were studied by rheology before and after gamma irradiation to understand the effect of radiation on the viscosity of the sample. The effect of ionizing radiation on polysaccharides is

TABLE 2: Crossover frequencies of guar and its various derivatives.

Radiation dose (KGy)	Crossover frequency (Hz)					
	CMG (D.S. = 0.1)	CMG (D.S. = 0.2)	HEG (M.S. = 0.1)	HEG (M.S. = 0.2)	MG (D.S. = 0.1)	MG (D.S. = 0.2)
0	Nil	0.299	0.161	0.134	Nil	0.167
1	2.101	6.539	2.717	0.5298	Nil	0.514
2.5	6.755	8.94	6.776	4.234	4.443	4.995
5	16.91	22.23	18.40	9.86	12.24	10.58
10	10.53	9.191	14.60	12.69	12.18	15.39
20	3.857	3.275	2.717	8.335	6.755	6.959
50	1.78	1.282	1.588	1.862	Nil	2.591

known to cause degradation with a consequent loss in the viscosity. The derivatives of polysaccharides like carboxymethyl derivative of cellulose are, however, known to result in formation of gel when their aqueous solutions are subjected to an appropriate radiation dose. The competition between radiation-induced chain scission and chain crosslinking influences the resulting viscosity or formation of gel by the polysaccharide after irradiation. Thus, polysaccharides can be modified using ionizing radiation provided there is a possibility of reducing chain scission and/or favoring chain crosslinking during irradiation.

The following sections discuss the rheology studies for each sample separately.

3.1. Study on Change in Storage Modulus and Loss Modulus. The plot obtained from the frequency sweep study gives the effect of change in frequency on the storage (elastic, G') and the loss (viscous, G'') moduli. The effect of radiation on the storage modulus and loss modulus was studied and the change in crossover frequencies of irradiated samples has been given in Table 2. The effect of radiation on the rheological behavior of each of the derivatives has been described separately.

3.1.1. Carboxymethyl Guar. The carboxymethyl guar (CMG) samples were irradiated and their viscoelastic behavior was studied as described in the previous section. The change in storage (elastic, G') and the loss (viscous, G'') moduli was studied over a range of frequency 1 to 30 Hz and the results obtained have been shown in Figures 2 and 3. Following inferences can be drawn from the data presented in Figures 2 and 3 and Table 2.

(i) As evident from the Figures 2 and 3, in case of CMG of D.S. 0.1, there is no crossover of the G' and G'' whereas in case of CMG of D.S. 0.2 crossover frequency is 0.29 Hz. This shows that the elastic nature predominates throughout in case of CMG of D.S. 0.1, whereas elasticity dominates only after frequency of 0.29 Hz in case of CMG of D.S. 0.2.

(ii) As shown in Table 2, the crossover frequency increases as the radiation dose increases up to 5 KGy and the value is higher in case of CMG of D.S. 0.2 than CMG of D.S. 0.1. This shows that the elastic nature is more dominating for CMG of D.S. 0.1 than CMG of D.S. 0.2. The explanation for this can be found in the fact that initially the chains in guar gum form H-bonding linkages in themselves leading to solid-like behavior. But as the carboxymethyl (-CH_2COONa) substitution is carried out on the guar chain, the distance between the chains increases leading to opening up of the chains as the H-bonding linkages get decreased. Thus solid-like behavior or elasticity decreases on increasing the substitution.

(iii) This can also be explained by saying that during frequency sweep study, the chains get pushed towards each other and form aggregates. The carboxymethyl group being an ionic group will cause repulsion amongst the chains due to repulsion between like-charged moieties. Thus, a greater frequency will be needed to push the chains together in case of a sample having higher ionic substituents, that is, CMG of D.S. = 0.2, to form aggregates leading to solid-like or elastic behavior.

(iv) Beyond 5 KGy radiation dose, the crossover frequency is very close for both CMGs because the molecular chain breaks on irradiation leading to formation of small oligomeric chains of similar size in both CMGs of D.S. of 0.1 and 0.2. Thus, the effect of frequency is felt equally by both these samples.

(v) The values of G' and G'' are seen to decrease with the increase in radiation dose, which indicates chain scissioning.

Another aspect is the effect of radiation on the carboxymethyl guar. As seen from Table 2, the crossover frequency increases with the increase in radiation dose up to 5 KGy and then it starts to decrease. Irradiation leads to breaking up of the chain leading to reduction in molecular weight. Up to 5 KGy radiation dose, chain scissioning is taking place leading to decrease in molecular weight and less interaction of the smaller chains formed. As the interaction between the chains decreases, viscous behavior predominates or in other words elasticity or the solid-like behavior gets decreased. Now, beyond 5 KGy radiation dose, it is observed that the crossover frequency gets decreased, which may signify formation of aggregates between the chains leading to more solid-like behavior at lower frequencies.

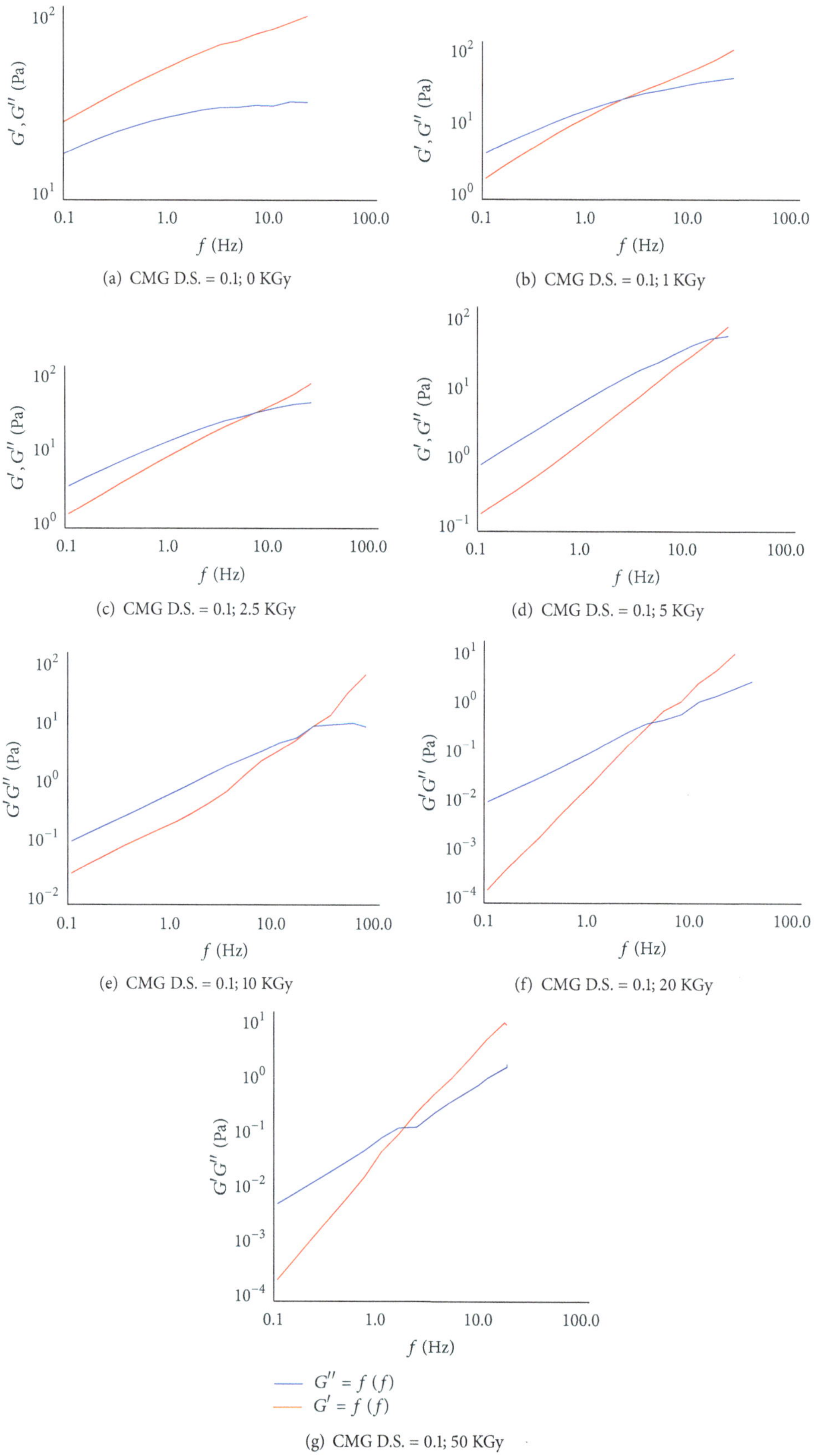

FIGURE 2: Effect of radiation processing on the viscoelastic behavior of carboxymethylated guar (CMG) of degree of substitution (D.S.) = 0.1: (a) unirradiated; (b) irradiated at 1 KGy; (c) irradiated at 2.5 KGy; (d) irradiated at 5 KGy; (e) irradiated at 10 KGy; (f) irradiated at 20 KGy; (g) irradiated at 50 KGy.

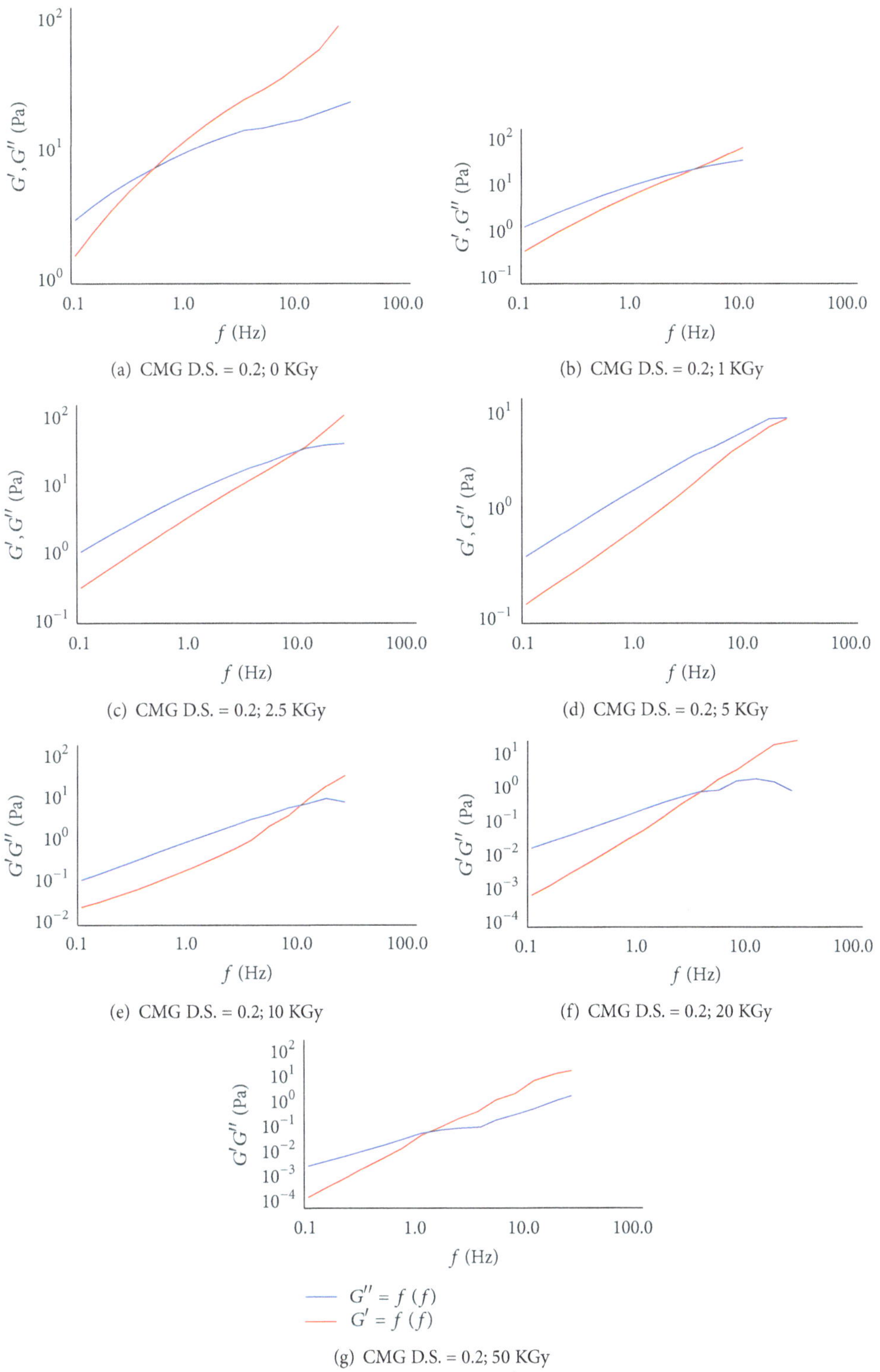

FIGURE 3: Effect of radiation processing on the viscoelastic behavior of carboxymethylated guar (CMG) of degree of substitution (D.S.) = 0.2: (a) unirradiated; (b) irradiated at 1 KGy; (c) irradiated at 2.5 KGy; (d) irradiated at 5 KGy; (e) irradiated at 10 KGy; (f) irradiated at 20 KGy; (g) irradiated at 50 KGy.

3.1.2. Hydroxyethyl Guar. The hydroxyethyl guar (HEG) samples were irradiated and their viscoelastic behavior was studied as described in the previous section. The change in storage (elastic, G') and the loss (viscous, G'') moduli was studied over a range of frequency 1 to 30 Hz and the results obtained have been shown in Figures 4 and 5. On the basis of the results obtained, the following observations were made.

(i) As evident from Figures 4 and 5, in case of HEG of M.S. 0.2, the crossover frequency of the G' and

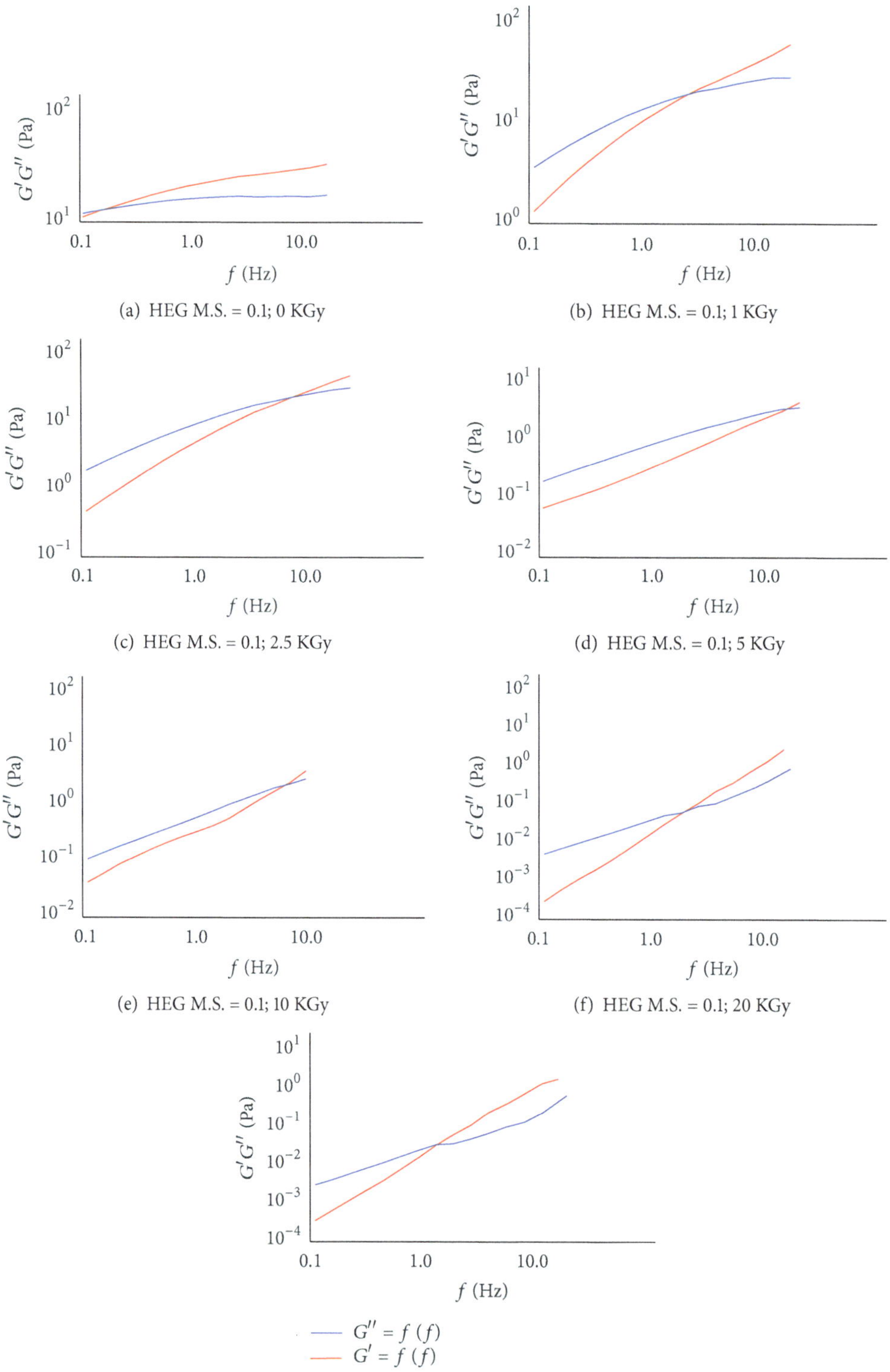

(a) HEG M.S. = 0.1; 0 KGy

(b) HEG M.S. = 0.1; 1 KGy

(c) HEG M.S. = 0.1; 2.5 KGy

(d) HEG M.S. = 0.1; 5 KGy

(e) HEG M.S. = 0.1; 10 KGy

(f) HEG M.S. = 0.1; 20 KGy

(g) HEG M.S. = 0.1; 50 KGy

FIGURE 4: Effect of radiation processing on the viscoelastic behavior of hydroxyethylated guar (HEG) of molar substitution (M.S.) = 0.1: (a) unirradiated; (b) irradiated at 1 KGy; (c) irradiated at 2.5 KGy; (d) irradiated at 5 KGy; (e) irradiated at 10 KGy; (f) irradiated at 20 KGy; (g) irradiated at 50 KGy.

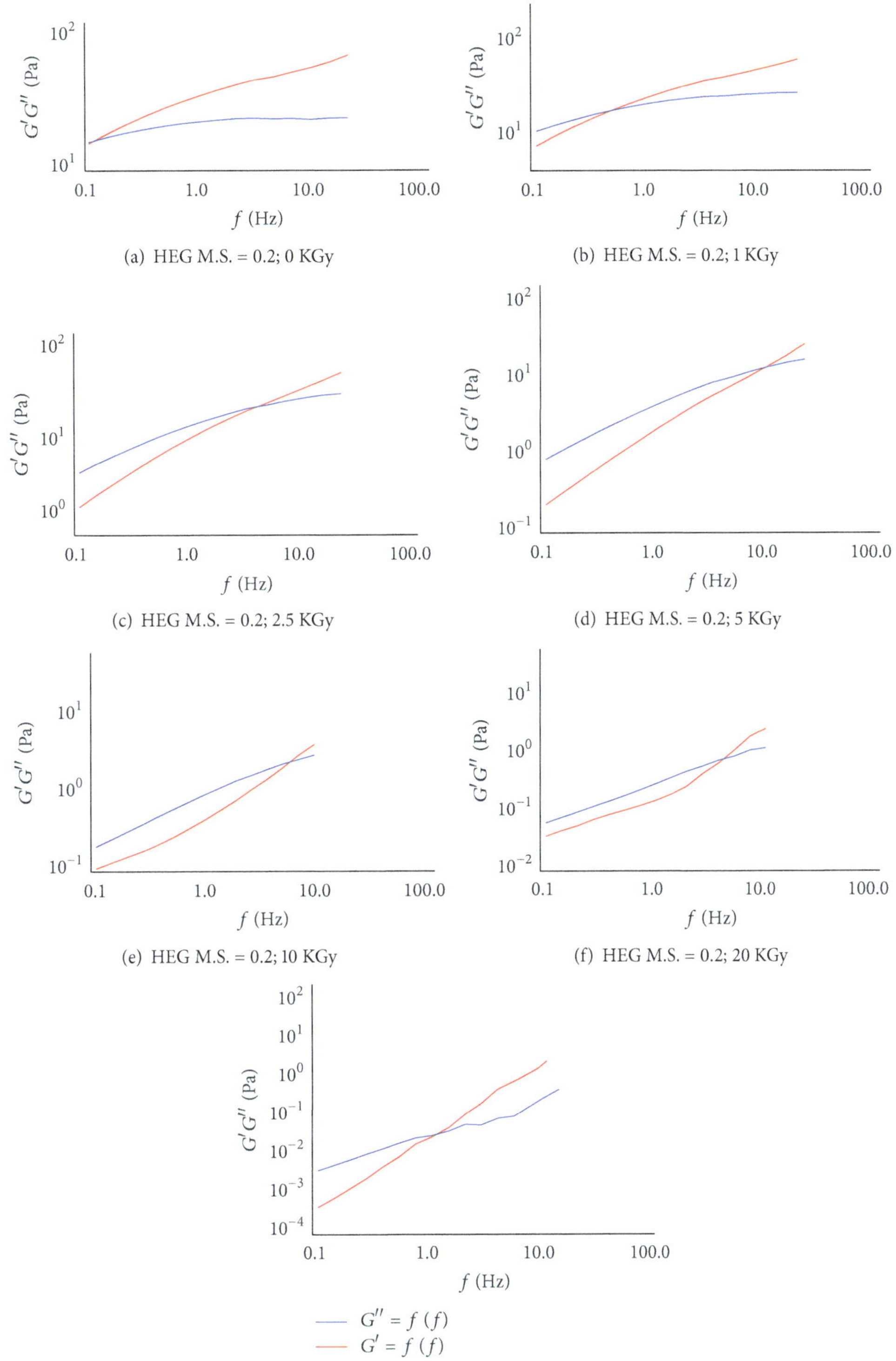

(a) HEG M.S. = 0.2; 0 KGy

(b) HEG M.S. = 0.2; 1 KGy

(c) HEG M.S. = 0.2; 2.5 KGy

(d) HEG M.S. = 0.2; 5 KGy

(e) HEG M.S. = 0.2; 10 KGy

(f) HEG M.S. = 0.2; 20 KGy

(g) HEG M.S. = 0.2; 50 KGy

FIGURE 5: Effect of radiation processing on viscoelastic properties of hydroxyethyl guar (HEG) having molar substitution 0.2: (a) unirradiated; (b) irradiated at 1 KGy; (c) irradiated at 2.5 KGy; (d) irradiated at 5 KGy; (e) irradiated at 10 KGy; (f) irradiated at 20 KGy; (g) irradiated at 50 KGy.

G'' is lesser than that of HEG of M.S. 0.1 till 20 KGy radiation dose. This means that elastic nature predominates in case of HEG of M.S. 0.2, which may be due to the presence of greater number of $-(CH_2-O)_n-$H groups leading to increase in H-bonding and thus more solid-like behavior.

(ii) On comparing the crossover frequencies of HEG of M.S. 0.1 after irradiation, it can be seen from the data that crossover frequency increases till 5 KGy dose and then it decreases, whereas in case of HEG of M.S. 0.2, the crossover frequency increases till 10 KGy dose and then it decreases. This can be explained on the basis of interactions taking place between the macromolecular chains in these two HEGs. In case of HEG of M.S. 0.1, intermolecular interactions are less as compared to HEG of M.S. 0.2 and thus less elastic nature.

(iii) Also, in case of HEG of M.S. 0.2, as the substitution increases, the coiling of macromolecular chain decreases due to steric hindrance, thereby causing more intramolecular interactions than intermolecular interactions. Less coiling of the chain causes more exposure to irradiation and thus greater chain scissioning and more viscous response of the macromolecule.

(iv) As the radiation dose increases, there is more chain scissioning leading to formation of oligomeric chains. These chains form intermolecular hydrogen bonding linkages leading to elastic behavior at lower frequencies.

(v) The G' and G'' values decrease on increasing the radiation dose, which signifies chain scissioning.

3.1.3. Methyl Guar. The methyl guar (MG) samples were irradiated and their viscoelastic behavior was studied as described in the previous section. The change in storage (elastic, G') and the loss (viscous, G'') moduli was studied over a range of frequency 1 to 30 Hz and the results obtained have been shown in Figures 6 and 7.

The following observations were made from the results.

In case of MG of D.S. 0.1, no crossover of G' and G'' curves is seen in unirradiated and irradiated sample at 1 KGy dose. This shows that the elasticity predominates. Methyl group is hydrophobic in nature and its substitution reduces the number of hydroxyl groups on guar molecule and opens up the guar chains. This leads to more number of intermolecular interactions than intramolecular interactions and thus increases elasticity.

(i) On increasing the radiation dose up to 10 KGy, the elastic component decreases. This can be attributed to the fact that chain scissioning takes place and now the H-bonding interactions take place between smaller chains, which reduces the elasticity and viscous behavior predominates. Due to this, the crossover frequency increases till 10 KGy dose.

(ii) Beyond 10 KGy dose, there is increase in the elastic component, which may be because of interactions in the smaller chains because of hydrophobic methyl group.

3.2. Study on Effect of Irradiation on Shear Rate. The effect of radiation on the change in viscosity with increase in shear rate (1 to 30/s) was studied and has been shown in Figures 8 to 13. To analyze this data, the minimum viscosity (viscosity obtained at shear rate value of 30/s) obtained for each of the samples was plotted against the radiation dose (Figure 14) and the shear rate at which shear rate versus viscosity curve reaches a plateau was plotted with respect to the increase in radiation dose (Figure 15) for all the samples (CMG, HEG and MG).

3.3. Radiation Dose versus Viscosity. As evident from Figure 14, the minimum viscosity attained by the 1% solution of various derivatives both before and after irradiation shows a decreasing trend with respect to increase in the radiation dose. The decrease is less pronounced in the case of samples having higher substitution, which shows that samples with greater substitution level exhibit higher resistance to depolymerisation.

3.4. Radiation Dose versus Shear Rate. Figure 15 shows the effect of increasing the radiation dose on shear rate needed to achieve nearly constant viscosity with respect to change in shear rate. As the radiation dose is increased, the shear rate value, at which nearly constant viscosity is achieved, decreases and at the same time the behavior of the three types of derivatives is quite different for each type.

In case of CMG, the shear rate required to achieve nearly constant viscosity decreases with the increase in radiation dose. This means that as the radiation dose is increasing, the shear stability of carboxymethyl guar is increasing. In other words, the oligomeric chains of carboxymethyl guar formed on its irradiation exhibit better shear stability. When we look at the results obtained in case of hydroxyethyl guar, then it becomes evident that nearly constant viscosity in this case is achieved at higher shear rate values, although a decreasing trend in shear rate values is observed in this case also. In case of methyl guar also, on increasing the radiation dose, the shear rate required to achieve nearly constant viscosity decreases. In case of methyl guar of D.S. 0.1, the shear rate value was found to be nearly constant from 2.5 to 50 KGy radiation dose.

4. Conclusion

The depolymerisation of polymers and polymeric materials by radiation processing is a dry technique. Further, for achieving the desired results, one does not have to use any of the additives. In other words, by radiation processing one can depolymerize guar powder as such without making its solution in water and without incorporation of additives.

Radiation processing of materials such as polymers, food products, precious stones, medical goods, has been widely adopted industrially since it is a continuous operation, which is highly precise, energy saving, and reproducible.

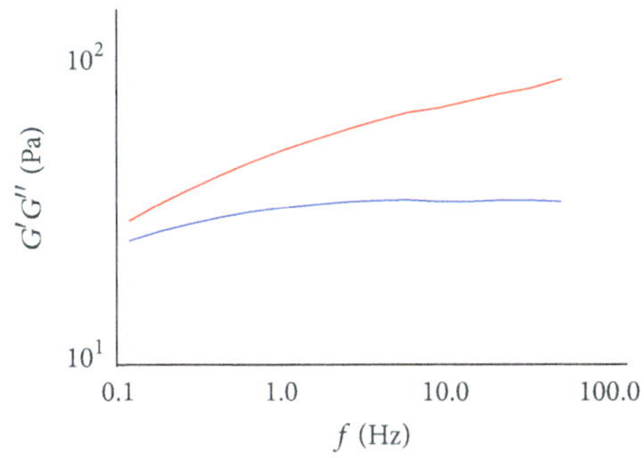

(a) MG D.S. = 0.1; 0 KGy

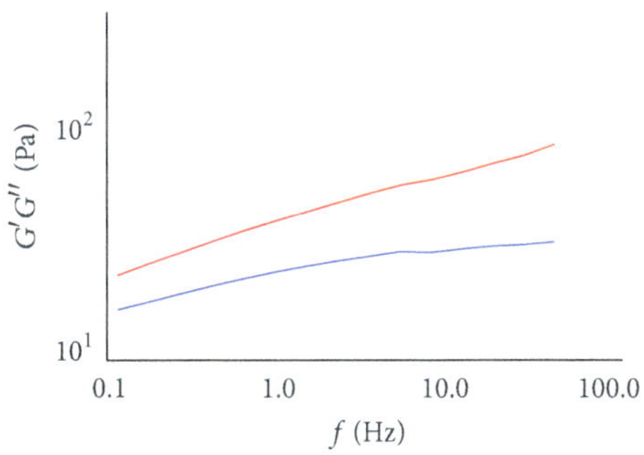

(b) MG D.S. = 0.1; 1 KGy

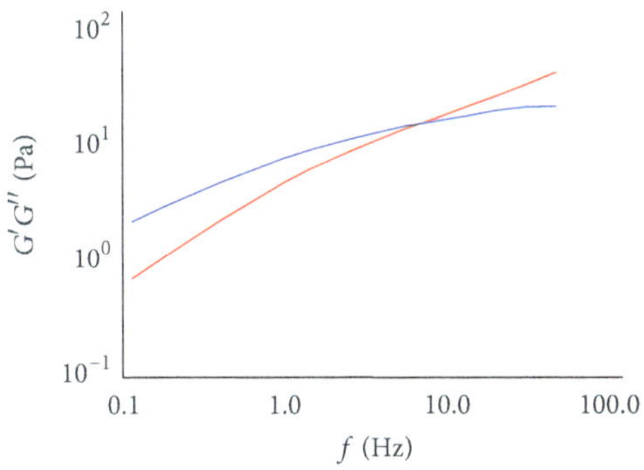

(c) MG D.S. = 0.1; 2.5 KGy

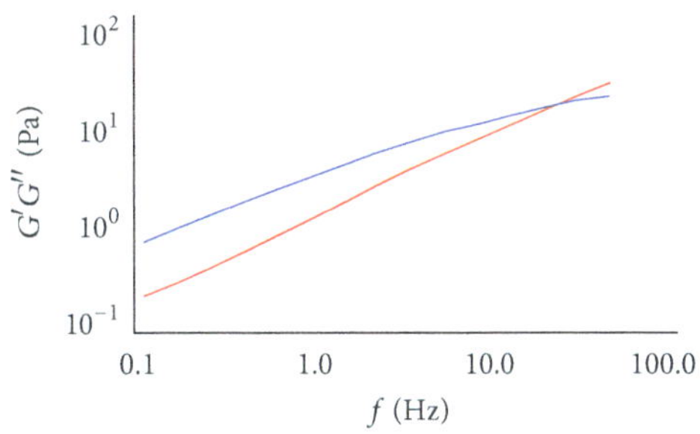

(d) MG D.S. = 0.1; 5 KGy

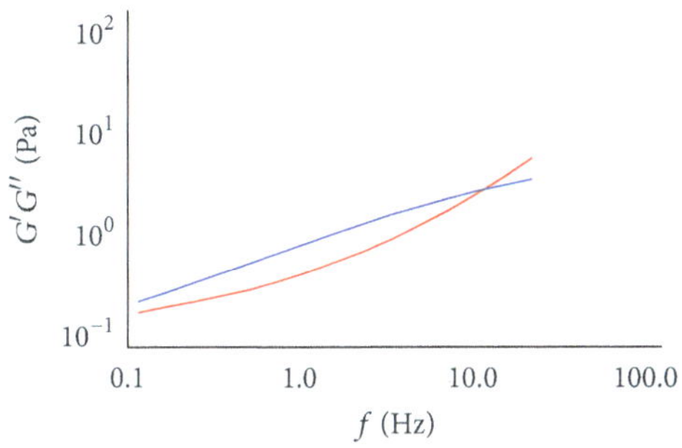

(e) MG D.S. = 0.1; 10 KGy

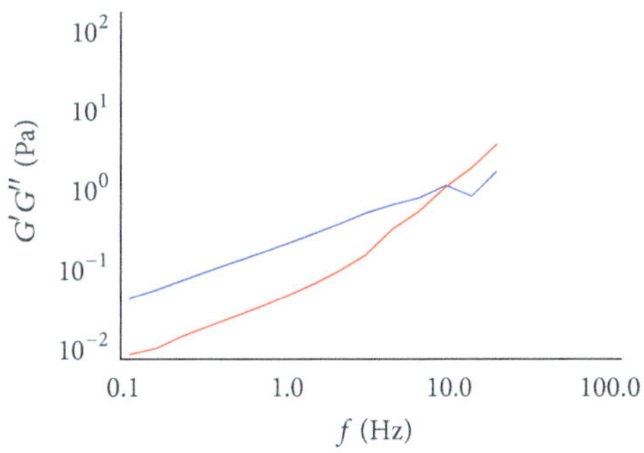

(f) MG D.S. = 0.1; 20 KGy

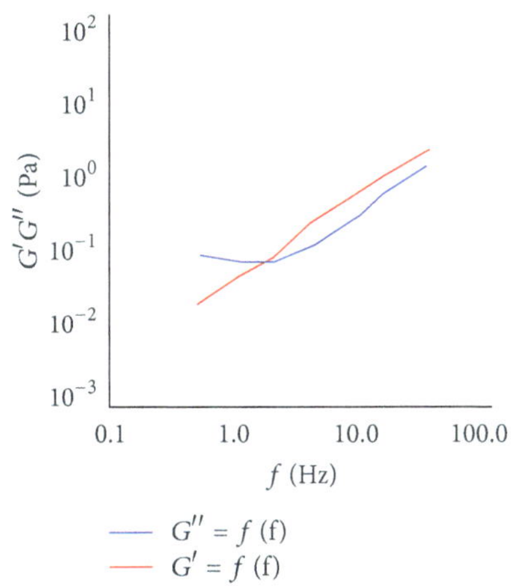

(g) MG D.S. = 0.1; 50 KGy

FIGURE 6: Effect of radiation processing on the viscoelastic behavior of methylated guar (MG) of degree of substitution (D.S.) = 0.1: (a) unirradiated; (b) irradiated at 1 KGy; (c) irradiated at 2.5 KGy; (d) irradiated at 5 KGy; (e) irradiated at 10 KGy; (f) irradiated at 20 KGy; (g) irradiated at 50 KGy.

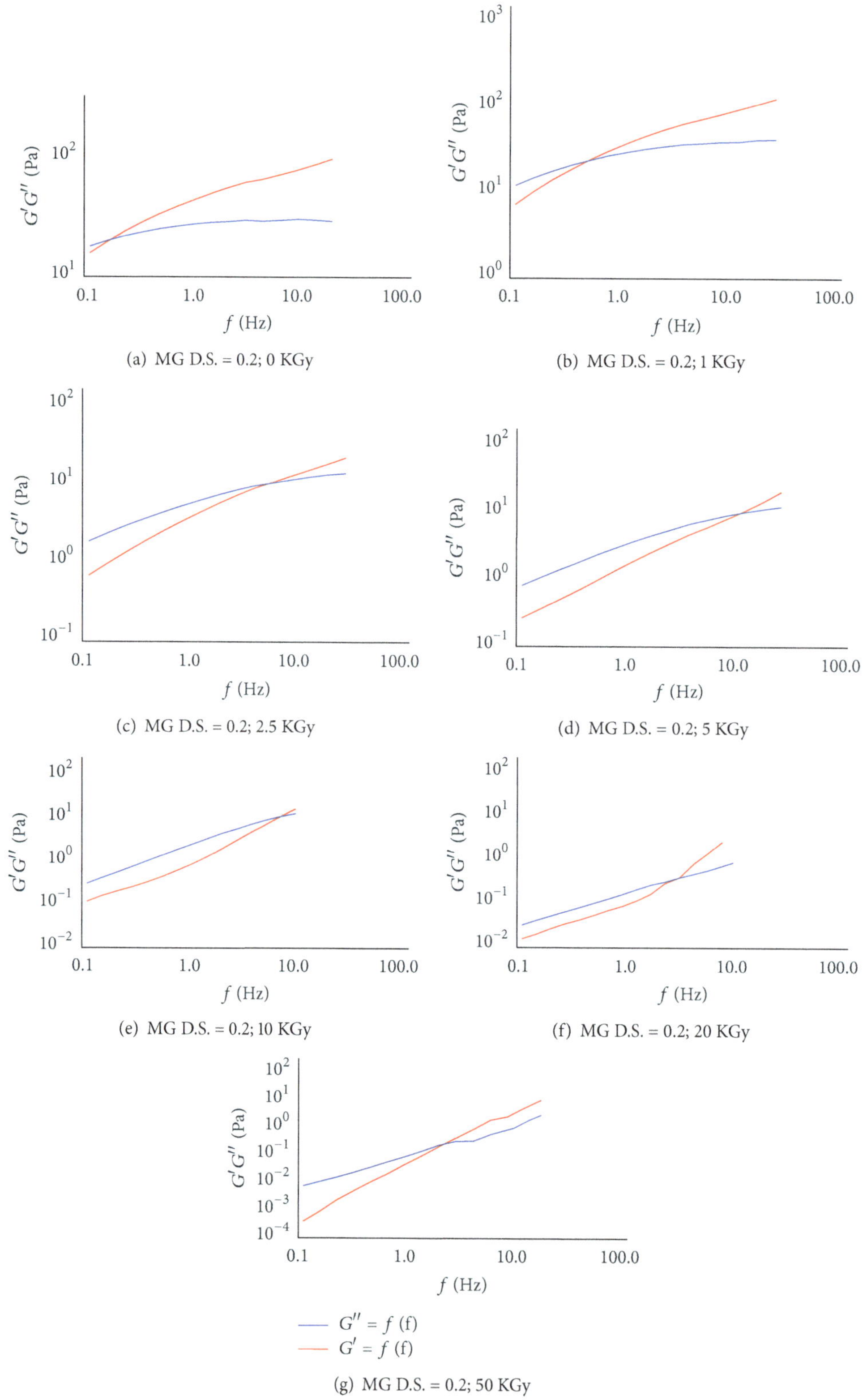

FIGURE 7: Effect of radiation processing on the viscoelastic behavior of methylated guar (MG) of degree of substitution (D.S.) = 0.2: (a) unirradiated; (b) irradiated at 1 KGy; (c) irradiated at 2.5 KGy; (d) irradiated at 5 KGy; (e) irradiated at 10 KGy; (f) irradiated at 20 KGy; (g) irradiated at 50 KGy.

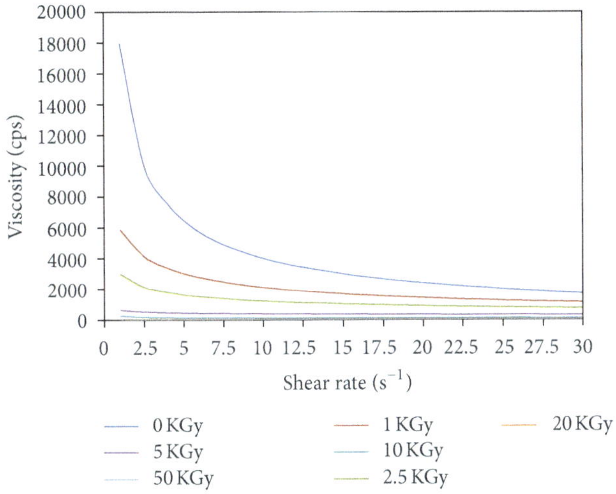

FIGURE 8: Effect of radiation processing on the shear stability of carboxymethylated guar (CMG) of degree of substitution (D.S.) = 0.1; unirradiated and irradiated.

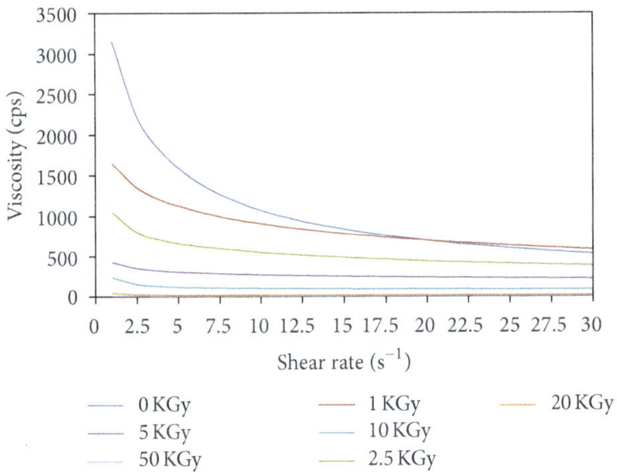

FIGURE 9: Effect of radiation processing on the shear stability of carboxymethylated guar (CMG) of degree of substitution (D.S.) = 0.2; unirradiated and irradiated.

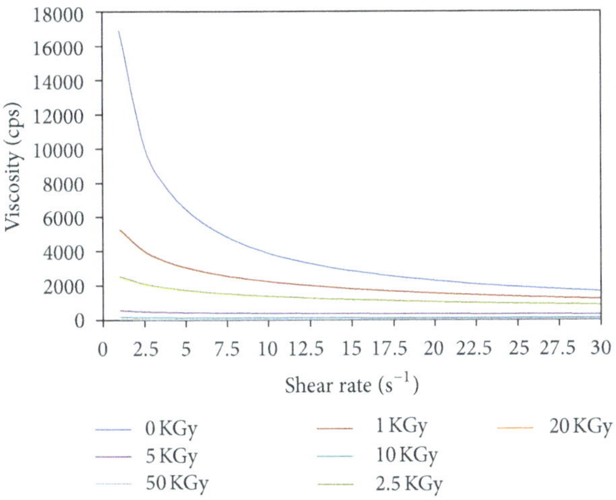

FIGURE 10: Effect of radiation processing on the shear stability of hydroxyethylated guar (HEG) of molar substitution (M.S.) = 0.1; unirradiated and irradiated.

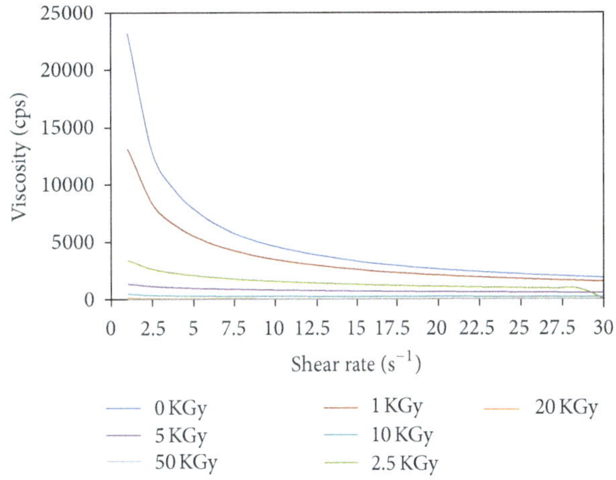

FIGURE 11: Effect of radiation processing on the shear stability of hydroxyethylated guar (HEG) of molar substitution (M.S.) = 0.2; unirradiated and irradiated.

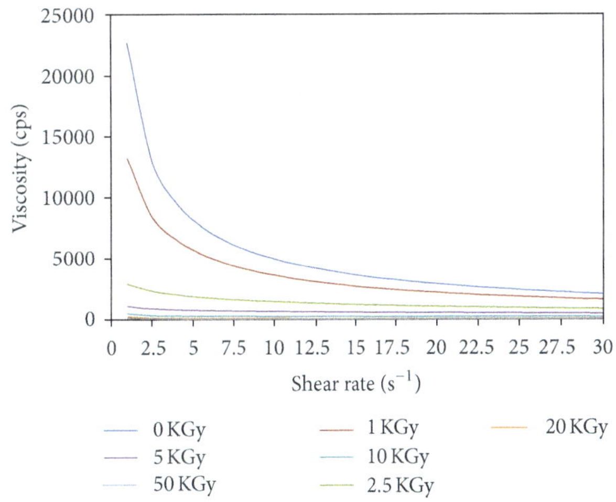

FIGURE 12: Effect of radiation processing on the shear stability of methylated guar (MG) of degree of substitution (D.S.) = 0.1; unirradiated and irradiated.

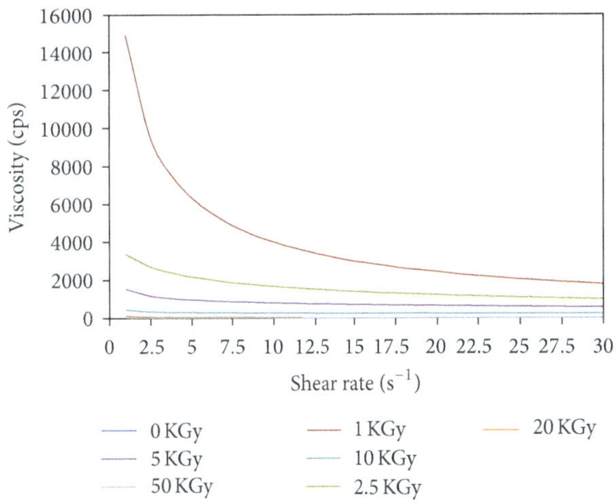

FIGURE 13: Effect of radiation processing on the shear stability of methylated guar (MG) of degree of substitution (D.S.) = 0.2; unirradiated and irradiated.

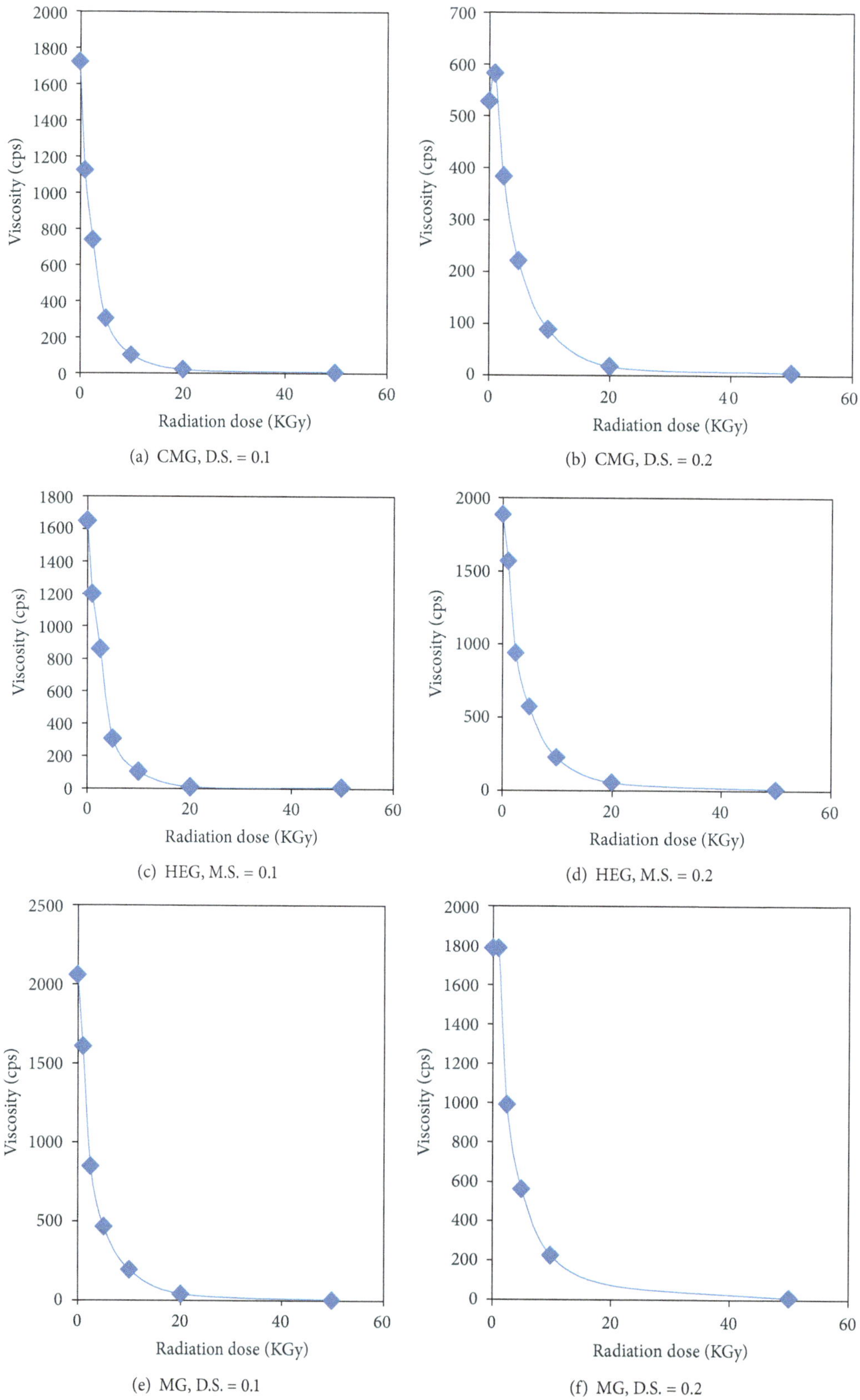

FIGURE 14: The plot of minimum viscosity (at shear rate 30/s) attained by the sample while subjecting it to shear from 1 to 30/s for (a) CMG D.S. = 0.1 and (b) 0.2, (c) HEG M.S. = 0.1 and (d) 0.2, and (e) MG D.S. = 0.1 and (f) 0.2 with respect to radiation dose.

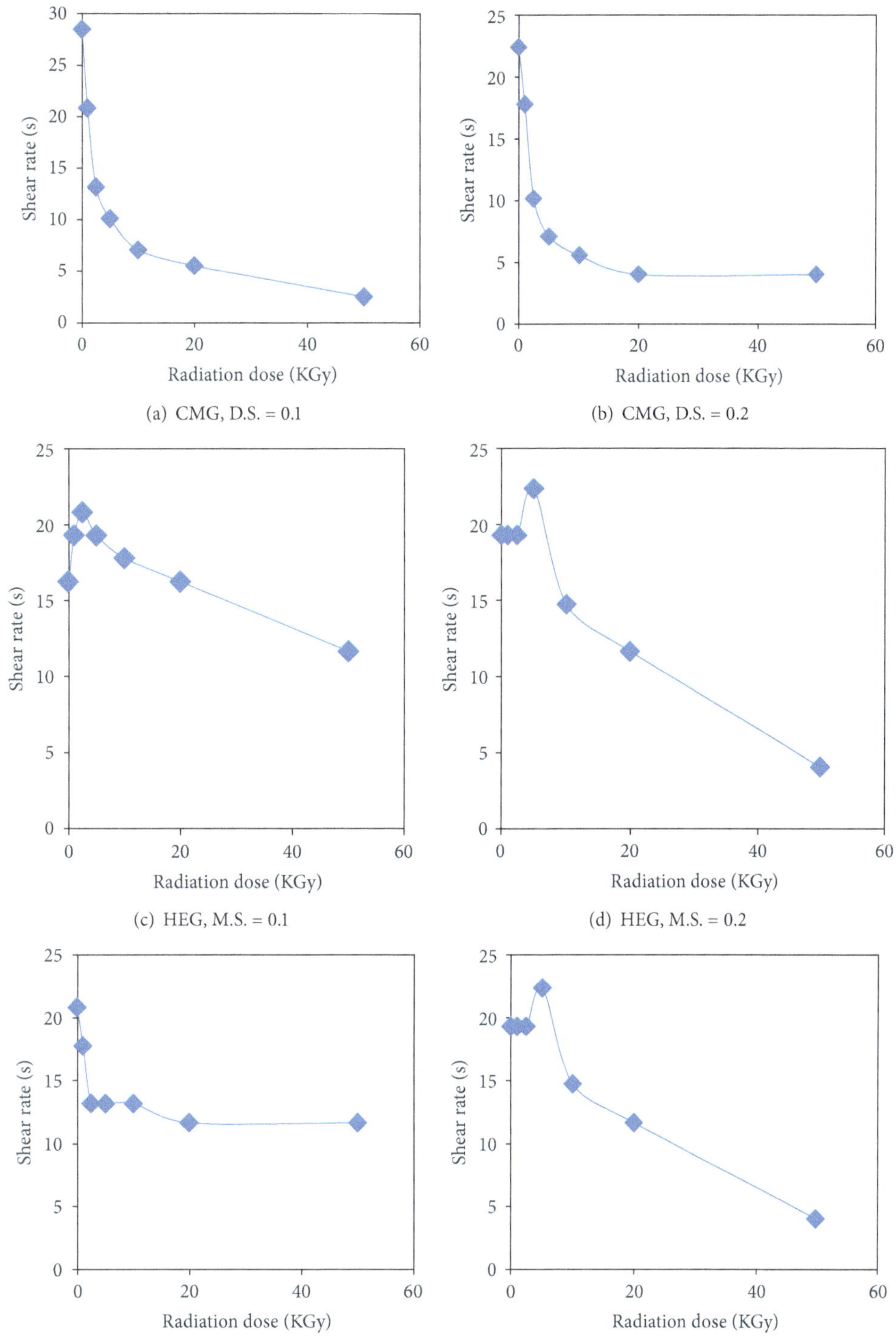

(a) CMG, D.S. = 0.1
(b) CMG, D.S. = 0.2
(c) HEG, M.S. = 0.1
(d) HEG, M.S. = 0.2
(e) MG, D.S. = 0.1
(f) MG, D.S. = 0.2

FIGURE 15: The plot of shear rate value versus radiation dose. The shear rate value was taken from the study carried out by increasing the shear rate and monitoring change in viscosity. The point at which the curve between shear rate and viscosity reaches plateau was taken for plotting radiation dose versus shear rate for samples (a) CMG D.S. = 0.1 and (b) 0.2, (c) HEG M.S. = 0.1 and (d) 0.2, and (e) MG D.S. = 0.1 and (f) 0.2 with respect to radiation dose.

The depolymerisation of synthetic polymers by radiation processing for recycling of monomers has been known for quite some time. The present paper is an attempt to initiate the use of this technology for radiation processing of modified natural polymers. The gamma irradiation facility used for the current studies is an industrial plant and it has been in operation for about 20 years now where large volumes of industrial products are irradiated every day.

From the results presented here, it is evident that the depolymerisation of guar is achieved with the radiation dose of 1 KGy and above, as seen from the observation of viscosity of 1% of guar derivatives getting reduced from as high a level as 10,000 cps (for unirradiated guar derivative) to as low a level as 1 cps at 25°C (for guar derivatives irradiated at 20–50 KGy).

Inspite of this knowledge, which can be easily used for scaling up of the process to commercial level, doubts are raised about the scalability of the process. Whether the process can be used for bulk material is the common apprehension in the minds of the processors of guar. While the present study would help clear many of the doubts regarding the suitability of radiation processing technology, but the data about the scalability of the process would actually eliminate all sorts of doubts.

Eventhough, the results presented in this paper pertain to the batch size of 200 g for each guar derivative at each dose, the same was found valid even for the bigger batch size (in tons). The results of large batches were the same as obtained for the small batches. Here, it must be mentioned that the dose optimization for the purpose would be necessary for the purpose of studying the scalability.

From the present study the following can be concluded.

(i) Radiation processing of guar derivatives leads to their chain scissioning.

(ii) Radiation processing leads to reduction in viscosity of aqueous solutions of guar derivatives.

(iii) Irradiation technique can be a good tool for tailor making the guar derivatives of desired rheological properties.

(iv) The minimum viscosity attained by the 1% solution of various derivatives shows a decreasing trend with respect to increase in the radiation dose.

(v) In case of carboxymethyl guar, the crossover frequency increases as the radiation dose increases up to 5 KGy and the value is higher in case of CMG of D.S. 0.2 than in that of CMG of D.S. 0.1. This shows that the elastic nature is more dominating for CMG of D.S. 0.1 than for CMG of D.S. 0.2.

(vi) In case of hydroxyethyl guar of M.S. 0.2, the crossover frequency of the G' and G'' is lesser than that of HEG of M.S. 0.1 till 20 KGy radiation dose. This means that elastic nature predominates in case of HEG of M.S. 0.2, which may be due to the presence of greater number of $-(CH_2-O)_n-H$ groups leading to increase in H-bonding and thus more solid-like behavior.

(vii) In case of methyl guar, on increasing the radiation dose up to 10 KGy, the elastic component decreases. This can be attributed to the fact that chain scissioning takes place and now the H-bonding interactions take place between smaller chains, which reduces the elasticity and viscous behavior predominates. Due to this, the crossover frequency increases till 10 KGy dose.

(viii) As the radiation dose is increased, the shear rate value at which nearly constant viscosity is achieved decreases.

Acknowledgment

The authors express sincere gratitude to the management of Shriram Institute for Industrial Research, Delhi, India, for the kind support.

References

[1] L. Wang and L. M. Zhang, "Viscoelastic characterization of a new guar gum derivative containing anionic carboxymethyl and cationic 2-hydroxy-3-(trimethylammonio)propyl substituents," *Industrial Crops and Products*, vol. 29, no. 2-3, pp. 524–529, 2009.

[2] C. Sandolo, P. Matricardi, F. Alhaique, and T. Coviello, "Effect of temperature and cross-linking density on rheology of chemical cross-linked guar gum at the gel point," *Food Hydrocolloids*, vol. 23, no. 1, pp. 210–220, 2009.

[3] H. N. Englyst, V. Anderson, and J. H. Cummings, "Starch and non-starch polysaccharides in some cereal foods," *Journal of the Science of Food and Agriculture*, vol. 34, no. 12, pp. 1434–1440, 1983.

[4] R. L. Feddersen and S. N. Thorp, *Industrial Gums*, Acaedemic Press, San Diego, Calif, USA, 1993.

[5] D. D. Roberts, J. S. Elmore, K. R. Langley, and J. Bakker, "Effects of sucrose, guar gum, and carboxymethylcellulose on the release of volatile flavor compounds under dynamic conditions," *Journal of Agricultural and Food Chemistry*, vol. 44, no. 5, pp. 1321–1326, 1996.

[6] D. R. Picout, S. B. Ross-Murphy, K. Jumel, and S. E. Harding, "Pressure cell assisted solution characterization of polysaccharides. 2. Locust bean gum and tara gum," *Biomacromolecules*, vol. 3, no. 4, pp. 761–767, 2002.

[7] R. S. Blackburn, "Natural polysaccharides and their interactions with dye molecules: applications in effluent treatment," *Environmental Science and Technology*, vol. 38, no. 18, pp. 4905–4909, 2004.

[8] M. Urdiaín, A. Doménech-Sánchez, S. Albertí, V. J. Benedí, and J. A. Rosselló, "New method of DNA isolation from two food additives suitable for authentication in polymerase chain reaction assays," *Journal of Agricultural and Food Chemistry*, vol. 53, no. 9, pp. 3345–3347, 2005.

[9] R. P. Singh, S. Pal, and D. Mal, "A high performance flocculating agent and viscosifiers based on cationic guar gum," *Macromolecular Symposia*, vol. 242, pp. 227–234, 2006.

[10] S. P. Zhao, D. Ma, and L. M. Zhang, "New semi-interpenetrating network hydrogels: synthesis, characterization and properties," *Macromolecular Bioscience*, vol. 6, no. 6, pp. 445–451, 2006.

[11] J. Z. Yi and L. M. Zhang, "Biodegradable blend films based on two polysaccharide derivatives and their use as Ibuprofen-releasing matrices," *Journal of Applied Polymer Science*, vol. 103, no. 6, pp. 3553–3559, 2007.

[12] S. Venkataiah and E. G. Mahadevan, "Rheological properties of hydroxypropyl and sodium carboxymethyl substituted guar gums in aqueous solution," *Journal of Applied Polymer Science*, vol. 27, no. 5, pp. 1533–1548, 1982.

[13] R. H. W. Wientjes, M. H. G. Duits, R. J. J. Jongschaap, and J. Mellema, "Linear rheology of guar gum solutions," *Macromolecules*, vol. 33, no. 26, pp. 9594–9605, 2000.

[14] T. Aubry and M. Moan, "Rheological behavior of a hydrophobically associating water soluble polymer," *Journal of Rheology*, vol. 38, no. 6, pp. 1681–1692, 1994.

[15] L. M. Zhang, T. Kong, and P. S. Hui, "Semi-dilute solutions of hydroxypropyl guar gum: viscosity behaviour and thixotropic properties," *Journal of the Science of Food and Agriculture*, vol. 87, no. 4, pp. 684–688, 2007.

[16] N. N. G. Swamy, T. S. Dharmarajan, and K. L. K. Paranjothi, "Derivatization of guar to various hydroxy alkyl derivatives and their characterization," *Indian Drugs*, vol. 43, no. 9, pp. 756–759, 2006.

[17] H. Gong, M. Liu, J. Chen, F. Han, C. Gao, and B. Zhang, "Synthesis and characterization of carboxymethyl guar gum and rheological properties of its solutions," *Carbohydrate Polymers*, vol. 88, no. 3, pp. 1015–1022, 2012.

12

High-Performance Anion-Exchange Chromatography Coupled with Pulsed Electrochemical Detection as a Powerful Tool to Evaluate Carbohydrates of Food Interest: Principles and Applications

Claudio Corradini, Antonella Cavazza, and Chiara Bignardi

Dipartimento di Chimica Generale de Inorganica, Chimica Analitica, Chimica Fisica, Università Degli Studi di Parma, Parco Area delle, Scienze 17/A, 43124 Parma, Italy

Correspondence should be addressed to Claudio Corradini, claudio.corradini@unipr.it

Academic Editor: Margarita Stoytcheva

Specific HPLC approaches are essential for carbohydrate characterization in food products. Carbohydrates are weak acids with pK_a values in the range 12–14 and, consequently, at high pH can be transformed into oxyanions, and can be readily separated using highly efficient anion-exchange columns. Electrochemical detection in HPLC has been proven to be a powerful analytical technique for the determination of compounds containing electroactive groups; pulsed amperometric detection of carbohydrates is favourably performed by taking advantage of their electrocatalytic oxidation mechanism at a gold working electrode in a basic media. High-performance Anion Exchange Chromatography (HPAEC) at high pH coupled with pulsed electrochemical detection (PED) is one of the most useful techniques for carbohydrate determination either for routine monitoring or research application. This technique has been of a great impact on the analysis of oligo- and polysaccharides. The compatibility of electrochemical detection with gradient elution, coupled with the high selectivity of the anion-exchange stationary phases, allows mixtures of simple sugars, oligo- and polysaccharides to be separated with high resolution in a single run. A few reviews have been written on HPAEC-PED of carbohydrates of food interest in the last years. In this paper the recent developments in this field are examined.

1. Introduction

The analysis of carbohydrates in food has extreme nutritional importance because they are a primary source of energy and have healthy beneficial effects resulting mainly from dietary fibers and other unavailable carbohydrates resistant to digestion. In food industry they are largely used as technological coadjuvants in order to obtain physicochemical and selected sensorial characteristics for some products. Moreover, they are suitable to be considered as markers for quality and authenticity control.

Several analytical techniques have been proposed for carbohydrates analysis; among them, high-performance liquid chromatography (HPLC) coupled to different detection systems. This technique offers the advantages of high resolution, fast analysis, direct injection of the sample without or with little pretreatment, and easy of automation. Different operation modes of HPLC have been applied to the analysis of carbohydrates of food interest.

HPLC of carbohydrates is frequently performed using cation-exchange resins based on copolymers in protonated or metal ion forms. Metal-loaded strong cation exchangers such as sulphonated styrene-divinylbenzene copolymers in calcium, lithium, sodium, lead, and silver form, have been extensively used in many separations of mono- and oligosaccharides. Loading of sulfonated resins with various cations causes substantial changes in retention and selectivity of neutral carbohydrates. By these approaches separations are obtained involving different mechanisms such as ion exchange, ion exclusion, size exclusion, ligand exchange, and counterions in which underivatized saccharides are usually chromatographed using plain water as the mobile phase [1, 2].

Hydrophilic interaction chromatographic systems using bonded-phase packaging-based silica columns and employing acetonitrile-water mixtures as the mobile phase, as well as reversed-phase chromatography, have also been used for carbohydrate separation.

However, these modes of carbohydrate HPLC can suffer several drawbacks such as instability and short lifetime of the bonded phases and poor column performance in selectivity and efficiency. Furthermore, sensitive detection of carbohydrates after HPLC separation represents an additional difficulty for their determination; in fact carbohydrates mainly lack the chromophores and fluorophores. This often results in attempts to detect carbohydrates using low-wavelength UV absorbance or refractive index detection (RI). As it is well known, neutral carbohydrates adsorb only in the region 190–210 nm, where the organic mobile phase modifiers such as acetonitrile also adsorb strongly, and consequently the UV detection is unsuitable for use in HPLC of underivatized carbohydrates in mobile phases rich in these solvents. A greater sensitivity of detection can be achieved by derivatization introducing either a chromophoric or a fluorescent group into carbohydrate analytes [3]. Moreover, post- and precolumn chemical derivatizations often enhance the selectivity of the detection since only reducing sugars can be modified in the presence of amino acids and lipids when carbonyl groups are derivatized. Either indirect UV absorbance or indirect fluorescence detection has also been proposed. After absorbance and fluorescence, refractive index (RI) detection is the third mode in order of importance in conventional HPLC, and it has been the most frequently used in HPLC of carbohydrates. However, the major disadvantage of RI detection is its sensitivity to changes in mobile phase composition, preventing its use when gradient elution is employed.

Consequently, the lack of selectivity and sensitivity of these detection modes is often not sufficient for the analysis of carbohydrates present in complex mixtures such as food. Alternatively, evaporative light-scattering detection (ELSD) has also been proposed. ELSD is insensitive to temperature fluctuations, allows gradient solvent systems, and is dependent on the mass of the vaporized analyte. On the other hand, ELSD requires the vaporization of all the components of the HPLC eluent and the nonvaporization of carbohydrates which can be detected. It is thus possible to use any gradient of eluents requiring the lowest possible temperature to prevent thermal degradation of carbohydrates. Consequently, the chromatographic mobile phase composition is not independent from the detection system. In conclusion, ELSD detection, which can be considered a universal detector, found scarce application in HPLC analysis of carbohydrates.

High-performance anion-exchange chromatography (HPAEC) is a powerful analytical tool in carbohydrate separation due its ability to separate all classes of alditols, aminosugars, mono-, oligo- and polysaccharides, according to structural features such as size, composition, anomericity and linkage isomerism. Moreover, detector insensitivity is overcome coupling HPAEC with pulsed amperometric detection. HPAEC-PED is presently being applied to a variety of routine monitoring and research applications. Moreover, this technique has had a major impact on the analysis of oligo-, and polysaccharides. The compatibility of electrochemical detection with gradient elution coupled with the high selectivity of the anion-exchange stationary phases allows mixtures of simple sugars, oligo- and polysaccharides to be separated with suitable resolution in a single run.

2. High-Performance Anion-Exchange Chromatography (HPAEC)

Anion exchange chromatography is not a technique commonly associated with the analysis of neutral carbohydrates. However, many carbohydrates are weak acids with pK_a values in the range 12–14, and, consequently, at high pH values their hydroxyl groups are partially or totally transformed into oxyanions, enabling this class of compounds to be selectively eluted as anions by high-performance anion-exchange chromatography in a single run.

Under alkaline conditions, carbohydrates are readily separated by quaternary-ammonium-bonded pellicular anion-exchange columns, where the order of increasing retention is correlated with decreasing pK_a value.

Anion exchange chromatography on high pH-resistant polymeric-based strong anion-exchange columns specifically tailored for carbohydrate analysis enables selective elution of carbohydrates, where the most important parameters influencing the separations are the number of hydroxyl groups, anomerism, positional isomerism and the polymerization degree. Monosaccharides possess several potentially ionisable hydroxyl groups having, taking glucose as reference, the following hierarchy of acidity: 1-OH > 2-OH ≥ 6-OH > 3-OH > 4-O.

aldoses exhibit greater retention than their reduced alditol forms. A study conducted by Paskach et al. [4] on the retention behavior of 93 alditols and carbohydrates on HPAEC has demonstrated that capacity factors for monosaccharides are lowest for the alditols and are higher and approximately similar for analogous aldoses and ketoses. There is a general trends, observed to be strongest with alditols, for capacity factors to increase with increasing numbers of carbon atoms. The most common columns employed in HPAEC-PED, specifically designed for carbohydrate anion-exchange chromatography, are manufactured by Dionex (Sunnyvale, CA, USA; now Thermo Fisher Scientific Inc., USA). As reported in Table 1, the CarboPac MA1 is a macroporous resin fully functionalized with a high-capacity, strong exchange column, whereas the others are prepared by proprietary processes, which permit the independent manufacture of polymeric nonporous sulphonated resins and anion-exchange latex particles functionalized with quaternary ammonium compounds that are brought together only at the last steps of preparing anion-exchanger columns.

Table 1: Most employed columns for HPAEC-PAD analysis*.

Column	Specifications	Samples	Notes
CarboPac MA1	7.5 μm diameter vinylbenzylchloride-divinylbenzene macroporous substrate fully functionalized with alkyl quaternary ammonium groups (15% crosslinked)	Reduced mono- and disaccharide alditol analyses	Other carbohydrates: fucose, N-acetyl-(D)-glucosamine, N-acetyl-galactosamine, mannose, galactose
CarboPac PA1	10 μm diameter substrate (polystyrene 2% crosslinked with divinylbenzene) agglomerated with MicroBead quaternary ammonium functionalized latex (5% crosslinked)	Mono- and disaccharides, oligosaccharides	Column well suited for the analysis of food for nutrition labelling
CarboPac PA10	10 μm diameter substrate (ethylvinylbenzene 55% cross linked with divinylbenzene) agglomerated with a 460 nm MicroBead difunctional quaternary ammonium ion (5% crosslinked)	Mono- and disaccharides, glucosamine, galactosamine; sialic acids	Ideal for the analysis of mono- and disaccharides in foods, drugs, and plants
CarboPac PA20	6.5 μm diameter substrate (ethylvinylbenzene 55% crosslinked with divinylbenzene) agglomerated with a 130 nm MicroBead difunctional quaternary ammonium ion (5% crosslinked)	Mono- and disaccharides	High resolution regarding the six monosaccharides commonly found in mammalian glycoproteins
CarboPac PA100	8.5 μm diameter substrate (ethylvinylbenzene 55% crosslinked with divinylbenzene) agglomerated with a 275 nm MicroBead quaternary amine functionalized latex (6% crosslinked)	Separation of closely related oligosaccharides, oligosaccharides released from glycoproteins which can differ in size, charge, branching, and linkage	Separation of neutral and charged oligosaccharides in the same run Separation of closely related oligosaccharides
CarboPac PA200	5.5 μm diameter substrate (ethylvinylbenzene 55% crosslinked with divinylbenzene) agglomerated with a 43 nm MicroBead quaternary amine functionalized latex (6% crosslinked)	Oligosaccharides	Separations based on size, charge, degree of branching, and linkage isomerism
CarboPac SA10	6.0 μm diameter substrate (ethylvinylbenzene 55% crosslinked with divinylbenzene) agglomerated with a 55 nm MicroBead quaternary ammonium ion (4.5% crosslinked)	Mono- and disaccharides	Fast analysis of mono- and disaccharides of food interest.

*As reported in technical note from Dionex (Thermo Fisher Inc.).

Monosaccharide separations are normally performed on CarboPac PA1 and CarboPac PA10 columns, which have high selectivity for mono- and disaccharide separations. These sugars are usually eluted isocratically using sodium hydroxide solutions at concentration ranging from 10 to 20 mM. Particular attention should be paid to the preparation of eluents in order to avoid the presence of carbon dioxide and subsequent production of carbonate in the mobile phase. Carbonate, being a divalent ion at pH 12, binds strongly to the anion exchangers and interferes with carbohydrate retention, causing shortened retention times, decrease in column selectivity, and loss in resolution. Efforts to solve this problem may rely on the use of an electrochemical generator of potassium hydroxide eluents which are not affected by carbonate interference. Cataldi and coworkers proposed the use of barium and strontium ions in the alkaline eluent for reducing the uptake of carbon dioxide [5].

The separation of closely related carbohydrates, such as glucose and mannose which differ only in the axial-equatorial configuration of their hydroxyl groups, can be enhanced significantly by lowering the pH of the alkaline eluent to a value that is comparable to the pK_a of the sugar molecules, namely pH 12. In this case, postcolumn addition to the eluent stream of a strong base through a tee connection can provide the alkaline environment necessary for carbohydrate detection at the gold working electrode of the detector.

Retention of oligosaccharides on HPAEC column is not directly predictable by extrapolating the hierarchy of monosaccharide hydroxyl acidities to oligosaccharides. For

TABLE 2: Gradient elution program employed for inulin analysis reported in Figure 1; flow = 0.50 mL/min.

Time (minutes)	% eluent A H_2O	% eluent B NaOH 600 mM 50%	% eluent C $NaNO_3$ 500 mM
0	83	16	1
10	80	16	4
10.1	80	16	4
40	60	16	24

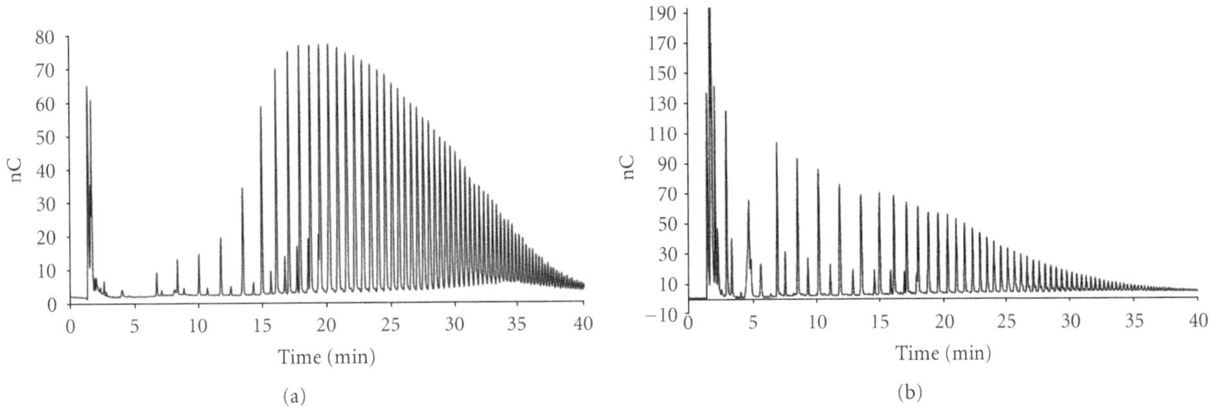

FIGURE 1: Chromatographic profiles obtained for Inulin Tex 0.5 mg/mL (a) and IQ 0.5 mg/mL (b) on a column CarboPac PA200 at the conditions of elution reported in Table 2.

members of homologous series of oligosaccharides, the capacity factors increase in a regular and predictable manner with chain length. However, beside chain length, other than saccharide composition, linkage positions are expected to affect the chromatographic retention of oligosaccharides.

Corradini et al. [6] showed that separation by HPAEC of twelve disaccharides was strongly affected, beside their acidity, by the accessibility of oxyanions to the functional groups of the anion exchanger column. This effect has been observed in particular for the glucobioses trehalose, isomaltose, gentiobiose, nigerose, and maltose, which are disaccharides composed of two D-glucosyl residues, differing only in the configuration of their glycosidic bonds.

The CarboPac PA100 and CarboPac PA200, strong anion exchange columns, are proposed for oligosaccharide characterization. To elute oligosaccharides and polysaccharide until DP 85, a stronger eluent than sodium hydroxide is needed. Sodium acetate solutions are usually employed for these separations that still require sodium hydroxide because a strongly basic environment is necessary for detection.

Most oligosaccharide separations are performed by sodium acetate gradient with a constant concentration of sodium hydroxide. The effects of acetate and nitrate as "pushing agent" have also been investigated. Wong and Jane [7] compared separations of debranched amylopectin achieved adding acetate or nitrate anions in the mobile phases. They found that, in comparison with the commonly used pushing agent, nitrate offered greater reproducibility, accuracy, and lower limit of detection.

More recently, we verified that an improvement in resolution with shorter chromatographic runs can be achieved using nitrate instead of acetate as the pushing agent in the mobile phase composition for the separation and detection of inulin by HPAEC-PED. The different chain-length distribution profiles of two commercial available inulins have been accomplished with the CarboPac PA200 column using nitrate as the "pushing agent" and gradient elution following the program reported in Table 2. Under the selected conditions, baseline separation was achieved in a short analysis time (see Figure 1).

3. Pulsed Electrochemical Detection (PED)

Electrochemical detection in HPLC has been proven to be a powerful analytical technique for the determination of compounds containing electroactive groups, and recently the progress in application of this detection strategy coupled to liquid chromatography has been reviewed by Trojanowicz [11].

Amperometric detection is typically performed at a solid anode, such as gold, platinum, and glassy carbon, under constant applied potential, which is denoted as DC amperometry. In constant potential amperometric mode-numerous aromatic compounds are easily detected by anodic reactions. These include phenols, aminophenols, catecholamines, and other metabolic amines.

The presence of π-resonance in aromatic compounds functions to stabilize intermediate oxidation products of anodic oxidation, and, as a consequence, the activation barrier for electrochemical reaction is significantly lowered. In dc amperometry, the electrode must be inert and able to accept or donate electrons to the analyte of interest. This is particularly true for glassy carbon electrodes which are particularly resistant to fouling. Glassy carbon is the most commonly used working electrode material in chromatographic amperometric detection, using a constant applied potential. In the case of carbohydrates, this system is not adequate since glassy carbon amperometric electrodes exhibit no response for aliphatic compounds. As expected, the absence of π-resonance for these compounds results in very low oxidation rates at an inert electrode, although the reactions may be favoured thermodynamically.

Considering that an inherent electronic stabilization mechanism does not exist for polar aliphatic compounds, the stabilization of the formed free-radical intermediates can be achieved with the use of catalytic electrode surfaces. Aldehyde and terminal alcohol moieties in carbohydrates can be oxidated at golden electrodes in alkaline media and at Pt electrodes under all pH conditions.

Electrocatalytic properties of gold electrodes are much less pronounced under acidic conditions, and only analytes with a strong affinity to gold, such as sulfur-containing species which are readily adsorbed on the gold surface, are oxidized and detected on gold electrodes under these conditions. However, intermediate oxidation products of analytes can adsorb on the surface of solid electrodes with direct participation of the electrode surface within the oxidation mechanisms. Unfortunately, a serious consequence of adsorption is the fouling of the electrode by accumulation of reaction products, resulting in the gradual loss of electrode response. The problem of electrode fouling can be solved by application of a pulsed potential waveform.

The application of alternated positive and negative potential pulses as a method in reactivating Au and Pt electrodes that have become fouled by adsorption of organic compounds has been employed since the early part of the last century [12].

Pulsed amperometric detection (PAD) applying a repetitive three-step waveform was proposed the first time in 1981 for the detection of simple alcohols at Pt electrodes in flow-injection systems [13, 14]. In these applications anodic detection was alternated with oxidative cleaning by stepping the potential to a greater positive value and reductive cleaning of adsorbed products for the reactivation of noble-metal electrodes to maintain a reproducibly high electrode activity. By utilizing a simple three-step potential waveform, reproducible detection of polar aliphatic compounds can be achieved. This technique had been popularly known as pulsed amperometric detection (PAD).

This detection approach was particularly attractive in combination with high-performance anion-exchange chromatography using quaternary ammonium polymer-based stationary phases and high pH.

In Figure 2 is reported a generic triple-pulse potential waveform designed for detection of carbohydrate at a gold electrode in alkaline media. As shown in the figure, detection occurs at a constant potential E_1, which is applied for the time t_1 with electronic integration of the current during the period t_{int}, following a delay of t_{del} to allow changing current to decay to a negligible value ($t_1 = t_{\text{del}} + t_{\text{det}}$). The potential of the electrode is then increased from E_1 to E_2 (with $E_2 \gg E_1$), during a period t_2 to fully oxidize any absorbed impurities and/or detect products on the electrode surface. This oxidation step generates the formation of a thin layer of oxide on the electrode surface. The applied voltage on the electrode is then reversed to a strongly reducing potential ($E_3 \ll E_1$) during the period t_3, which convert the noble-metal oxide layer at the surface of the electrode back to native metal, thus renewing the electrode surface. Following a delay during which charging current resulting from the step from the final potential (E_3) back to the initial potential (E_1) decays, analyte oxidation current is measured by integration for a fixed duration, t_{det} depicted in Figure 2. The resulting measured charge, expressed in micro-Coulomb, obtained from the analyte integration current, is proportional to the rate of the oxidation reaction.

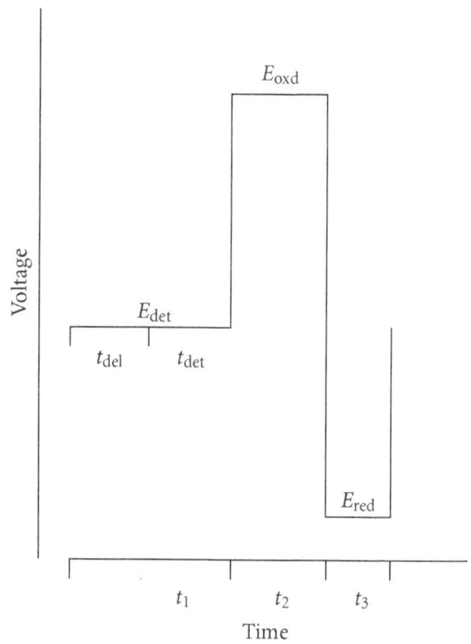

FIGURE 2: Triple-pulsed potential waveform designed for detection of carbohydrates. Qualitative waveform, where E_{det} is the constant potential E_1, applied for the time t_1 E_{oxd} is referred to potential E_2 (fully oxidation on the working gold electrode), applied for the time t_2; E_{red} is the negative potential applied during the time t_3 to convert the gold oxide back to native metal.

PAD detection can be defined with the more generic name of "pulsed electrochemical detection" (PED). The measurement of the current in PED instrumentation is due by electronic integration of electrode current during the period t_{int} (see Figure 2). Hence, the output voltage signal can be directly proportional to the average current or to the current integral [12].

The former style of signal presentation has resulted in the common name "pulsed amperometric detection" (PAD), whereas the names of "pulsed coulometric detection" and "integrated pulsed amperometric detection" (IPAD) have been suggested for the later style of the signal presentation [15].

Sensitivity of PED detection of alditols, monosaccharides, oligosaccharides, and polysaccharides at gold electrodes is maximized under alkaline conditions (pH > 12), which are also the chromatographic conditions requested in HPAEC.

Pulsed potential waveform optimization to determine the appropriate detection potential in HPAEC-PED is usually performed by cyclic voltammetry; the choice of the potential and time parameters in PED waveforms have been extensively reviewed [12, 16].

Improvement of long-term reproducibility of pulsed amperometric detection of carbohydrate has been achieved applying a quadruple-potential waveform to the working gold electrode in the flow-through detector cell instead of the triple-pulse potential waveform [17]. With the new selected waveform, a negative cleaning and a positive activation potentials maintain a clean and active gold electrode surface without causing electrode corrosion, with its consequent recession. Recession of the working gold electrode increases the volume between the working electrode and the counter electrode, decreasing the linear velocity of flow over the surface of the working electrode. This causes decreased transport of analyte to the electrode and consequent decreased detector response.

The proposed quadruple-potential waveform maximizes signal-to-noise ratio, showing similar minimum detection limits obtained employing with the triple-potential waveform, but greatly improving long-term reproducibility.

Since its introduction, pulsed electrochemical detection has considerably improved the sensitivity and selectivity of the separation of carbohydrates by HPAEC using alkaline mobile phases. The alkaline conditions are appropriate for the direct detection of carbohydrate by PED at gold electrode.

HPAEC coupled with PED detection has proven to be a very useful method for the analysis of the carbohydrates, including those of food interest. Applications regarding HPAEC-PED of carbohydrates determination in several food products such as fruit juices, milk, soluble instant coffee, infant formulas, dietetic candy, chocolate, honey, wine, wheat, rice, barley, sweet potato, corn syrups, and so on have been extensively reviewed by Cataldi and coworkers in the year 2000 [18] and by El Rassi in 2002 [19].

4. Recent Applications of HPAEC-PAD in Food Analysis

Food carbohydrates come in two principal groups: simple carbohydrates and complex carbohydrates. Simple carbohydrates consist of any soluble, digestible, and absorbable sugar such as monosaccharides, disaccharides, and sugar alcohols or alditols. Complex carbohydrates comprise the oligosaccharides and polysaccharides, such as starch and fiber components.

The analysis of carbohydrates in food is of great importance for many studies related to several sectors of investigation such as compositional evaluation, estimation of the nutritional value of a food product, food technology and food processing, control of food origin, authenticity, and eventual adulterations, monitoring of a product during storage.

In Table 3 a list of applications of HPAEC-PAD for the analysis of food products performed in the late years is reported, with the related references [8–10, 20–56]. As can be seen, the analytes most studied have been oligosaccharides rather than simple sugars.

While the main impact of carbohydrates occurrence in a food preparation is linked to the sweetener effect related to the presence of mono- and disaccharides, the occurrence of oligo- and polysaccharides gives to the food different properties depending on the type of linkage between the monomeric units that may be attacked by the human enzymes or not.

Fructooligosaccharides (FOSs) and inulin can be defined as nondigestible food ingredients and are effective prebiotics, which beneficially affect the host by selectively stimulating the growth and/or activity of the microbial groups. They occur naturally in many foods of vegetable origin, such as onions, Jerusalem artichokes, asparagus, leeks, garlic, and chicory.

FOSs are composed by chains of fructose moieties linked by β-$(2 \rightarrow 1)$-glycosidic bonds with or without a terminal glucose unit. FOS has a degree of polymerization (DP) of 2–9 and is produced from inulin by controlled hydrolysis. Inulin is extracted from chicory roots and contains fructose chains ranging in DP from 3 to 80.

FOS and inulin pass the stomach and small intestine unaltered, since they cannot be degraded or absorbed, and reach the colon where they are fermented by the intestinal microbiota exhibiting favorable effects on health promoting the growth of enteral bacteria, the intestinal motility, and the stimulation of the immune response [57].

HPAEC-PED has been demonstrated by Corradini et al. [58] to be a powerful tool to evaluate changing in chain length distribution of FOS and inulin fermented by bifidobacteria, providing significant information about their prebiotic capabilities. Fermentation of oligo- and polysaccharides in the colon is the result of a complex microbial metabolic activity, and long inulin chains are hydrolysed by extracellular enzymes, supplying FOS to bifidobacteria for selective growth [59].

Besides, inulin is widely used in food formulation for its technological properties on texture: inulin can develop a gel-like structure when mixed with aqueous liquid, forming a gel with creamy appearance. Therefore, it can be easily incorporated into food as sugar and fat replacer, and it is often used as ingredient in the so-called functional foods.

The traditional and the official analytical methods for inulin and oligosaccharides are based on enzymatic assays. Those systems are long-time consuming and give results related only to the total sugar content, by quantification of the monosaccharidic units, without providing any information about chains distribution and DP. On the other hand, HPAEC-PAD has been reported to show the advantage of giving a signal for any single chain of different DP [60] and provide useful information on chain length distribution [4], a parameter of great importance for the physicochemical properties of these compounds when used as food ingredients.

The ability of commercial available inulins with different oligo- and polysaccharide profile to form gel was recently investigated by Chiavaro et al. [61]. The knowledge of

Table 3: Recent applications of HPAEC-PAD in food analysis.

Food sample	Analytes	Column	Eluent	Ref.
	Galactooligosaccharides (GOSs)	CarboPac PA1	NaOH, CH_3COONa	[20]
	Mixture of neutral oligosaccharides	CarboPac PA100	NaOH, CH_3COONa	[21]
Agave	Fructooligosaccharides (FOSs)	CarboPac PA100	NaOH, CH_3COONa	[22]
Agave americana	Fructans	CarboPac PA10, CarboPac PA200	NaOH, CH_3COONa	[23]
Agave tequilana	Water-soluble carbohydrate composition	CarboPac PA100	NaOH, CH_3COONa	[24]
Artichoke	Inulooligosaccharides (IOSs)	CarboPac PA100	NaOH, CH_3COONa	[25]
Banana	Fructooligosaccharides (FOSs)	CarboPac PA1	NaOH, CH3COONa	[26]
Barley	Oligosaccharides	CarboPac PA1	NaOH, CH3COONa	[27]
Brewer's spent grain	Oligosaccharides	CarboPac PA1	NaOH, CH_3COONa	[28]
Dietary supplements	Maltose	CarboPac PA1	NaOH, CH_3COONa	[29]
Fermented beverage of plant extract	Oligosaccharides	Carbo Pac PA1	NaOH, CH_3COONa	[30]
Fermented beverage of plant extract	Fructopyranose oligosaccharides	Carbo Pac PA1	NaOH, CH_3COONa	[31]
Fermented beverage of plant extract	Oligosaccharides	CarboPac PA1	NaOH, CH_3COONa	[32]
Fermented milk	Galactooligosaccharides (GOSs)	CarboPac PA1	NaOH, CH_3COONa	[33]
Fermented milk	Fructooligosaccharides (FOSs), inulooligosaccharides (IOSs)	CarboPac PA100	NaOH, CH_3COONa	[8]
Fermented milk, Cooked Ham, Food ingredients	Fructooligosaccharides (FOSs) and inulins	CarboPac PA100 CarboPac PA200	NaOH, CH_3COONa NaOH, $NaNO_3$	[34]
Fermented milk (synbiotic)	Fructooligosaccharides (FOSs), inulooligosaccharides (IOSs)	CarboPac PA200	NaOH, $NaNO_3$	[35]
Fruit juice	Sucralose, fructooligosaccharides (FOSs)	CarboPac PA1	NaOH, CH_3COONa	[36]
Grape	Simple sugars, fructooligosaccharides (FOSs)	CarboPac PA1	NaOH, CH_3COONa	[37]
Honey	Oligosaccharides	CarboPac PA100	NaOH, CH_3COONa	[38]
Honey	Oligosaccharides	CarboPac PA100	NaOH, CH_3COONa	[9]
Honey	Monosaccharides, Oligosaccharides	CarboPac PA1	NaOH	[39]
Honey	Polysaccharides	CarboPac PA100	NaOH, CH_3COONa	[40]
Instant coffee	Glucose, xylose	CarboPac PA1	NaOH	[41]
Konjac	Glucomannan oligosaccharides	CarboPac PA1	NaOH, CH_3COONa	[42]

TABLE 3: Continued.

Food sample	Analytes	Column	Eluent	Ref.
Onions	Fructooligosaccharides (FOSs)	CarboPac PA1	NaOH, CH_3COONa	[43]
Prebiotic mixture	Galactooligosaccharides (GOSs)	CarboPac PA1	NaOH, CH_3COONa	[44]
Rice husk	Maltodextrins	CarboPac PA1	$NaNO_3$	[45]
Rice bran (defatted)	Oligosaccharides	CarboPac PA1, PA10	NaOH, $NaNO_3$	[46]
Rice husk	Xylooligosaccharides (XOSs)	CarboPac PA1	NaOH, CH_3COONa	[47]
Rice husk	Xylooligosaccharides (XOSs),	CarboPac PA1	NaOH, CH_3COONa	[48]
Soybean	Glucose, fructose, sucrose, raffinose, stachyose	CarboPac PA10	NaOH	[49]
Soybean flour (defatted)	Monosaccharides	CarboPac PA1	NaOH	[50]
Sugar beet	Arabinooligosaccharides (AOSs)	CarboPac PA1	NaOH	[51]
Sugar beet	Monosaccharides, arabinooligosaccharides (AOSs), oligosaccharides	CarboPac PA1	NaOH, CH_3COONa	[52]
Sugar beet	Oligosaccharides	CarboPac PA1	NaOH, CH_3COONa	[53]
Sugar Beet	Arabinose and arabinooligosaccharides (AOSs)	CarboPac PA1	NaOH, CH_3COONa	[54]
Wheat bran, chicory root	Arabinoxylooligosaccharides (AOSs), xylooligosaccharides (XOSs), fructooligosaccharides (FOSs)	CarboPac PA100	NaOH, CH_3COONa	[55]
Wheat flour	Arabinoxylans	CarboPac PA1	NaOH	[56]
Wheat milling fractions	Fructans	CarboPac PA1	NaOH, CH_3COONa	[10]

physicochemical and mechanical properties as well as their stability is of pivotal importance for the formulation of new products. In this paper oligo- and polysaccharide distribution of three different inulins has been characterized by HPAEC-PED and has showed that chemical structure of the tested inulins significantly affected the stability, the texture, and the thermal properties of the gels.

Inulin is usually derived from chicory roots. Other promising crops for this production seem to be the Jerusalem artichoke and cereals of C_3 type. The metabolism of FOS during the maturation of wheat kernels has been investigated and characterized by HPAEC-, and some biological effects have been studied [62]. FOS and inulin differ for the degree of polymerization and molecular weight, depending on the source, the harvest time, and the processing conditions. Commercial inulin and FOS are extracted by vegetables and are composed by a polydisperse mixture of chains characterized by differences in ramification and chain length. The employment of inulin in food industry requires a preliminary fractionation to obtain products with a defined range of DP. This treatment is usually realized using specific enzymes. Artichoke is a common source of inulin with an average DP around 80; its composition has been recently studied by Ronkart et al. [25] who isolated and purified several fractions with the aim of producing oligomer standards for their identification in a complex inulin chromatogram.

Despite the advantages of HPAEC-PED, the main limit of this technique is related to the lack of commercial standards for chains with high DP. Therefore the peak identification is obviously difficult to be performed, and peak assignment is often based on a generally accepted assumption that the retention time of a homologous series of carbohydrates increases as the DP increases. It means each peak eluted represents a chain with one more unit than the previous peak. Mass spectrometry could be of help in the attribution of the DP of a fraction, and MALDI-TOF has been used successfully for molecular sizing of carbohydrates and to determine chain length distribution of commercial FOS and inulin largely employed as food ingredients [34]. Comparing the two techniques, it has been shown that HPAEC-PED

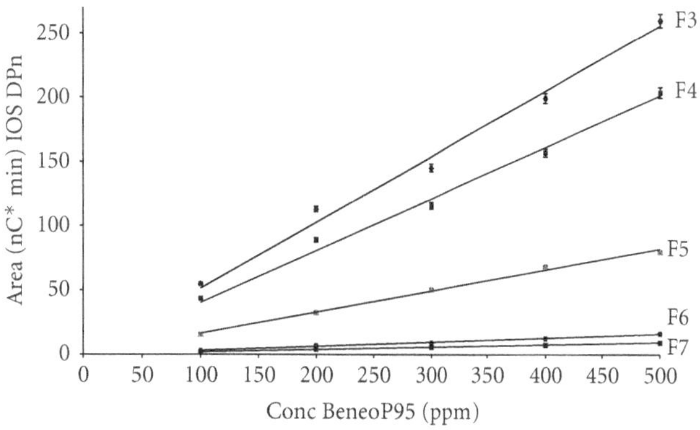

FIGURE 3: Calibration curves built by injecting different concentrations of the commercial mixture of oligosaccharides Beneo (from [8]).

allows the record of a profile showing a single peak for each FOS at any DP (even for branched isomers) and the simultaneous analysis of simple sugars and fructans at the same time. On the other hand, MALDI-TOF permits the characterization of molar mass giving details on maximum DP and on every chain present in the samples, although no information on branched and linear isomers can be obtained. Therefore the two techniques can be considered complementary.

As for the quantitative approach, HPAEC-PED is limited to studies involving the use of lab-synthesized standards or methods employing a common external standards to be used for all peaks. This last system does not overcome the inconvenient represented by the variation of detector response with the increasing retention time above discussed. We studied this matter evaluating the differences of slope for calibration curves built by injecting different concentration of a commercial mixture of oligosaccharides containing compounds having different DP [35]. In Figure 3 the different calibration curves obtained for peaks representing chains with DP between 3 and 7 are reported, and the difference in the slope value is evident. A mathematical formula has been proposed [8] permitting to calculate the amount of each compound taking into account the different value of detector response that can be calculated with the aim of an internal standard. Another approach has been proposed by Abballe and coworkers [60], based on the complementary employment of both HPAEC-PED and capillary electrophoresis (CEs) coupled with UV detection and has been applied to verify the reliability of HPAEC-PED for quantitative evaluation of dextran mixtures.

In a recent review from Raessler et al. [63] on the determination of carbohydrates in plant samples, the role of HPAEC-PAD in the analysis of different classes of sugars has been pointed out, and several applications in that field have been reported. The same review presents an overview of the common sample pretreatments, among which hydrolysis of oligo- and polysaccharides that can be performed by acid treatment or by employment of enzymes. The use of enzymes can be useful for conformational studies since it permits a cleavage of selective linkages as in the case of branched oligosaccharides.

As presented in Table 3, many s of foods containing sugars have been analysed by HPAEC-PED. One of the main sugar sources is undoubtedly sugar beet, rich in sucrose but also containing many other carbohydrates. The presence of arabinooligosaccharides with different degrees of polymerization, together with the monosaccharidic profile, has been studied by Zaidel et al. [51] who investigated the presence of branched and linear forms. Sugar beet pulp is a byproduct of the sugar industry abundant in many countries, and it is very rich in pectin, a complex polysaccharide composed by homogalacturonan, and rhamnogalacturonans. It can be a good source for prebiotic production after a controlled enzymatic treatment. Martínez et al. [53] investigated this possibility analyzing the profile of the raw material and of the product of its hydrolysis. Westphal et al. [54] characterized the structure of the chains with the help of NMR studies, while Holck et al. [52] evaluated the in vitro prebiotic activity of different fractions of beet pulp extracts on human intestinal microbiota.

The stability of FOS, arabinoxylooligosaccharides, and xylooligosaccharides during shelf-life has also been investigated by Courtin et al. [55] considering different temperatures and different pH for different time periods and observing the decomposition of the chains due to the cleavage of the units under the extreme conditions.

Some authors investigated the composition of arabinoxylan, the predominant cell-wall polysaccharides occurring in wheat flour. A comparison of different wheat flours has also been proposed evaluating the effect of processing the content and the structure of water-extractable arabinoxylans [56]. A study on the fructan content in milling fractions and outer part of wheat in two different cultivars has been object of analysis by Haskå et al. [10]. The chromatograms reported in Figure 5 show several differences observed between germ and wheat grain extracts. Isolation, identification, and determination of molecular weight of fructans were evaluated by MALDI-TOF. Using the complementarity of HPLC-PAD and MALDI-TOF, as previously reported [34], the DP of the latest fructan peak was estimated to be 19. Another study was focused on the monosaccharide composition of soybean defatted flour, after acid and enzymatic hydrolysis, and differences between the considered varieties were found [50].

Onion bulbs represent another common vegetable rich in fructooligosaccharides; the profile of those compounds, analysed by HPAEC-PAD, has been found variable among different cultivars, dependent on the bulb size and on storage conditions and time. An interesting study has been developed on onions var. Tropeana harvested in different moments along bulb maturation, and a different pattern of FOS has been found [43]. Similarly, Ghedini Der Agopian et al. [26] analyzed FOS in banana of different cultivars and at different ripening stages, and another study of that kind has been conducted on Agave Americana by Arrizon et al. [24]. Fructans from Agave have been also analyzed by Ávila-Fernández [22] and Ravenscroft et al. [23], who made structural studies using also NMR and ESI-MS techniques.

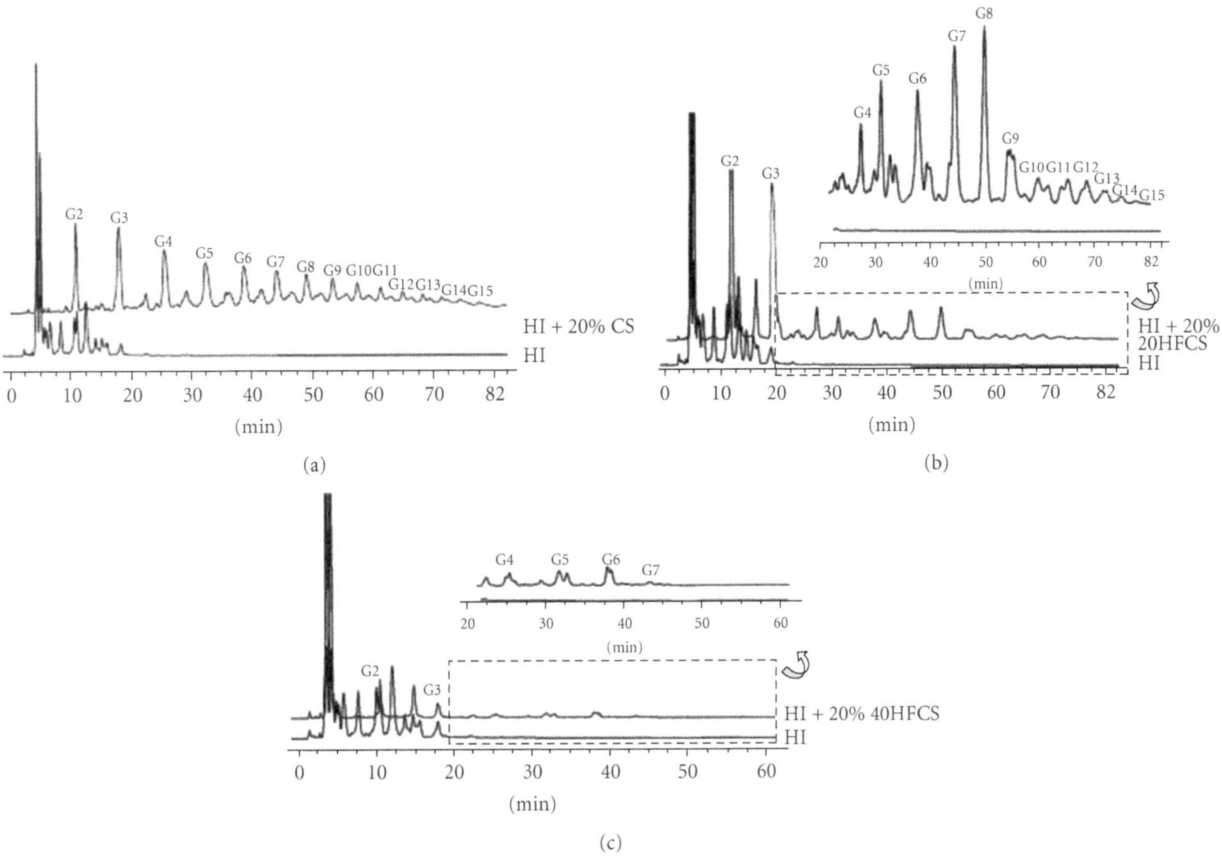

Figure 4: HPAEC-PED chromatographic profiles of the oligosaccharides fraction of a honey sample (H1) and the same sample adulterated with 20% of (a) corn syrups (CS); (b) high fructose corn syrups with 20% of isomerization (20HFCS); (c) high fructose corn syrups with 20% of isomerization (40HFCS). (From [9].)

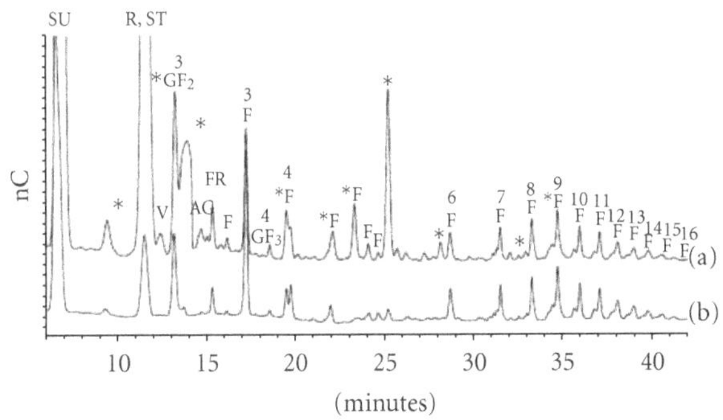

Figure 5: HPAEC-PAD profiles of extracts from (a) germ and (b) wheat grain. SU, sucrose; R, raffinose; ST, stachyose; V, verbascose; GF2, 1-kestose; AG, unidentified a-galactoside; FR, fructosylraffinose; GF3, 1,1-kestotetraose; F, unidentified fructan; *, unidentified non-fructan compound. Numbers indicate DP of fructans. (From [10]).

In the field of food quality control, the origin of a product can be investigated by the analysis of the carbohydrate profile. An example can be honey characterization: the analysis of carbohydrates in honey represents an extremely complicated case for HPAEC-PED analysis because of the number of oligosaccharides present, their relatively low concentration, and their very similar structures, especially regarding the di- and trisaccharide fractions useful for floral origin characterization [64]. A recent study has been proposed by Ouchemoukh et al. [39] who analyzed several honey samples of different botanical origin finding differences in the sugar profiles. A similar approach, based on the examination of the carbohydrate profile, has been shown to be of great help for discovering food adulteration in the case of addition of corn syrup in honey [9, 38, 39]. In Figure 4 the chromatograms obtained by Morales et al. [9], for a honey sample adulterated with corn syrups are reported.

The detection of honey adulteration by addition of cane or beet sugars and/or sugars obtained from starch hydrolysis is particularly difficult because the variations in honey carbohydrates and the similarities with sugar syrups composition make it difficult to detect these adulterations.

Carbohydrate analysis has been demonstrated to be useful in monitoring the eventual changes observed during food storage. An example is the study of the effect of a preliminary treatment of grape berries with CO_2, found to influence the sugar profile of grape stored at low temperature [37]. Similarly, synbiotic milks, containing prebiotics and probiotics, have been studied during their shelf-life; in particular, a parallel study on the simultaneous variations of the fructooligosaccaridic profile and probiotics microorganisms count has been carried out [35]. The obtained data permitted to evaluate the microbial activity occurring in the product. A similar study on fermented milks has been conducted on galactooligosaccharides (GOSs) [33], where prebiotic GOS content has been determined by HPAEC-PED by using an external calibration curve with raffinose as standard reference.

Recently, N-acetylneuraminic acid and N-glycolylneuraminic acid have been determined by HPAEC-PED in yogurt, milk, and powdered milk [65]. N-acetylneuraminic acid, also known as sialic acid, has been the recipient of considerable attention in recent years regarding the possibility to be added to different kinds of food products for improving health.

In conclusion, as widely reported, pulsed amperometric detection combined with anion exchange chromatography provides a sensitive and selective means of determining simple sugars and more complex carbohydrates in food matrices.

This review of papers published in the last years proves wide application of HPAEC-PED determination of carbohydrates of food interest and opens large possibilities of applications in routine analysis as well as in research studies.

References

[1] C. G. Huber and G. K. Bonn, "HPLC of carbohydrates with cation and anion-exchange silica and resin-based stationary phases," in *Carbohydrate Analysis, Journal of Chromatography Library* , Z. El Rassi, Ed., vol. 58, pp. 147–180, Elsevier, Amsterdam, The Netherlands, 1995.

[2] W. R. LaCourse, "Ion chromatography in food analysis," in *Handbook of Food Analysis Instruments*, S. Otles, Ed., pp. 161–196, CRC press, Taylor and Francis Group, 2009.

[3] F. N. Lamari, R. Kuhn, and N. K. Karamanos, "Derivatization of carbohydrates for chromatographic, electrophoretic and mass spectrometric structure analysis," *Journal of Chromatography B*, vol. 793, no. 1, pp. 15–36, 2003.

[4] T. J. Paskach, H. P. Lieker, P. J. Reilly, and K. Thielecke, "High-performance anion-exchange chromatography of sugars and sugar alcohols on quarternary ammonium resins under alkaline conditions," *Carbohydrate Research*, vol. 215, no. 1, pp. 1–14, 1991.

[5] T. R. I. Cataldi, G. Margiotta, and C. G. Zambonin, "Determination of sugars and alditols in food samples by HPAEC with integrated pulsed amperometric detection using alkaline eluents containing barium or strontium ions," *Food Chemistry*, vol. 62, no. 1, pp. 109–115, 1998.

[6] C. Corradini, D. Corradini, C. G. Huber, and G. K. Bonn, "Synthesis of a polymeric-based stationary phase for carbohydrate separation by high-pH anion-exchange chromatography with pulsed amperometric detection," *Journal of Chromatography A*, vol. 685, no. 2, pp. 213–220, 1994.

[7] K. S. Wong and J. Jane, "Effects of pushing agents on the separation and detection of debranched amylopectin by high-performance anion-exchange chromatography with pulsed amperometric detection," *Journal of Liquid Chromatography*, vol. 18, no. 1, pp. 63–80, 1995.

[8] C. Borromei, A. Cavazza, C. Merusi, and C. Corradini, "Characterization and quantitation of short-chain fructooligosaccharides and inulooligosaccharides in fermented milks by high-performance anion-exchange chromatography with pulsed amperometric detection," *Journal of Separation Science*, vol. 32, no. 21, pp. 3635–3642, 2009.

[9] V. Morales, N. Corzo, and M. L. Sanz, "HPAEC-PAD oligosaccharide analysis to detect adulterations of honey with sugar syrups," *Food Chemistry*, vol. 107, no. 2, pp. 922–928, 2008.

[10] L. Haskå, M. Nyman, and R. Andersson, "Distribution and characterisation of fructan in wheat milling fractions," *Journal of Cereal Science*, vol. 48, no. 3, pp. 768–774, 2008.

[11] M. Trojanowicz, "Recent developments in electrochemical flow detections—a review—part II. Liquid chromatography," *Analytica Chimica Acta*, vol. 688, no. 1, pp. 8–35, 2011.

[12] W. R. LaCourse, "Origins of pulsed potential cleaning," in *Pulsed Electrochemical Detection in High Performance Liquid Chromatography*, pp. 6–8, John Wiley & Sons, New York, NY, USA, 1997.

[13] S. Hughes, P. Lawrence Meschi, and D. C. Johnson, "Amperometric detection of simple alcohols in aqueous solutions by application of a triple-pulse potential waveform at platinum electrodes," *Analytica Chimica Acta*, vol. 132, no. C, pp. 1–10, 1981.

[14] S. Hughes and D. C. Johnson, "Amperometric detection of simple carbohydrates at platinum electrodes in alkaline solutions by application of a triple-pulse potential waveform," *Analytica Chimica Acta*, vol. 132, no. C, pp. 11–22, 1981.

[15] D. C. Johnson and W. R. LaCourse, "Pulsed electrochemical detection of carbohydrates at gold electrodes following liquid chromatographic separation," in *Carbohydrate Analysis, Journal of Chromatography Library* , Z. El Rassi, Ed., vol. 58, pp. 391–429, Elsevier, Amsterdam, The Netherlands, 1995.

[16] D. C. Johnson and W. R. LaCourse, "Pulsed electrochemical detection at noble metal electrodes in liquid chromatography," *Electroanalysis*, vol. 4, pp. 367–380, 1992.

[17] R. D. Rocklin, A. P. Clarke, and M. Weitzhandler, "Improved long-term reproducibility for pulsed amperometric detection of carbohydrates via a new quadruple-potential waveform," *Analytical Chemistry*, vol. 70, no. 8, pp. 1496–1501, 1998.

[18] T. R. I. Cataldi, C. Campa, and G. E. De Benedetto, "Carbohydrate analysis by high-performance anion-exchange chromatography with pulsed amperometric detection: the potential is still growing," *Fresenius' Journal of Analytical Chemistry*, vol. 368, no. 8, pp. 739–758, 2000.

[19] Z. El Rassi, "Carbohydrate analysis by modern chromatography and electrophoresis," in *Journal of Chromatography Library*, Z. El Rassi, Ed., vol. 66, Elsevier Science, Amsterdam, The Netherlands, 2002.

[20] A. Cardelle-Cobas, N. Corzo, M. Villamiel, and A. Olano, "Isomerization of lactose-derived oligosaccharides: a case study using sodium aluminate," *Journal of Agricultural and Food Chemistry*, vol. 56, no. 22, pp. 10954–10959, 2008.

[21] M. Brokl, O. Hernández-Hernández, A. C. Soria, and M. L. Sanz, "Evaluation of different operation modes of high performance liquid chromatography for the analysis of complex mixtures of neutral oligosaccharides," *Journal of Chromatography A*, vol. 1218, no. 42, pp. 7697–7703, 2011.

[22] A. Ávila-Fernández, N. Galicia-Lagunas, M. E. Rodríguez-Alegría, C. Olvera, and A. López-Munguía, "Production of functional oligosaccharides through limited acid hydrolysis of agave fructans," *Food Chemistry*, vol. 129, no. 2, pp. 380–386, 2011.

[23] N. Ravenscroft, P. Cescutti, M. A. Hearshaw, R. Ramsout, R. Rizzo, and E. M. Timme, "Structural analysis of fructans from agave americana grown in South Africa for spirit production,"

Journal of Agricultural and Food Chemistry, vol. 57, no. 10, pp. 3995–4003, 2009.

[24] J. Arrizon, S. Morel, A. Gschaedler, and P. Monsan, "Comparison of the water-soluble carbohydrate composition and fructan structures of Agave tequilana plants of different ages," *Food Chemistry*, vol. 122, no. 1, pp. 123–130, 2010.

[25] S. N. Ronkart, C. S. Blecker, H. Fourmanoir et al., "Isolation and identification of inulooligosaccharides resulting from inulin hydrolysis," *Analytica Chimica Acta*, vol. 604, no. 1, pp. 81–87, 2007.

[26] R. G. Der Agopian, C. A. Soares, E. Purgatto, B. R. Cordenunsi, and F. M. Lajolo, "Identification of fructooligosaccharides in different banana cultivars," *Journal of Agricultural and Food Chemistry*, vol. 56, no. 9, pp. 3305–3310, 2008.

[27] D. H. Yoo, B. H. Lee, P. S. Chang, G. L. Hyeon, and S. H. Yoo, "Improved quantitative analysis of oligosaccharides from lichenase-hydrolyzed water-soluble barley β-glucans by high-performance anion-exchange chromatography," *Journal of Agricultural and Food Chemistry*, vol. 55, no. 5, pp. 1656–1662, 2007.

[28] J. Treimo, B. Westereng, S. J. Horn et al., "Enzymatic solubilization of brewers' spent grain by combined action of carbohydrases and peptidases," *Journal of Agricultural and Food Chemistry*, vol. 57, no. 8, pp. 3316–3324, 2009.

[29] M. Mosca, C. Boniglia, B. Carratù, S. Giammarioli, V. Nera, and E. Sanzini, "Determination of α-amylase inhibitor activity of phaseolamin from kidney bean (*Phaseolus vulgaris*) in dietary supplements by HPAEC-PAD," *Analytica Chimica Acta*, vol. 617, no. 1-2, pp. 192–195, 2008.

[30] N. Kawazoe, H. Okada, E. Fukushi et al., "Two novel oligosaccharides isolated from a beverage produced by fermentation of a plant extract," *Carbohydrate Research*, vol. 343, no. 3, pp. 549–554, 2008.

[31] H. Okada, E. Fukushi, A. Yamamori et al., "Novel fructopyranose oligosaccharides isolated from fermented beverage of plant extract," *Carbohydrate Research*, vol. 345, no. 3, pp. 414–418, 2010.

[32] H. Okada, E. Fukushi, A. Yamamori et al., "Isolation and structural confirmation of the oligosaccharides containing α-d-fructofuranoside linkages isolated from fermented beverage of plant extracts," *Carbohydrate Research*, vol. 346, no. 16, pp. 2633–2637, 2011.

[33] C. Martínez-Villaluenga, A. Cardelle-Cobas, N. Corzo, and A. Olano, "Study of galactooligosaccharide composition in commercial fermented milks," *Journal of Food Composition and Analysis*, vol. 21, no. 7, pp. 540–544, 2008.

[34] C. Borromei, M. Careri, A. Cavazza et al., "Evaluation of fructooligosaccharides and inulins as potentially health benefiting food ingredients by HPAEC-PED and MALDI-TOF MS," *International Journal of Analytical Chemistry*, vol. 2009, Article ID 530639, 9 pages, 2009.

[35] C. Borromei, A. Cavazza, C. Corradini et al., "Validated HPAEC-PAD method for prebiotics determination in synbiotic fermented milks during shelf life," *Analytical and Bioanalytical Chemistry*, vol. 397, no. 1, pp. 127–135, 2010.

[36] M. Feinberg, J. San-Redon, and A. Assié, "Determination of complex polysaccharides by HPAE-PAD in foods: validation using accuracy profile," *Journal of Chromatography B*, vol. 877, no. 23, pp. 2388–2395, 2009.

[37] M. Blanch, M. T. Sanchez-Ballesta, M. I. Escribano, and C. Merodio, "Fructo-oligosaccharides in table grapes and response to storage," *Food Chemistry*, vol. 129, no. 3, pp. 724–730, 2011.

[38] V. Morales, M. L. Sanz, A. Olano, and N. Corzo, "Rapid separation on activated charcoal of high oligosaccharides in honey," *Chromatographia*, vol. 64, no. 3-4, pp. 233–238, 2006.

[39] S. Ouchemoukh, P. Schweitzer, M. Bachir Bey, H. Djoudad-Kadji, and H. Louaileche, "HPLC sugar profiles of Algerian honeys," *Food Chemistry*, vol. 121, no. 2, pp. 561–568, 2010.

[40] M. Megherbi, B. Herbreteau, R. Faure, and A. Salvador, "Polysaccharides as a marker for detection of corn sugar syrup addition in honey," *Journal of Agricultural and Food Chemistry*, vol. 57, no. 6, pp. 2105–2111, 2009.

[41] P. Girard, P. Stöber, M. Blanc, and J. Prodolliet, "Carbohydrate specification limits for the authenticity assessment of soluble (Instant) Coffee: statistical approach," *Journal of the Association of Official Agricultural Chemists*, vol. 89, no. 4, pp. 999–1003, 2006.

[42] S. Albrecht, G. C. J. Van Muiswinkel, H. A. Schols, A. G. J. Voragen, and H. Gruppen, "Introducing capillary electrophoresis with laser-induced fluorescence detection (CE-LIF) for the characterization of konjac glucomannan oligosaccharides and their in vitro fermentation behavior," *Journal of Agricultural and Food Chemistry*, vol. 57, no. 9, pp. 3867–3876, 2009.

[43] A. Cavazza, C. Corradini, M. Giannetto, R. Greco, G. Mori, and T. Vicchio, "Studio del profilo di fruttooligosaccaridi in cipolle di diversa cultivar mediante HPAEC-PAD: valutazione chemiometrica," in *Proceedings of the VIII Congresso Nazionale di Chimica degli Alimenti*, pp. 550–553, Marsala, Italy, 2010.

[44] L. Coulier, J. Timmermans, B. Richard et al., "In-depth characterization of prebiotic galactooligosaccharides by a combination of analytical techniques," *Journal of Agricultural and Food Chemistry*, vol. 57, no. 18, pp. 8488–8495, 2009.

[45] J. Patindol, Y. J. Wang, and J. L. Jane, "Structure-functionality changes in starch following rough rice storage," *Starch/Staerke*, vol. 57, no. 5, pp. 197–207, 2005.

[46] J. Patindol, L. Wang, and Y. J. Wang, "Cellulase-assisted extraction of oligosaccharides from defatted rice bran," *Journal of Food Science*, vol. 72, no. 9, pp. C516–C521, 2007.

[47] P. Gullón, P. Moura, M. P. Esteves, F. M. Girio, H. Domínguez, and J. C. Parajó, "Assessment on the fermentability of xylooligosaccharides from rice husks by probiotic bacteria," *Journal of Agricultural and Food Chemistry*, vol. 56, no. 16, pp. 7482–7487, 2008.

[48] P. Gullón, M. J. González-Muñoz, M. P. van Gool et al., "Production, refining, structural characterization and fermentability of rice husk xylooligosaccharides," *Journal of Agricultural and Food Chemistry*, vol. 58, no. 6, pp. 3632–3641, 2010.

[49] E. Giannoccaro, Y. J. Wang, and P. Chen, "Comparison of two HPLC systems and an enzymatic method for quantification of soybean sugars," *Food Chemistry*, vol. 106, no. 1, pp. 324–330, 2008.

[50] E. M. Bainy, S. M. Tosh, M. Corredig, V. Poysa, and L. Woodrow, "Varietal differences of carbohydrates in defatted soybean flour and soy protein isolate by-products," *Carbohydrate Polymers*, vol. 72, no. 4, pp. 664–672, 2008.

[51] D. N. A. Zaidel, A. Arnous, J. Holck, and A. S. Meyer, "Kinetics of enzyme-catalyzed cross-linking of feruloylated arabinan from sugar beet," *Journal of Agricultural and Food Chemistry*, vol. 59, no. 21, pp. 11598–11607, 2011.

[52] J. Holck, A. Lorentzen, L. K. Vigsnæs, T. R. Licht, J. D. Mikkelsen, and A. S. Meyer, "Feruloylated and nonferuloylated arabino-oligosaccharides from sugar beet pectin selectively stimulate the growth of bifidobacterium spp. In human fecal in vitro fermentations," *Journal of Agricultural and Food Chemistry*, vol. 59, no. 12, pp. 6511–6519, 2011.

[53] M. Martínez, B. Gullón, R. Yáñez, J. L. Alonso, and J. C. Parajó, "Direct enzymatic production of oligosaccharide mixtures from sugar beet pulp: experimental evaluation and mathematical modeling," *Journal of Agricultural and Food Chemistry*, vol. 57, no. 12, pp. 5510–5517, 2009.

[54] Y. Westphal, S. Kühnel, P. de Waard et al., "Branched arabino-oligosaccharides isolated from sugar beet arabinan," *Carbohydrate Research*, vol. 345, no. 9, pp. 1180–1189, 2010.

[55] C. M. Courtin, K. Swennen, P. Verjans, and J. A. Delcour, "Heat and pH stability of prebiotic arabinoxylooligosaccharides, xylooligosaccharides and fructooligosaccharides," *Food Chemistry*, vol. 112, no. 4, pp. 831–837, 2009.

[56] L. Virkki, H. N. Maina, L. Johansson, and M. Tenkanen, "New enzyme-based method for analysis of water-soluble wheat arabinoxylans," *Carbohydrate Research*, vol. 343, no. 3, pp. 521–529, 2008.

[57] S. Kolida, K. Tuohy, and G. R. Gibson, "Prebiotic effects of inulin and oligofructose," *British Journal of Nutrition*, vol. 87, no. 2, pp. S193–S197, 2002.

[58] C. Corradini, F. Bianchi, D. Matteuzzi, A. Amoretti, M. Rossi, and S. Zanoni, "High-performance anion-exchange chromatography coupled with pulsed amperometric detection and capillary zone electrophoresis with indirect ultra violet detection as powerful tools to evaluate prebiotic properties of fructooligosaccharides and inulin," *Journal of Chromatography A*, vol. 1054, no. 1-2, pp. 165–173, 2004.

[59] M. Rossi, C. Corradini, A. Amaretti et al., "Fermentation of fructooligosaccharides and inulin by bifidobacteria: a comparative study of pure and fecal cultures," *Applied and Environmental Microbiology*, vol. 71, no. 10, pp. 6150–6158, 2005.

[60] F. Aballe, M. Toppazzini, C. Campa, F. Uggeri, and S. Paoletti, "Study of molar response of dextrans in electrochemical detection," *Journal of Chromatography A*, vol. 1149, no. 1, pp. 38–45, 2007.

[61] E. Chiavaro, E. Vittadini, and C. Corradini, "Physicochemical characterization and stability of inulin gels," *European Food Research and Technology*, vol. 225, no. 1, pp. 85–94, 2007.

[62] N. Merendino, M. D'Aquino, R. Molinari et al., "Chemical characterization and biological effects of immature durum wheat in rats," *Journal of Cereal Science*, vol. 43, no. 2, pp. 129–136, 2006.

[63] M. Raessler, B. Wissuwa, A. Breul, W. Unger, and T. Grimm, "Determination of water-extractable nonstructural carbohydrates, including inulin, in grass samples with high-performance anion exchange chromatography and pulsed amperometric detection," *Journal of Agricultural and Food Chemistry*, vol. 56, no. 17, pp. 7649–7654, 2008.

[64] C. Corradini, G. Canali, and I. Nicoletti, "Application of HPAEC-PAD to carbohydrate analysis in food products and fruit juices," *Seminars in Food Analysis*, vol. 2, pp. 99–111, 1997.

[65] K. T. Tang, L. N. Liang, Y. Q. Cai, and S. F. Mou, "Determination of sialic acid in milk and products using high performance anion-exchange chromatography coupled with pulsed amperometric detection," *Fenxi Huaxue/ Chinese Journal of Analytical Chemistry*, vol. 36, no. 11, pp. 1535–1538, 2008.

13

Analysis by Vibrational Spectroscopy of Seaweed Polysaccharides with Potential Use in Food, Pharmaceutical, and Cosmetic Industries

Leonel Pereira,[1] Saly F. Gheda,[2] and Paulo J. A. Ribeiro-Claro[3]

[1] *IMAR-CMA, Department of Life Sciences, FCTUC, University of Coimbra, 3004-516 Coimbra, Portugal*
[2] *Phycology Lab., Botany Department, Faculty of Science, Tanta University, Tanta 31527, Egypt*
[3] *Department of Chemistry, CICECO, University of Aveiro, 3810-193 Aveiro, Portugal*

Correspondence should be addressed to Leonel Pereira; leonel@bot.uc.pt

Academic Editor: Otto Holst

Polysaccharides present in several seaweeds (*Kappaphycus alvarezii*, *Calliblepharis jubata*, and *Chondrus crispus*—Gigartinales, Rhodophyta; *Gelidium corneum* and *Pterocladiella capillacea*—Gelidiales, Rhodophyta; *Laurencia obtusa*—Ceramiales, Rhodophyta; *Himanthalia elongata*, *Undaria pinnatifida*, *Saccorhiza polyschides*, *Sargassum vulgare*, and *Padina pavonica*—Phaeophyceae, Ochrophyta) are analyzed by spectroscopic techniques. The nature of the polysaccharides (with extraction and without any type of extraction) present in these seaweeds was determined with FTIR-ATR and FT-Raman analysis of extracted phycocolloids and ground dry seaweed.

1. Introduction

Many species of seaweed (marine macroalgae) are used as food and they have also found use in traditional medicine because of their perceived health benefits. Seaweeds are rich sources of sulphated polysaccharides, including some that have become valuable additives in the food industry because of their rheological properties as gelling and thickening agents (e.g., alginates, agar, and carrageenan). Sulphated polysaccharides are recognized to possess a number of biological activities including anticoagulant, antiviral, antitumor, anti-inflammatory, and immunostimulating activities that might find relevance in nutraceutical/functional food, cosmetic, and pharmaceutical applications [1].

Some seaweeds produce hydrocolloids, associated with the cell wall and intercellular spaces. Members of the red algae (Rhodophyta) produce galactans (e.g., carrageenans and agars) and the brown algae (Heterokontophyta, Phaeophyceae) produce uronates (alginates) and other sulphated polysaccharides (e.g., fucoidan and laminaran) [2–8].

The different phycocolloids used in food industry as natural additives are (European codes of phycocolloids)

(i) alginic acid—E400,
(ii) sodium alginate—E401,
(iii) potassium alginate—E402,
(iv) ammonium alginate—E403,
(v) calcium alginate—E404,
(vi) propylene glycol alginate—E405,
(vii) agar—E406,
(viii) carrageenan—E407,
(ix) semirefined carrageenan or "processed *Eucheuma* seaweed"—E407A.

Carrageenan and agar (Figure 1) are the principal sulphated polysaccharides produced by red seaweeds (Rhodophyta);

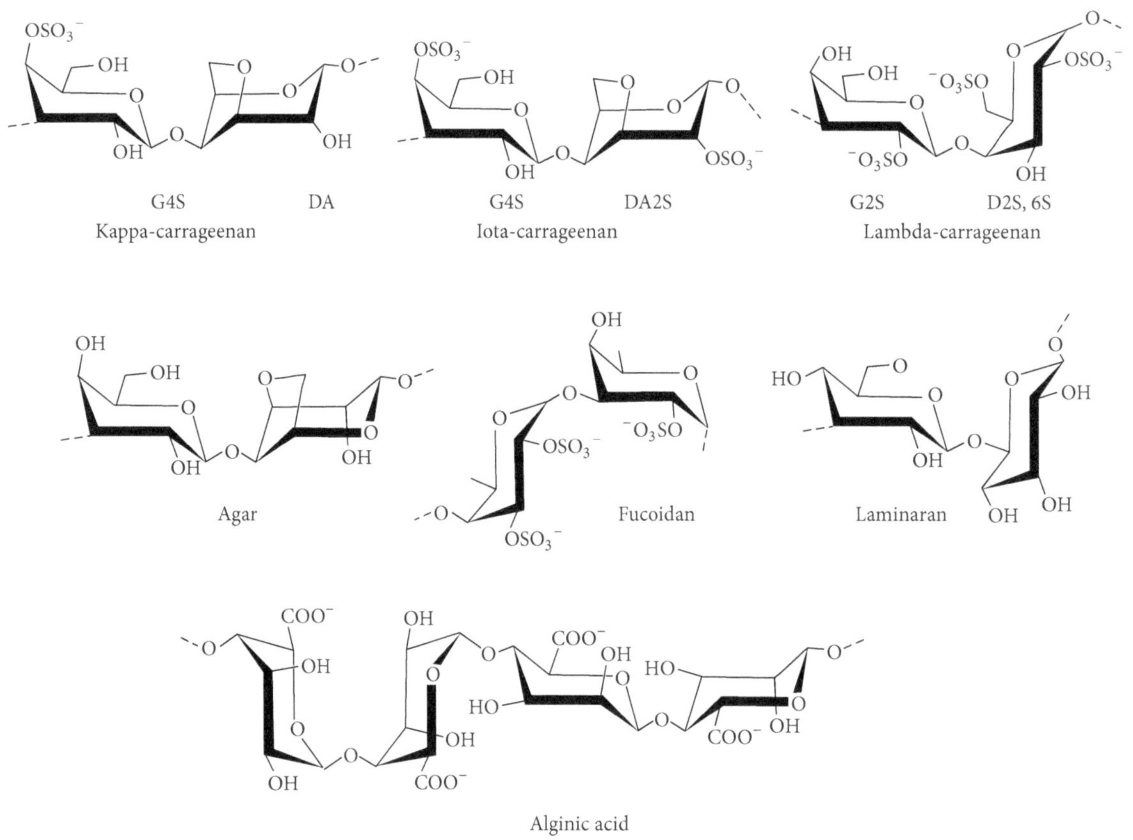

Figure 1: Idealized structures of the chemical units of kappa-, iota-, and lambda-carrageenan, agar, alginic acid (M = mannuronic acid and G = guluronic acid), fucoidan and laminaran [13, 14].

the main difference between the highly sulphated carrageenans from the less sulphated agars is the presence of D-galactose and anhydro-D-galactose in carrageenans and of D-galactose, L-galactose, or anhydro-L-galactose in agars.

The structure of the various types of carrageenans is defined by the number and position of sulphate groups, the presence of 3,6-anhydro-D-galactose, and conformation of the pyranose ring. There are about fifteen idealized carrageenan structures traditionally identified by Greek letters [9].

The commercial carrageenans are normally divided into three main types: kappa-, iota-, and lambda-carrageenans. Generally, seaweeds do not produce these idealized and pure carrageenans but more likely a range of hybrid structures. The precursors (mu and nu), when exposed to alkali conditions, are modified into kappa and iota, respectively, through formation of the 3,6-anhydrogalactose bridge [10].

Different types of carrageenan are obtained from different species of the Gigartinales (Rhodophyta). Kappa-carrageenan is predominantly obtained by extraction from the cultivated tropical seaweed *Kappaphycus alvarezii* (known in the trade as "cottonii"). *Eucheuma denticulatum* (trade name "spinosum") is the main species for the production of iota-carrageenan. Lambda-carrageenan is obtained from different species from the genera *Gigartina* and *Chondrus* (trade name "Irish moss") [11].

The rheological properties of the gelling carrageenans (e.g., kappa and iota) are quite distinct: the kappa type forms gels that are hard, strong, and brittle, whereas iota-carrageenan forms soft and weak gels. The common feature of these carrageenans is the anhydrogalactose bridge of the 4-linked galactose residue, DA and DA2S, respectively, which adopts the $^1C_4$-chair conformation. This conformation is crucial for the formation of the helical structure and, thereby, for the ability to form a gel. Lambda-carrageenan and the precursors mu- and nu-carrageenan lack the 3,6-anhydrobridge and, therefore, the 4-linked residue adopts the $^4C_1$-chair conformation, which disturbs the helical conformation. Thus, lambda-carrageenan acts simply as a thickening agent [11].

Agar (Figure 1) was the first colloid to be developed and it has applications as a gelling agent for food and also as an inert support medium for microbial culture. This polysaccharide is the dried hydrophilic, colloidal substance extracted commercially from certain marine algae of the phylum Rhodophyta. The most important commercial agarophyte genera are *Gelidium*, *Pterocladiella*, *Gelidiella*, and *Gracilaria*. Agar has also been found in species of *Ceramium*, *Phyllophora*, *Ahnfeltia*, *Campylaephora*, *Acanthopholis*, and *Gracilariopsis*. It is a polysaccharide, consisting primarily of D- and L-galactose units. About every tenth D-galactopyranose unit contains a sulphate ester group. Calcium, magnesium, potassium, or sodium cations are also associated with the polysaccharide. Agar may be separated into two fractions. One is a neutral polymer, agarose, composed of repeating units, referred to as agarobiose, and of alternating 1,3-linked β-D-galactopyranose and 1,4-linked 3,6-anhydro-α-L-galactopyranose. The second fraction has agaropectin, a more complicated structure. It contains residues of sulphuric, pyruvic, and uronic acids, in addition to D-galactose and 3,6-anhydro-L-galactose [10, 12].

Alginic acid (Figure 1) was discovered in 1883 by E. C. C. Stanford, a British pharmacist who called it algin. In seaweeds, algin is extracted as a mixed salt of sodium and/or potassium, calcium, and magnesium. The exact composition varies with algal species. Since Stanford discovered algin, the name has been applied to a number of substances, such as alginic acid and all alginates, derived from alginic acid. The extraction process is based on the conversion of an insoluble mixture of alginic acid salts of the cell wall in a soluble salt (alginate) which is appropriate for the water extraction [13–15]. This polysaccharide is derived from several genera of brown algae (e.g., mixed Fucales and Laminariales) that are utilized as raw materials by commercial alginate producers; these include *Macrocystis, Laminaria, Lessonia, Ascophyllum, Alaria, Ecklonia, Eisenia, Nereocystis, Sargassum, Cystoseira* and *Fucus*, with *Macrocystis pyrifera* and *Ascophyllum nodosum* being the principal sources of the world's alginate supply. The intercellular mucilage in these seaweeds has been regarded as the principal site of algin, although it has also been found to occur in the cell walls. Alginic acid is a complex organic compound composed of D-mannuronic acid and L-guluronic acid monomers [15–17].

Fucans (Figure 1) are sulphated polysaccharides that are composed of a fucose backbone. One of the best studied fucans from brown algae is fucoidan, which was first isolated by Kylin in 1913 [18]. The fucoidan from *Fucus vesiculosus* has been available commercially for decades (Sigma-Aldrich Chemical Company, St. Louis, MO, USA). Early work on its structure showed that it contained primarily $(1 \rightarrow 2)$ linked 4-*O*-sulphated fucopyranose residues. However, 3-linked fucose with 4-sulphated groups were subsequently reported to be present on some of the fucose residues. Additionally, it was determined to contain branches every 2-3 fucose residues. Subsequently, Chevolot and colleagues reported that the fucoidan from *F. vesiculosus* and *Ascophyllum nodosum* contains a predominant disaccharide motif containing sulphate at the 2-position of the 3-linked fucose and sulphate groups on the 2- and 3-positions of the 4-linked fucose [19].

Laminaran (Figure 1) is a small glucan present in either soluble or insoluble forms. The first form is characterized by complete solubility in cold water, while the other is only soluble in hot water [20, 21]. This polysaccharide is composed of D-glucose with b-(1,3) linkages, with b-(1,6) intrachain branching [6–8].

In this work, a combined FTIR-ATR and FT-Raman spectroscopy analysis was used to identify the main seaweed polysaccharides, namely, alginate, fucoidan, laminaran, agar, and kappa-, iota-, and lambda-carrageenans. Therefore, vibrational spectroscopy (FTIR-ATR and FT-Raman) is proposed as a useful tool for the cosmetic, pharmaceutical, and food industries to check the phycocolloid quality of a raw seaweed material by a quick and nondestructive method [5, 16].

2. Materials and Methods

2.1. Algal Material and Standard Samples of Hydrocolloids. Specimens of red algae (Rhodophyta) *Kappaphycus alvarezii* (Gigartinales) were collected in the Philippines. Specimens of red algae (Rhodophyta) *Calliblepharis jubata, Chondrus crispus* (Gigartinales), and *Gelidium corneum* (Gelidiales) and the brown algae (Phaeophyceae) *Saccorhiza polyschides* (Tilopteridales) were collected in the central zone of the western coast of Portugal. *Pterocladiella capillacea* (Gelidiales, Rhodophyta), *Laurencia obtusa* (Ceramiales, Rhodophyta), *Padina pavonica* (Dictyotales, Phaeophyceae), and *Sargassum vulgare* (Fucales, Phaeophyceae) were collected in the Mediterranean (Egypt). The edible brown seaweeds (Phaeophyceae) *Himanthalia elongata* (Fucales), and *Undaria pinnatifida* (Laminariales) were obtained from Algamar (Galicia, Spain).

Standard samples were obtained from Sigma (kappa-carrageenan, type III, C-1263; iota-carrageenan, type V, C-4014; alginic acid, A0682), TAAB Laboratories (agar, A010), and CP Kelco (pure lambda-carrageenan).

2.2. Preparation of Ground Seaweed Samples for FTIR-ATR and FT-Raman. The seaweed samples were rinsed in distilled freshwater to eliminate salt and debris from the thallus surface and dried to constant weight at 60°C. The dried seaweeds were finely ground in order to render the samples uniform. For FTIR analysis the samples do not need additional treatment. The analysis by FT-Raman requires that these are without pigmentation. The lack of pigmentation can be achieved by sun drying (process used by collectors/producers of commercial seaweeds) or by pigment elimination in the laboratory by the addition of calcium hypochlorite solution (4%, 30/60 s, 4°C) [5, 16, 22].

2.3. FTIR-ATR and FT-Raman Analysis. The FTIR spectra of sample materials (ground-dried seaweed, native and alkali-modified carrageenan) were recorded on an IFS 55 spectrometer, using a Golden Gate single reflection diamond ATR system, with no need for sample preparation. All spectra are the average of two independent measurements with 128 scans, each at a resolution of 2 cm^{-1}.

The corresponding FT-Raman spectra were recorded on an RFS-100 Bruker FT-spectrometer using an Nd:YAG laser with an excitation wavelength of 1064 nm. Each spectrum was the average of two repeated measurements, with 150 scans at a resolution of 2 cm^{-1}.

3. Results and Discussion

The assignments of the IR spectra were mostly based on previous works [4, 5, 9, 16, 23]. The Raman spectra were assigned based on the IR information and on the comparison between samples of known composition.

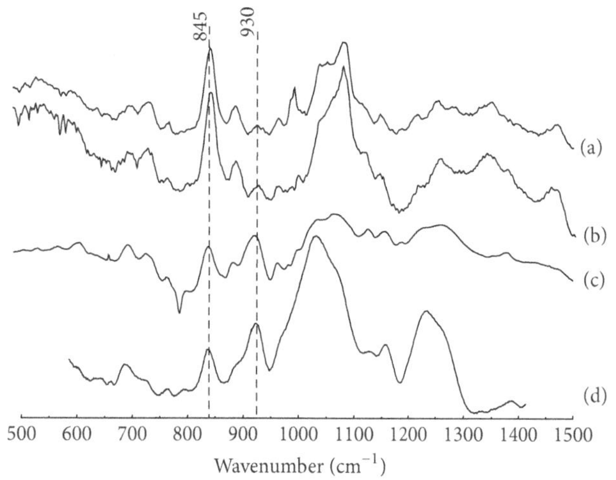

FIGURE 2: Spectra of commercial kappa-carrageenan and of ground seaweed sample (*Kappaphycus alvarezii*): FT-Raman ((a) and (b), resp.) and FTIR ((c) and (d), resp.).

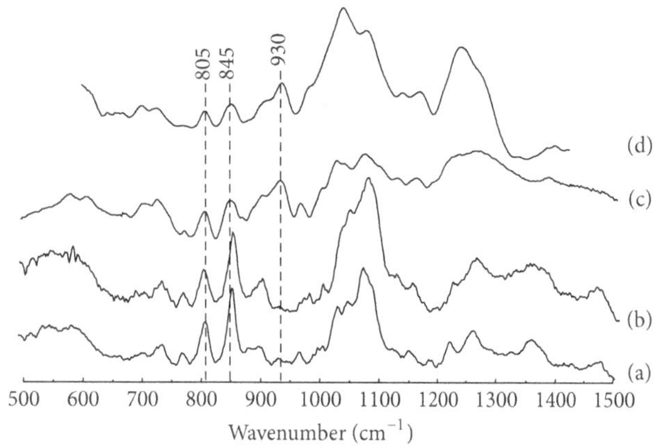

FIGURE 3: Spectra of commercial iota-carrageenan and of ground seaweed sample (*Calliblepharis jubata*): FT-Raman ((a) and (b), resp.) and FTIR ((c) and (d), resp.).

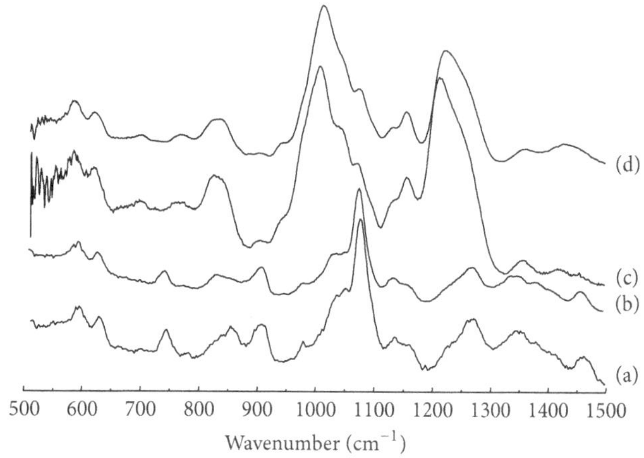

FIGURE 4: Spectra of commercial lambda-carrageenan and of ground seaweed sample (*Chondrus crispus* tetrasporophyte): FT-Raman ((a) and (b), resp.) and FTIR ((c) and (d), resp.).

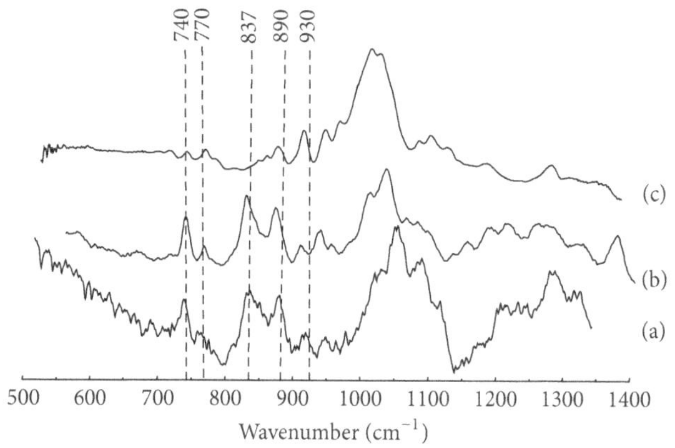

FIGURE 5: FT-Raman spectra of ground seaweed *Gelidium corneum* (a) and commercial agar (b) andFTIR-ATR spectra of ground seaweed *Gelidium corneum* (c).

3.1. Identification of Carrageenan. The FTIR-ATR and FT-Raman spectra of *Kappaphycus alvarezii* were compared with those of commercial kappa-carrageenan in Figure 2. The spectra of the ground seaweed show the main features of commercial kappa-carrageenan: a strong Raman band at approximately 845 cm^{-1} (with moderate intensity in the IR spectrum), which is assigned to D-galactose-4-sulphate (G4S) and a relatively strong band at approximately 930 cm^{-1} in the FTIR-ATR spectra, weak in FT-Raman spectrum, indicating the presence of 3,6-anhydro-D-galactose (DA) [5, 16].

Figure 3 presents the FTIR-ATR and FT-Raman spectra of iota-carrageenan and of *Calliblepharis jubata*. The spectra of these samples also show the bands at approximately 930 and 845 cm^{-1}, with the same intensity pattern as in kappa-carrageenan. However, an additional well-defined feature is visible in both IR and Raman spectra, around 805 cm^{-1}, indicating the presence of sulphate ester in the 2-position of the anhydro-D-galactose residues (DA2S), a characteristic band of the iota-carrageenan [5, 16].

The FTIR-ATR and FT-Raman spectra of lambda-carrageenan and ground *Chondrus crispus* tetrasporophytes are shown in Figure 4. These samples present high sulphate content as indicated by the broad band between 820 and 830 cm^{-1} in FTIR-ATR spectra. The *C. crispus* and lambda-carrageenan FT-Raman spectra show the two combined weak bands between 815 and 830 cm^{-1} [5, 16].

3.2. Identification of Agar. Agars differ from carrageenans as they have the L-configuration for the 4-linked galactose residue; nevertheless, they have some structural similarities with carrageenans. The characteristic broad band of sulphate esters in general [9] between 1210 to 1260 cm^{-1} (Figures 2, 3, and 4) was much stronger in carrageenan than in agar (Figures 5 and 6). Especially in the anomeric region (700–950 cm^{-1}) agar and carrageenan showed several similar bands. Thus, the strong IR band at 930 cm^{-1} (Figures 5(c) and

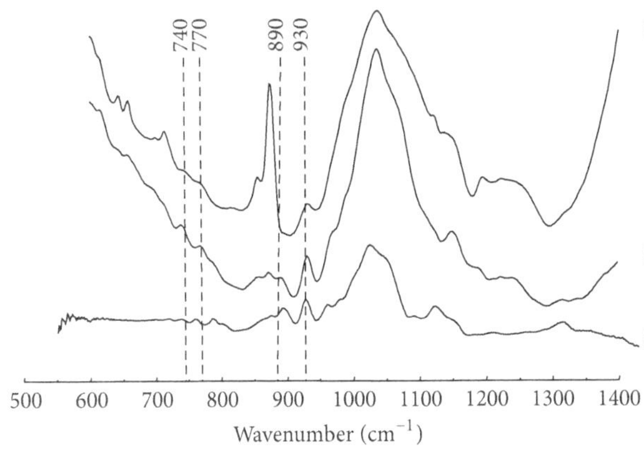

FIGURE 6: FTIR-ATR spectra of commercial agar (a), ground seaweed *Pterocladiella capillacea* (b), and *Laurencia obtusa* (c).

6(a)–6(c)) assigned to the presence of 3,6-anhydrogalactose was common to agar and carrageenans; the band at 890 cm^{-1} (Figures 5(a), 5(b), and 6) corresponded to anomeric CH of β-galactopyranosyl residues and Raman bands at 770, 740 (strong in the FT-Raman spectra and weak in the FTIR-ATR) are assigned to the skeleton bending of pyranose ring [23–25] both in agar and carrageenans. Also, the bands at 1010–1030 cm^{-1} (Figures 5(c) and 6(d)–6(f)) may be assigned to C–O and C–C stretching vibrations of pyranose ring common to all polysaccharides. So, the main polysaccharide composition of *Gelidium corneum* and of *Pterocladiella capillacea* is agar [23, 26, 27].

The FT-Raman spectra (Figures 5(a) and 5(b)) show a strong band centred at 837 cm^{-1}, which is absent in the FTIR spectra. After Matsuhiro [24], this band is associated with the CH vibration coupled with C–OH related modes of α residues. Moreover, the spectral feature at 890 cm^{-1}, also particularly intense in the FT-Raman spectra, is mainly associated with vibrational modes of the β-galactose residues.

Laurencia obtusa (Figure 6(f)) presents a complex agar-like sulphated galactan. These polysaccharides belong to the agar group, being agarose derivatives with a rather high content sulfate groups and with a reduced amount of 3,6-anhydro-L-galactose residues (700–950 cm^{-1}) [28].

3.3. Polysaccharides from Brown Algae. The main polysaccharide found in studied brown seaweeds (Phaeophyceae) was alginate, a linear copolymer of mannuronic (M) and guluronic acid (G). Different types of alginic acid present different proportions and/or different alternating patterns of guluronic (G) and mannuronic (M) units. The presence of these acids can be identified from their characteristic bands in the vibrational spectra; in accordance with Mackie [25] these phycocolloids show two characteristic bands in IR spectra: 808 cm^{-1}, assigned to M units, and 787 cm^{-1}, assigned to G units. However, Matsuhiro and coworkers, in work with specimens of *Lessonia* genus, assign both bands to G units [29, 30]. Filipov and Kohn [31] propose that M/G ratios of the different samples can be estimated from

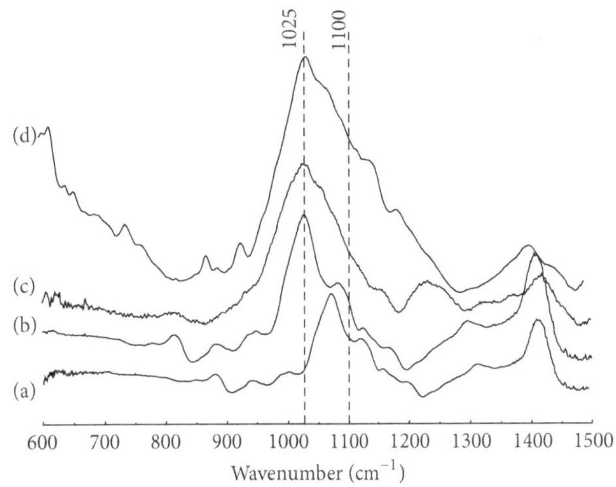

FIGURE 7: FT-Raman spectra of commercial alginate (a) and FTIR spectra of commercial alginate (b) and *Saccorhiza polyschides* (c). FTIR-ATR spectra of ground seaweed *Himanthalia elongata* (d).

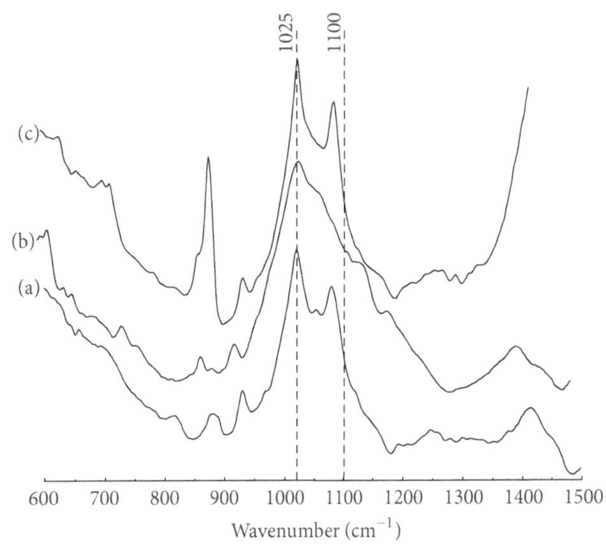

FIGURE 8: FTIR-ATR spectra of ground seaweed *Undaria pinnatifida* (a), *Padina pavonica* (b), and *Sargassum vulgare* (c).

the ratio of absorbance of the bands at 1320 and 1290 cm^{-1} in FTIR spectra. According to Sakugawa and collaborators [32] the M/G concentration ratio characterizing a certain alginate sample can be inferred from the relative intensity ratio of the two bands 1030/1080 cm^{-1}, in calcium alginate and 1019/1025 cm^{-1}, in manganese alginate. In accordance with the same authors, the absorbance at 1030 cm^{-1} directly reflects the change of mannuronate concentration of calcium alginate and the 1025 cm^{-1} is attributed to the OH bending of guluronate [32].

Alginate M/G ratio was tentatively estimated from the 1030/1080 cm^{-1} band ratio in infrared spectra, suggesting higher values of mannuronic than guluronic acid blocks (M/G > 1) in *Himanthalia elongata* (Figure 7(d)). However, the FTIR spectra of *Saccorhiza polyschides* (Figure 7(c)) show an intense broad band centred at 1025 cm^{-1}, indicating that the samples considered are particularly rich in guluronic acid. According to our spectrum on *U. pinnatifida* (old adult thallus, Figure 8(a)), the relative amounts of both mannuronate and guluronate residues were similar, and result in accordance with other published works [33].

Our spectra suggesting higher values of guluronic than mannuronic acid blocks in *Padina pavonica* (Figure 8(b)); *Sargassum vulgare* spectra (Figure 8(c)) suggesting the presence of similar amounts of the mannuronate and guluronate residues, results in accordance with other published works [33–35].

Some brown algae (see Figures 7(c) and 8(a), *Saccorhiza polyschides* and *Undaria pinnatifida,* resp.) also exhibited a broad band around 1220–1260 cm^{-1}, assigned to the presence of sulphate ester groups (S=O) which is a characteristic component in fucoidan and sulphated polysaccharides other than alginate in brown seaweeds. *Padina pavonica* and *Sargassum vulgare* (see Figures 8(b) and 8(c)) also exhibited a broad band (around 1195–1237 cm^{-1} for *Padina* and 1210–1280 cm^{-1} for *Sargassum*) assigned to (S=O) which is characteristic component in fucoidan (main water soluble sulphated polysaccharide); however, *Sargassum vulgare* contains more amounts of fucoidan than in *Padina pavonica,* whereas, both seaweed species contain little amounts of laminaran [2, 6–8, 20, 21, 36].

4. Conclusion

Vibrational FTIR-ATR spectroscopy (FTIR-ATR and FT-Raman) is a useful tool in the food, pharmaceutical, and cosmetics industries for a preliminary identification of the main sulphated polysaccharides (namely, alginate, laminaran, fucoidan, agar, and carrageenan) produced by edible brown and red seaweeds, by a rapid and non-destructive method.

Acknowledgments

The authors acknowledge financial support from the Portuguese Foundation for Science and Technology—*IMAR-CMA (Institute of Marine Research) and Laboratório Associado CICECO.*

References

[1] L. Pereira, "A Review of the nutrient composition of selected edible seaweeds," in *Seaweed: Ecology, Nutrient Composition and Medicinal Uses*, V. H. Pomin, Ed., pp. 15–47, Nova Science, New York, NY, USA, 2011.

[2] G. Jiao, G. Yu, J. Zhang, and H. S. Ewart, "Chemical structures and bioactivities of sulfated polysaccharides from marine algae," *Marine Drugs*, vol. 9, no. 2, pp. 196–233, 2011.

[3] B. Rudolph, "Seaweed products: red algae of economic significance," in *Marine & Freshwater Products Handbook*, R. E. Martin, Ed., pp. 515–529, Technomic, Lancaster, UK, 2000.

[4] L. Pereira and F. van de Velde, "Portuguese carrageenophytes: carrageenan composition and geographic distribution of eight species (Gigartinales, Rhodophyta)," *Carbohydrate Polymers*, vol. 84, no. 1, pp. 614–623, 2011.

[5] L. Pereira, A. M. Amado, A. T. Critchley, F. van de Velde, and P. J. A. Ribeiro-Claro, "Identification of selected seaweed polysaccharides (phycocolloids) by vibrational spectroscopy (FTIR-ATR and FT-Raman)," *Food Hydrocolloids*, vol. 23, no. 7, pp. 1903–1909, 2009.

[6] L.-E. Rioux, S. L. Turgeon, and M. Beaulieu, "Structural characterization of laminaran and galactofucan extracted from the brown seaweed *Saccharina longicruris*," *Phytochemistry*, vol. 71, no. 13, pp. 1586–1595, 2010.

[7] V. Barry, "Constitution of laminarin—isolation of 2, 4, 6-trimethylglucopyranose," *Scientific Proceedings of Royal Dublin Society*, no. 22, pp. 59–67, 1939.

[8] S. Peat, W. J. Whelan, and H. G. Lawley, "The structure of laminarin. Part I. The main polymeric linkage," *Journal of the Chemical Society*, pp. 724–728, 1958.

[9] T. Chopin, B. F. Kerin, and R. Mazerolle, "Phycocolloid chemistry as a taxonomic indicator of phylogeny in the Gigartinales, Rhodophyceae: a review and current developments using Fourier transform infrared diffuse reflectance spectroscopy," *Phycological Research*, vol. 47, no. 3, pp. 167–188, 1999.

[10] H. J. Bixler and H. Porse, "A decade of change in the seaweed hydrocolloids industry," *Journal of Applied Phycology*, vol. 23, no. 3, pp. 321–335, 2011.

[11] F. van de Velde and G. A. de Ruiter, "Carrageenan," in *Biopolymers V. 6. Polysaccharides II, Polysaccharides from Eukaryotes*, E. J. Vandamme, S. D. Baets, and A. Steinbèuchel, Eds., pp. 245–274, Wiley-VCH, Chichester, UK, 2002.

[12] M. Lahaye, "Developments on gelling algal galactans, their structure and physico-chemistry," *Journal of Applied Phycology*, vol. 13, no. 2, pp. 173–184, 2001.

[13] C. S. Lobban, D. J. Chapman, and B. P. Kremer, *Experimental Phycology: A Laboratory Manual*, Phycological Society of America, Cambridge University Press, 1988.

[14] M. Lahaye, "Chemistry and physico-chemistry of phycocolloids," *Cahiers de Biologie Marine*, vol. 42, no. 1-2, pp. 137–157, 2001.

[15] M. Rinaudo, "Alginates and carrageenans," *Actualite Chimique*, no. 11-12, pp. 35–38, 2002.

[16] L. Pereira, *Estudos em macroalgas carragenófitas (Gigartinales, Rhodophyceae) da costa portuguesa—aspectos ecológicos, bioquímicos e citológicos [Ph.D. thesis]*, FCTUC, University of Coimbra, 2004.

[17] C. Sartori, D. S. Finch, B. Ralph, and K. Gilding, "Determination of the cation content of alginate thin films by FTi.r. Spectroscopy," *Polymer*, vol. 38, no. 1, pp. 43–51, 1997.

[18] H. Kylin, "Biochemistry of sea algae," *Zeitschrift für Physikalische Chemie*, vol. 83, pp. 171–197, 1913.

[19] L. Chevolot, B. Mulloy, J. Ratiskol, A. Foucault, and S. Colliec-Jouault, "A disaccharide repeat unit is the major structure in fucoidans from two species of brown algae," *Carbohydrate Research*, vol. 330, no. 4, pp. 529–535, 2001.

[20] A. O. Chizhov, A. Dell, H. R. Morris et al., "Structural analysis of laminarans by MALDI and FAB mass spectrometry," *Carbohydrate Research*, vol. 310, no. 3, pp. 203–210, 1998.

[21] T. E. Nelson and B. A. Lewis, "Separation and characterization of the soluble and insoluble components of insoluble laminaran," *Carbohydrate Research*, vol. 33, no. 1, pp. 63–74, 1974.

[22] M. Zinoun and J. Cosson, "Seasonal variation in growth and carrageenan content of *Calliblepharis jubata* (Rhodophyceae, Gigartinales) from the Normandy coast, France," *Journal of Applied Phycology*, vol. 8, no. 1, pp. 29–34, 1996.

[23] L. Pereira, A. Sousa, H. Coelho, A. M. Amado, and P. J. A. Ribeiro-Claro, "Use of FTIR, FT-Raman and ^{13}C-NMR spectroscopy for identification of some seaweed phycocolloids," *Biomolecular Engineering*, vol. 20, no. 4–6, pp. 223–228, 2003.

[24] B. Matsuhiro, "Vibrational spectroscopy of seaweed galactans," *Hydrobiologia*, vol. 326-327, pp. 481–489, 1996.

[25] W. Mackie, "Semi-quantitative estimation of the composition of alginates by infra-red spectroscopy," *Carbohydrate Research*, vol. 20, no. 2, pp. 413–415, 1971.

[26] E. Gómez-Ordóñez and P. Rupérez, "FTIR-ATR spectroscopy as a tool for polysaccharide identification in edible brown and red seaweeds," *Food Hydrocolloids*, vol. 25, no. 6, pp. 1514–1520, 2011.

[27] M. Sekkal, P. Legrand, J. P. Huvenne, and M. C. Verdus, "The use of FTIR microspectrometry as a new tool for the identification in situ of polygalactanes in red seaweeds," *Journal of Molecular Structure*, vol. 294, pp. 227–230, 1993.

[28] A. I. Usov and M. Y. Elashvili, "Polysaccharides of algae. 44. Investigation of sulfated galactan from *Laurencia nipponica* Yamada (Rhodophyta, Rhodomelaceae) using partial reductive hydrolysis," *Botanica Marina*, no. 34, pp. 553–560, 1991.

[29] N. P. Chandia, B. Matsuhiro, and A. E. Vasquez, "Alginic acids in Lessonia trabeculata: characterization by formic acid hydrolysis and FT-IR spectroscopy," *Carbohydrate Polymers*, vol. 46, no. 1, pp. 81–87, 2001.

[30] N. P. Chandía, B. Matsuhiro, E. Mejías, and A. Moenne, "Alginic acids in Lessonia vadosa: partial hydrolysis and elicitor properties of the polymannuronic acid fraction," *Journal of Applied Phycology*, vol. 16, no. 2, pp. 127–133, 2004.

[31] M. P. Filippov and R. Kohn, "Determination of composition of alginates by infra-red spectroscopic methods," *Chemické Zvesti*, no. 28, p. 817, 1974.

[32] K. Sakugawa, A. Ikeda, A. Takemura, and H. Ono, "Simplified method for estimation of composition of alginates by FTIR," *Journal of Applied Polymer Science*, vol. 93, no. 3, pp. 1372–1377, 2004.

[33] A. Skriptsova, V. Khomenko, and I. Isakov, "Seasonal changes in growth rate, morphology and alginate content in Undaria pinnatifida at the northern limit in the Sea of Japan (Russia)," *Journal of Applied Phycology*, vol. 16, no. 1, pp. 17–21, 2004.

[34] M. R. Torres, A. P. A. Sousa, E. A. T. Silva Filho et al., "Extraction and physicochemical characterization of Sargassum vulgare alginate from Brazil," *Carbohydrate Research*, vol. 342, no. 14, pp. 2067–2074, 2007.

[35] K. Sahayaraj, S. Rajesh, and J. M. Rathi, "Silver nanoparticles biosynthesis using marine alga *Padina pavonica* (Linn.) and its microbicidal activity," *Digest Journal of Nanomaterials and Biostructures*, no. 7, pp. 1557–1567, 2012.

[36] N. P. Chandía and B. Matsuhiro, "Characterization of a fucoidan from Lessonia vadosa (Phaeophyta) and its anticoagulant and elicitor properties," *International Journal of Biological Macromolecules*, vol. 42, no. 3, pp. 235–240, 2008.

Latent Fingerprint Enhancement Using Tripolyphosphate-Chitosan Microparticles

Issa M. A. Il Dueik and Gordon A. Morris

Department of Chemical and Biological Sciences, School of Applied Sciences, University of Huddersfield, Queensgate, Huddersfield, West Yorkshire HD1 3DH, UK

Correspondence should be addressed to Gordon A. Morris; g.morris@hud.ac.uk

Academic Editor: Thomas J. Heinze

Chitosan has been widely used in the preparation of microparticles for drug delivery; however, it has not been considered in forensic applications. Tripolyphosphate- (TPP-) chitosan microparticles were formed using ionotropic gelation in the presence of a coloured dye and deposited onto latent fingerprints enabling fingerprint identification.

1. Introduction

Chitosan is the generic name for a family of strongly polycationic derivatives of poly-N-acetyl-D-glucosamine (chitin) extracted from the shells of crustaceans or from the mycelia of fungi [1]. In chitosan the N-acetyl group is replaced either fully or partially by NH_2, and therefore the degree of acetylation can vary from DA = 0 (fully deacetylated) to DA = 1 (fully acetylated, i.e., chitin). The long carbon chains of chitosan molecules render them lipophilic. Furthermore, chitosan is the second most abundant polymer on earth (after cellulose) and it is the only known naturally occurring polycationic polysaccharide; therefore, chitosan and its derivatives, including microparticles, have received a great deal of attention from the food, cosmetic, and pharmaceutical industries [2–4]. Microparticles can be prepared by the electrostatic interaction and the resultant ionotropic gelation between chitosan and the tripolyphosphate (TPP) (Figure 1) polyanion [2–4]. Size can be controlled by varying the chitosan : TPP ratio, pH, and the molar mass of the chitosan.

Fingerprint detection is probably the oldest and most common method of identification used in forensic science. Fingerprints, therefore, present a perfect method for personal recognition; they are traces of an impression from the friction ridges on a person's fingertips. Fingerprinting is used in the tracking and identification of criminals, and because they are unique (identical twins have different fingerprints), fingerprints can provide a clear and positive proof of identity.

Recently, there has been great interest in the use of nanotechnology in the design of novel fingerprint detection systems. This is due to the fact that microparticles can provide improved latent fingerprint detection by using dye-functionalized microparticles (the dye or fluorophore may also be encapsulated within the microparticle) which can therefore provide an opportunity for improved visualisation.

In this study, TPP-chitosan microparticles (loaded with red dye for visualisation purposes) have been used to attach to the lipid residues present in the latent fingerprint. In traditional fingerprinting techniques (e.g., ninhydrin), reagents react with salt, lipids, proteins, or amino acids present in the fingerprint residue. Although other polysaccharide-based systems may be more suitable (e.g., lipophilic polysaccharide esters [6]), the potential of chitosan for latent fingerprint development has been demonstrated previously [7, 8].

2. Materials and Methods

2.1. Materials. All chemicals were purchased from Sigma-Aldrich (Gillingham, UK) and used without further purification.

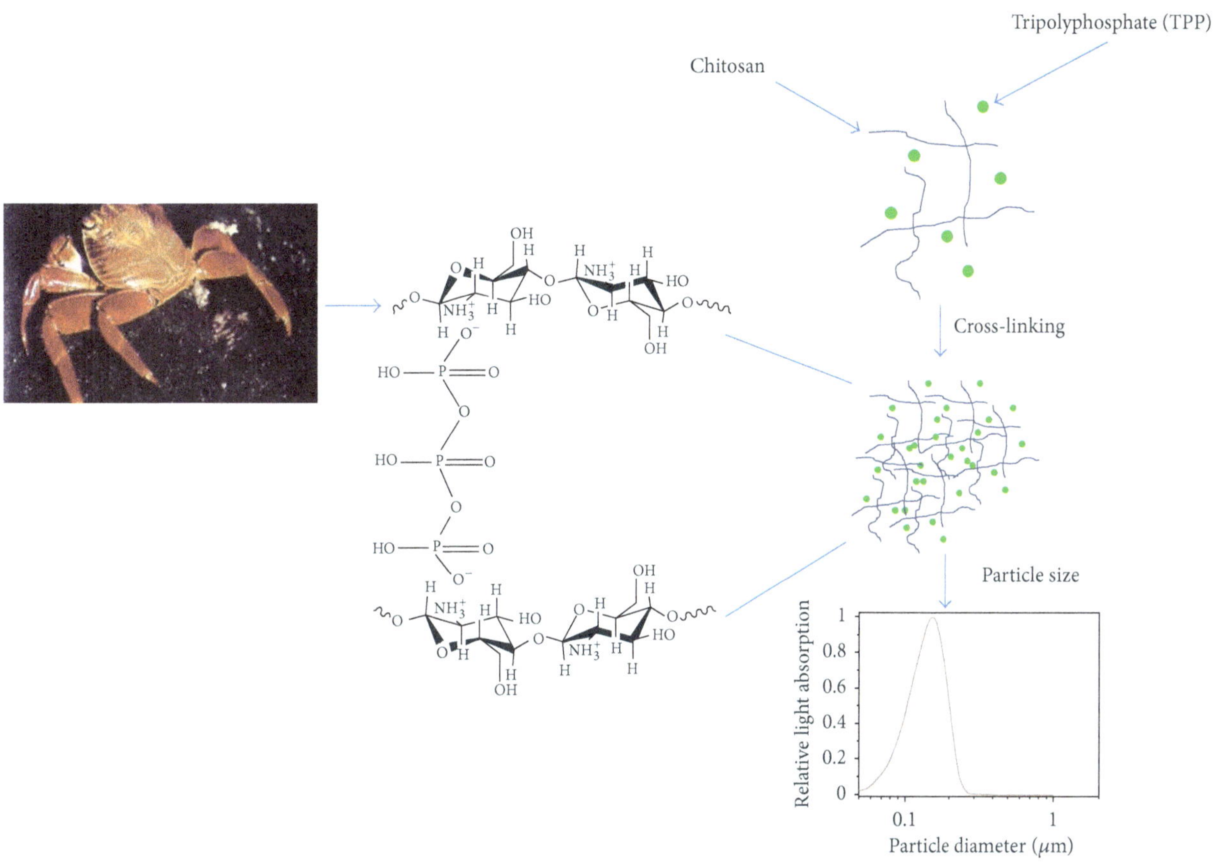

FIGURE 1: Formation of the tripolyphosphate-chitosan complex by ionotropic gelation [4, 5].

2.2. Sample Preparation. Chitosan (2.0 mg/mL) and tripolyphosphate pentasodium (0.84 mg/mL) were prepared in acetate buffer (0.2 M pH 4.3) as described in [2] and [4]. The resultant solutions were then mixed to give TPP : chitosan ratios of 1 : 6, 1 : 4, 1 : 2, 1 : 1, 2 : 1, 4 : 1, and 6 : 1, respectively, and the particle size distributions of the resultant microparticles were measured directly using a Malvern Mastersizer 2000 (Malvern Instruments Ltd., Malvern, UK) and under an optical microscope Leica DM 500 (Leica Microsystems, Milton Keynes, UK).

2.3. Fingerprint Enhancement. Latent fingerprint enhancement was investigated using the following protocol: the 7 different nanoparticle dispersions were centrifuged (Eppendorf UK, Stevenage, UK) at 4000 rpm for 90 minutes. After centrifugation, the supernatant was removed and the remaining of solid deposit was freeze-dried for 24 hours (Edwards High Vacuum International, Crawley, UK) after which the solid material was grounded with a pestle and mortar to produce powder suitable for fingerprinting. Fingerprints were then left on a glass slide as before and dusted with the TPP-chitosan powder.

3. Results and Discussion

When freshly prepared, the diameters of the TPP-chitosan particles were in the range 1–1000 μm (Figure 2) which is considerably larger than has been demonstrated in previous studies [4, 9–15] and may be explained in part by the molar mass and solubility of the chitosan. In so much as higher molecular weight chitosans produce larger particles [4, 15–18].

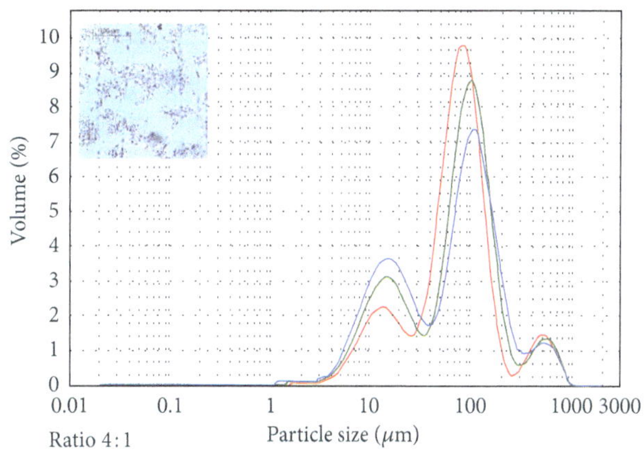

FIGURE 2: Particle size distribution (in triplicate) for TPP-chitosan in the ratio of 4 : 1. Inset a micrograph for the same sample.

It would appear that during sample preparation, the TPP-chitosan microparticles have aggregated. At this stage we have no apparent explanation; although we do not expect particle size to have a significant influence with respect to fingerprint

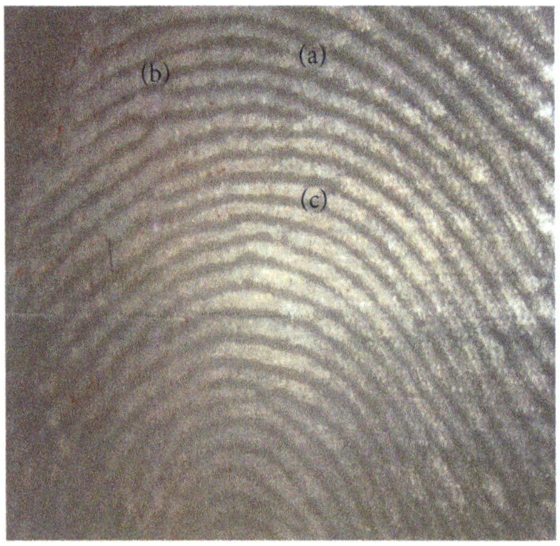

FIGURE 3: Latent fingerprint treated with TPP-chitosan (4 : 1). Some common features used for identification have been indicated: (a) ridge ending, (b) bifurcation, and (c) crossover.

enhancement. However, in any future applications the use of a stabiliser may be beneficial.

In the enhancement of latent fingerprints, it is only the sample with a TPP : chitosan ratio of 4 : 1 which gave satisfactory results (Figure 3). This is expected to be due to the effective charge on the particles [10], that is, as we increase the ratio of TPP, the particles will tend to lose their positive charge and it is expected that particles with little or no charge will interact to a greater extent with the lipids in fingerprint residues.

In Figure 3, the fingerprint details such as bifurcations and crossovers are clearly visible. As an alternative approach, fingerprints were left on a glass slides (nonporous surface) then immersed in the 7 different microparticle dispersions for 1 hour. The slides were then placed in drying oven for 45 minutes at temperature 80°C. However, this approach has not yet yielded any satisfactory results.

4. Conclusions

The use of TPP-chitosan in latent fingerprint enhancement was significantly affected by TPP : chitosan ratio, it was also expected that the storage temperature [4], concentration [18], molar mass [4, 15–18], and levels of aggregation, charge, and degree of deacetylation (DD) of chitosan will be of importance. Furthermore, it may be possible to form the microparticles directly on fingerprint in a 2-stage process. However, this new technique has the potential to be developed as a novel method for fingerprint enhancement.

Conflict of Interests

The authors declare that they have no conflict of interests.

References

[1] M. Rinaudo, "Chitin and chitosan: properties and applications," *Progress in Polymer Science*, vol. 31, no. 7, pp. 603–632, 2006.

[2] A. M. Dyer, M. Hinchcliffe, P. Watts et al., "Nasal delivery of insulin using novel chitosan based formulations: a comparative study in two animal models between simple chitosan formulations and chitosan nanoparticles," *Pharmaceutical Research*, vol. 19, no. 7, pp. 998–1008, 2002.

[3] G. A. Morris, M. S. Kök, S. E. Harding, and G. G. Adams, "Polysaccharide drug delivery systems based on pectin and chitosan," *Biotechnology and Genetic Engineering Reviews*, vol. 27, pp. 257–284, 2010.

[4] G. A. Morris, J. Castile, A. Smith, G. G. Adams, and S. E. Harding, "The effect of prolonged storage at different temperatures on the particle size distribution of tripolyphosphate (TPP)-chitosan nanoparticles," *Carbohydrate Polymers*, vol. 84, no. 4, pp. 1430–1434, 2011.

[5] L. E. Chávez de Paz, A. Resin, K.A. Howard, D.S. Sutherland, and P.L. Wejse, "Antimicrobial effect of chitosan nanoparticles on *Streptococcus mutans* biofilms," *Applied and Environmental Microbiology*, vol. 77, no. 11, pp. 3892–3895.

[6] S. Hornig and T. Heinze, "Efficient approach to design stable water-dispersible nanoparticles of hydrophobic cellulose esters," *Biomacromolecules*, vol. 9, no. 5, pp. 1487–1492, 2008.

[7] N. Ul Islam, K. F. Ahmed, A. Sugunan, and J. Dutta, "Forensic fingerprint enhancement using bioadhesive chitosan and gold nanoparticles," in *Proceedings of the 2nd IEEE International Conference on Nano/Micro Engineered and Molecular Systems (IEEE NEMS '07)*, pp. 411–415, January 2007.

[8] J. Dilag, H. Kobus, and A. V. Ellis, "Cadmium sulfide quantum dot/chitosan nanocomposites for latent fingermark detection," *Forensic Science International*, vol. 187, no. 1–3, pp. 97–102, 2009.

[9] A. Anitha, N. Deepa, K. P. Chennazhi, S. V. Nair, H. Tamura, and R. Jayakumar, "Development of mucoadhesive thiolated chitosan nanoparticles for biomedical applications," *Carbohydrate Polymers*, vol. 83, no. 1, pp. 66–73, 2011.

[10] A. Nasti, N. M. Zaki, P. de Leonardis et al., "Chitosan/TPP and chitosan/TPP-hyaluronic acid nanoparticles: systematic optimisation of the preparative process and preliminary biological evaluation," *Pharmaceutical Research*, vol. 26, pp. 1918–1930, 2009.

[11] Q. Gan, T. Wang, C. Cochrane, and P. McCarron, "Modulation of surface charge, particle size and morphological properties of chitosan-TPP nanoparticles intended for gene delivery," *Colloids and Surfaces B*, vol. 44, no. 2-3, pp. 65–73, 2005.

[12] B. Hu, C. Pan, Y. Sun et al., "Optimization of fabrication parameters to produce chitosan-tripolyphosphate nanoparticles for delivery of tea catechins," *Journal of Agricultural and Food Chemistry*, vol. 56, pp. 7451–7458, 2008.

[13] Y. Xu and Y. Du, "Effect of molecular structure of chitosan on protein delivery properties of chitosan nanoparticles," *International Journal of Pharmaceutics*, vol. 250, pp. 215–226, 2005.

[14] H. Zhang, M. Oh, C. Allen, and E. Kumacheva, "Monodisperse chitosan nanoparticles for mucosal drug delivery," *Biomacromolecules*, vol. 5, no. 6, pp. 2461–2468, 2004.

[15] M. L. Tsai, R. H. Chen, S. W. Bai, and W. Y. Chen, "The storage stability of chitosan/tripolyphosphate nanoparticles in a phosphate buffer," *Carbohydrate Polymers*, vol. 84, no. 2, pp. 756–761, 2011.

[16] M. L. Tsai, S. W. Bai, and R. H. Chen, "Cavitation effects versus stretch effects resulted in different size and polydispersity of ionotropic gelation chitosan-sodium tripolyphosphate nanoparticle," *Carbohydrate Polymers*, vol. 71, pp. 448–457, 2008.

[17] M. Luangtana-anan, P. Opanasopit, T. Ngawhirunpat et al., "Effect of chitosan salts and molecular weight on a nanoparticulate carrier for therapeutic protein," *Pharmaceutical Development and Technology*, vol. 10, no. 2, pp. 189–196, 2005.

[18] E. S. K. Tang, M. Huang, and L. Y. Lim, "Ultrasonication of chitosan and chitosan nanoparticles," *International Journal of Pharmaceutics*, vol. 265, pp. 103–114, 2003.

15

Production Methods for Hyaluronan

Carmen G. Boeriu,[1,2] Jan Springer,[1] Floor K. Kooy,[1,3] Lambertus A. M. van den Broek,[1] and Gerrit Eggink[1]

[1] *Wageningen UR Food & Biobased Research, P.O. Box 17, 6700 AA Wageningen, The Netherlands*
[2] *Universitatea "Aurel Vlaicu," Str. E. Dragoi nr. 2, 310330 Arad, Romania*
[3] *Genencor International B.V., P.O. Box 218, 2300 AE Leiden, The Netherlands*

Correspondence should be addressed to Carmen G. Boeriu; carmen.boeriu@wur.nl

Academic Editor: Thomas J. Heinze

Hyaluronan is a polysaccharide with multiple functions in the human body being involved in creating flexible and protective layers in tissues and in many signalling pathways during embryonic development, wound healing, inflammation, and cancer. Hyaluronan is an important component of active pharmaceutical ingredients for treatment of, for example, arthritis and osteoarthritis, and its commercial value far exceeds that of other microbial extracellular polysaccharides. Traditionally hyaluronan is extracted from animal waste which is a well-established process now. However, biotechnological synthesis of biopolymers provides a wealth of new possibilities. Therefore, genetic/metabolic engineering has been applied in the area of tailor-made hyaluronan synthesis. Another approach is the controlled artificial (*in vitro*) synthesis of hyaluronan by enzymes. Advantage of using microbial and enzymatic synthesis for hyaluronan production is the simpler downstream processing and a reduced risk of viral contamination. In this paper an overview of the different methods used to produce hyaluronan is presented. Emphasis is on the advancements made in the field of the synthesis of bioengineered hyaluronan.

1. Introduction

Hyaluronic acid, also known as hyaluronan, is a linear polysaccharide composed of a repeating disaccharide unit of β(1,4)-glucuronic acid (GlcUA)-β(1,3)-*N*-acetylglucosamine (GlcNAc) (Figure 1). Both individual carbohydrate residues in hyaluronan adopt the stable chair conformation which determines the conformation of the polymer in solution that is described as an overall random coil structure that may have also highly flexible regions. Nevertheless, in terms of chemical structure, hyaluronan is a simple linear polymer with high molecular mass and exceptional rheological properties. Hyaluronan is a member of the glycosaminoglycans family that includes chondroitin/dermatan sulfate, keratan sulfate, and heparin/heparan sulfate, each with a characteristic disaccharide-repeating structure of an amino sugar, either glucosamine or galactosamine, and a hexose, either galactose, glucuronic acid, or iduronic acid, which can be carboxylated or sulfated [1]. Hyaluronan is the only glycosaminoglycan member that is not sulfated and is not covalently bound to a proteoglycan core protein.

Research on hyaluronan expands over more than one century (Table 1). The first report that can be linked to hyaluronan dates from 1880, when the French chemist Portes observed that the mucin in the vitreous body, which he named "hyalomucine," behaved differently from other mucoids in cornea and cartilage. Fifty-four years later Meyer and Palmer described a procedure for isolating a novel glycosaminoglycan from the vitreous of bovine eyes, which they named hyaluronic acid based on hyaloid (vitreous) and uronic acid [3]. Since *in vivo* the polymer exists in an ionized form as polyanion, it is generally referred to as hyaluronan.

In living organisms, hyaluronan is produced by hyaluronan synthase enzymes, which synthesize large, linear polymers of the repeating disaccharide β(1,4)-GlcUA-β(1,3)-GlcNAc by alternate addition of GlcUA and GlcNAc to the growing chain using their activated nucleotide sugars (UDP-glucuronic acid and UDP-*N*-acetylglucosamine) as

TABLE 1: Important events in research on hyaluronan products.

Time	Event
1880	Portes reported that mucin from the vitreous body differs from other mucoids in cornea and cartilage and named it hyalomucine [2].
1934	Meyer and Palmer isolated and identified the polysaccharide from the vitreous body and named it hyaluronic acid [3].
30s–50s	Hyaluronan from many different tissues of vertebrates was isolated, identified, and characterized. A few pathogenic bacteria were found that produce hyaluronan and use it to encapsulate their cells.
50s	The chemical structure of hyaluronan was elucidated by Karl Meyer and his team. They used hyaluronidase to produce overlapping oligosaccharides that were structurally analyzed by conventional techniques [4]. Interest emerged to use hyaluronan in eye surgery as a substitute of the vitreous body.
40s–70s	Extraction processes from animal tissues were optimized to remove protein and to minimize hyaluronan degradation. First studies on hyaluronan production through bacterial fermentation and chemical synthesis were initiated.
1979	First patent on ultrapure hyaluronan isolated from rooster combs [5]. This was the starting of the industrial manufacturing of hyaluronan from animal sources for human applications. In 1980, using the methods of Balasz Pharmacia (Sweden) introduced Healon, a product used in cataract surgery.
90s–00s	Revival of studies on bacterial fermentation to produce hyaluronan of high molecular weight. Emphasis on controlling polymer size and polydispersity.
1993	The gene encoding for a single enzyme that polymerizes UDP-GlcNAc and UDP-GlcUA into hyaluronan is isolated by DeAngelis and coworkers from *Streptococcus pyogenes*. Hyaluronan synthases from other microorganisms were identified and characterized [6, 7].
1996	The largest hyaluronan fragment, an octamer, was chemically synthesized through controlled addition of disaccharide units [8].
2003	Research on the enzymatic synthesis of hyaluronan and monodisperse hyaluronan oligosaccharides with defined length [9, 10].

substrates. The number of repeat disaccharides, n, in a completed hyaluronan molecule is approximately 10,000 but chains of hyaluronan with up to 25,000 disaccharide units have been reported. The size of the polymers (M_w from 5,000 Da to 20 million Da) *in vivo* varies with the type of tissue. For example, hyaluronan in the human umbilical cord has a molecular mass of 3-4 million Da, whilst the molecular

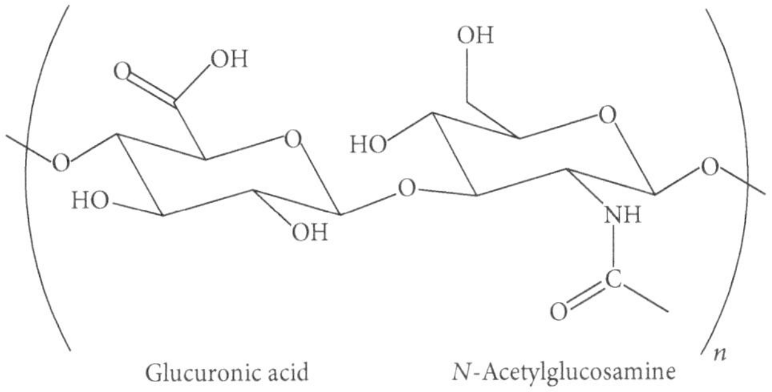

FIGURE 1: Repeating unit of hyaluronan.

mass of hyaluronan from human synovial fluid is 6 million Da. However, the reported molecular weight of isolated hyaluronan polymers may be underestimated, depending on the isolation and analysis method used. Hyaluronan is not monodisperse in molecular weight and the experimentally determined polydispersity (M_w/M_n) of the polymer depends as well on both the method of extraction and the method of analysis used. For example, the values of the polydispersity of a hyaluronan sample determined by high-performance liquid chromatography varied with the type of column, the detection method, and the flow rate used [11–13]. This results in a big variation of the values reported for the M_w and polydispersity of hyaluronan between laboratories and shows the importance for standardization of the methodology for characterization and reporting of hyaluronan molecular properties.

Hyaluronan is present in all vertebrates. High concentrations are found in the umbilical cord, the synovial fluid between joints, skin, and the vitreous body of the eye. It was estimated that in the body of a person of 70 kg, about 15 g hyaluronan is found in different tissues, of which one third is turned over every day. The human skin contains over 50% of the hyaluronan in the body [14, 15]. Rooster combs contain the highest concentrations of hyaluronan ($7.5\ mg\ g^{-1}$) ever reported for animal tissues [16]. Hyaluronan is also present in the capsule of a small number of microbial pathogens such as *Pasteurella multocida* and group A and C streptococci among which *Streptococcus pyogenes* (a human pathogen) and the animal pathogens *Streptococcus equi* and *Streptococcus uberis*, that quite likely pirated the enzymatic machinery from vertebrate hosts for its synthesis. These microorganisms use hyaluronan to encapsulate their cells, forming a perfect disguise against the animal defense system and facilitating the adhesion and colonization of the bacterial cells [17]. Since the hyaluronan polymer produced in animals and the aforementioned bacteria is identical, the host immune defense is not triggered to repel the pathogenic bacteria contrary to other bacteria with a different extracellular capsule. Intriguingly, the nonimmunogenic and noninflammatory properties of bacterial hyaluronan benefit mammals currently too, since bacterial hyaluronan forms an excellent source for medical grade hyaluronan. Initially hyaluronan was believed to be an inert compound having no specific interaction with other macromolecules. However,

Table 2: Biomedical and pharmacologic applications of hyaluronan.

Area	Application	Example
Biomedical	Orthopedic surgery	Arthritis
	Rheumatology	Osteoarthritis
	Ophthamology	Eye surgery
	Otolaryngology	Treatment of the vocal fold
	Dermatology and plastic surgery	Dermal filler
	Wound healing and dressing	Diabetic ulcer, skin burns
	Tissue engineering	
Pharmacology	Drug delivery	

from the discovery of the interaction of hyaluronan with cartilage proteoglycans by Hardingham and Muir [18] a large number of reports have been published on the role of hyaluronan in cellular activity, migration, mitosis, inflammation, cancer, angiogenesis, fertilization, and so on (see for more information [19, 20]).

In addition, hyaluronan and its derivatives have been largely studied and applied in the biomedical field (Table 2) [21–23]. Due to the high level of biocompatibility, hyaluronan is extensively used in viscosurgery, in the treatment of arthritis to supplement the lubrication of arthritic joints, as microcapsule for targeted drug delivery and in cosmetics as a hydrating and antiaging material.

The biological functions of hyaluronan are strongly depending upon its size. High molecular weight hyaluronan polymers ($M_w > 5 \times 10^5$ Da) are space filling, antiangiogenic, and immunosuppressive; medium size hyaluronan chains (M_w between 2×10^4–10^5 Da) are involved in ovulation, embryogenesis, and wound repair; oligosaccharides with 15–50 repeating disaccharide units (M_w between 6×10^3–2×10^4 Da) are inflammatory, immuno-stimulatory, and angiogenic while small hyaluronan oligomers (M_w from 400 to 4000 Da) are anti-apoptotic and inducers of heat shock proteins [24]. Lower molecular weight hyaluronan and small oligosaccharides are produced by controlled depolymerization of high molecular weight hyaluronan using physical treatment (thermal treatment, pressure), irradiation (electron beam, gamma ray, microwave), acid treatment, ozonolysis, metal catalyzed radical oxidation, and enzymatic hydrolysis with hyaluronidase (EC 3.2.1.35).

The commercial value of hyaluronan far exceeds that of other microbial extracellular polysaccharides. With an estimated world market value of $US 500 million, it is sold for up to $US 100,000 per kilogram [25]. On top of this, several groups have reported effects of hyaluronan oligosaccharides on cellular behavior that could imply the use of these molecules in cancer treatment or in wound healing [26]. It is clear that hyaluronan is much more than a space-filling inert component of the extracellular matrix and will become more and more important as a pharmaceutical component and indeed is a multifunctional megadalton stealth molecule [27].

In this paper the different methods to produce hyaluronan will be discussed as depicted in Figure 2.

2. Biosynthetic Pathways for Hyaluronan in Living Organisms

The overall reaction of hyaluronan synthesis is as follows:

$$n\,\text{UDP-GlcUA} + n\,\text{UDP-GlcNAc} \longrightarrow 2n\,\text{UDP} + [\text{GlcUA} + \text{GlcNAc}]_n. \quad (1)$$

The reaction is catalyzed by a single enzyme utilizing both sugar substrates to synthesize hyaluronan [28]. In contrast to other glycoaminoglycans which are synthesized in the Golgi network, hyaluronan is synthesized at the plasma membrane [29, 30]. Hyaluronan synthases (EC 2.4.1.212) are predicted to have multiple membrane spanning domains with a large intracellular loop on the plasma membrane's inner face. The only known exception is the *P. multocida* hyaluronan synthase which has a membrane attachment domain near the carboxyl terminus. Hyaluronan molecules are extruded into the extracellular matrix in coordination with their synthesis [31], but they also exist intracellularly. In the extracellular matrix hyaluronan exists in a number of different forms. For example, in vertebrates it can be intercalated within a proteoglycan complex, referred to glycocalyx if it is a more delicate pericellular matrix, or it can be bound to membrane receptors of the cell surface. It can interact with binding proteins so-called hyaladherins, which is also the case for intracellular hyaluronan [15]. The mechanisms regulating extracellular versus intracellular location of hyaluronan are not known yet. Recently, the presence of hyaluronan cables has been reported that was the result of the incorporation of the heavy chain of interalpha-trypsin inhibitor with hyaluronan [32]. Depending on the type of tissue, different hyaluronan concentrations and molecular weight evoke different cell responses, for example, during embryonic development, healing processes, inflammation, and cancer. How the cell response is influenced by the different molecular weight of hyaluronan is still an intriguing question [15, 24].

In 1996 the first reports were published concerning the cloning of mammalian hyaluronan synthases [33, 34]. Since then in human cells three hyaluronan synthase genes have been found encoding enzymes having distinct enzymatic properties with respect to hyaluronan size and size distribution [35] which are differentially inducible by growth factors and cytokines [36, 37].

In 1993 the first gene encoding the enzyme responsible for hyaluronan synthesis, denoted HA synthase or HAS, was identified from group A streptococci [6, 7]. This gene is part of an operon containing the *has*A gene encoding hyaluronan synthase, the *has*B gene encoding UDP-Glucose dehydrogenase, and the *has*C gene encoding UDP-Glucose-pyrophosphorylase [38–40]. A hydropathy plot of the *has*A gene predicts three transmembrane segments which is in accordance with the early findings of Markovitz and coworkers who showed that the hyaluronan synthase activity is located in the cell membrane of streptococci [41]. Cloning of

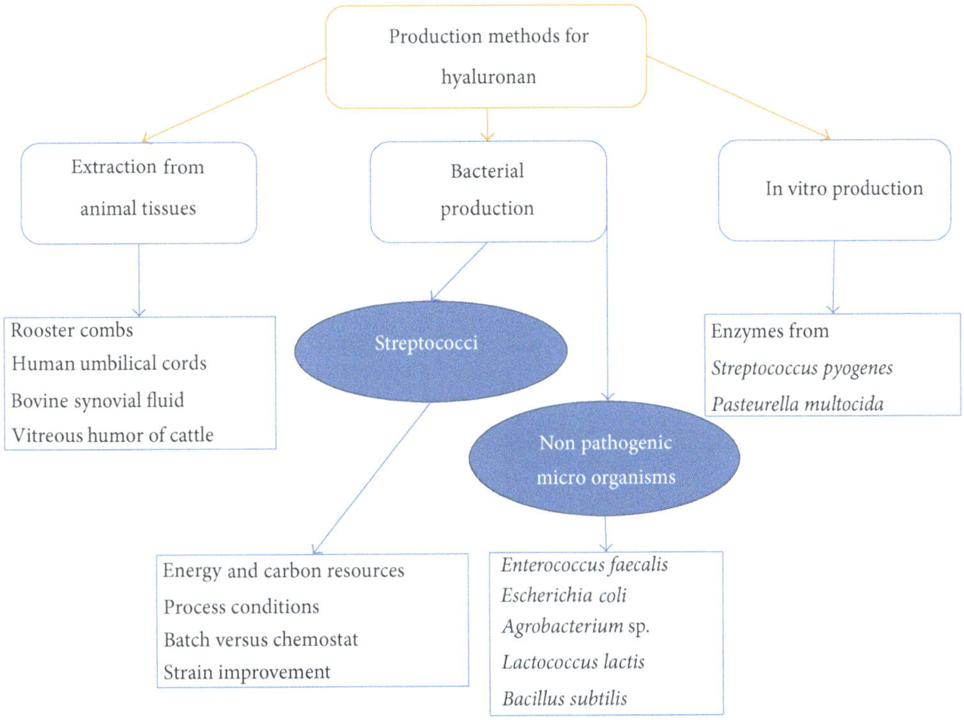

FIGURE 2: Production methods for hyaluronan.

the group A and later the group C streptococcal hyaluronan synthase genes [42] allowed researchers to confirm that only one gene product (the HA-synthase protein) is required for hyaluronan biosynthesis.

The first vertebrate gene identified as being HA synthase was the *Xenopus laevis* gene DG42 [43]. Since then more vertebrate HA synthase genes were identified [44] and in addition an HA synthase from an algal virus was discovered [17, 45]. The streptococcal, vertebrate, and viral HA synthases show protein sequence similarity and appear to have a single glycosyl transferase 2 (GT2) family module [46] with some similarities to chitin synthase. The last extension of organisms in which a HA synthase gene was identified is *Cryptococcus neoformans*. The CPS1 gene from this pathogenic yeast encodes a HA synthase with high homology to the previously mentioned HA synthase proteins [47]. An HA synthase that differs in protein sequence and in predicted topology from all other HA synthases was identified and cloned from type A *P. multocida*, an animal pathogen [48].

Unlike the other HA synthases (Table 3; grouped as Class I HA synthases) which are integral membrane proteins and elongate hyaluronan chains at the reducing end (Class I-R) or at the nonreducing end [50], the *P. multocida* HA synthase has a membrane attachment domain near the carboxyl terminus, has two GT2 modules, and elongates the HA chains at the nonreducing end. It is therefore called a Class II HA synthase [17]. So far the *P. multocida* HA synthase is the only known member of the Class II HA synthases [49, 51, 52].

The regulation of hyaluronan synthesis is much more complex in vertebrates than in bacteria because hyaluronan fulfills various functions in the mammalian body depending on the tissue involved and the hyaluronan chain length required. For the three mammalian HA synthase isozymes different expression patterns are observed between adult and embryonic tissues [54, 55]. Several regulatory factors are known to be involved in controlling HA synthase transcription, such as morphogenesis, cytokines, growth factors, and antisense mRNA [56, 57]. Furthermore, hyaluronan synthesis seems to be controlled through translation regulation, since a latent pool of HA synthase exits within the cell interior, in the endoplasmic reticulum-Golgi compartments, that upon insertion in the cell membrane becomes activated [53]. HA synthase activity can be modulated by posttranslational modification, such as phosphorylation and *N*-glycosylation [58]. Intermediate and small oligosaccharides are probably provided through hyaluronan cleavage by hyaluronidases, since mammalian HA synthase isozymes synthesize hyaluronan

TABLE 3: Hyaluronan synthase classification system as proposed by Weigel and DeAngelis [49].

	Class I		Class II
	Reducing	Nonreducing	
Members	*Streptococcus pyogenes*, *S. equisimilis*, *S. uberis*, humans, and mice	*Xenopus laevis*, algal virus	*Pasteurella multocida*
Hyaluronan chain growth	Reducing end	Nonreducing end	Nonreducing end

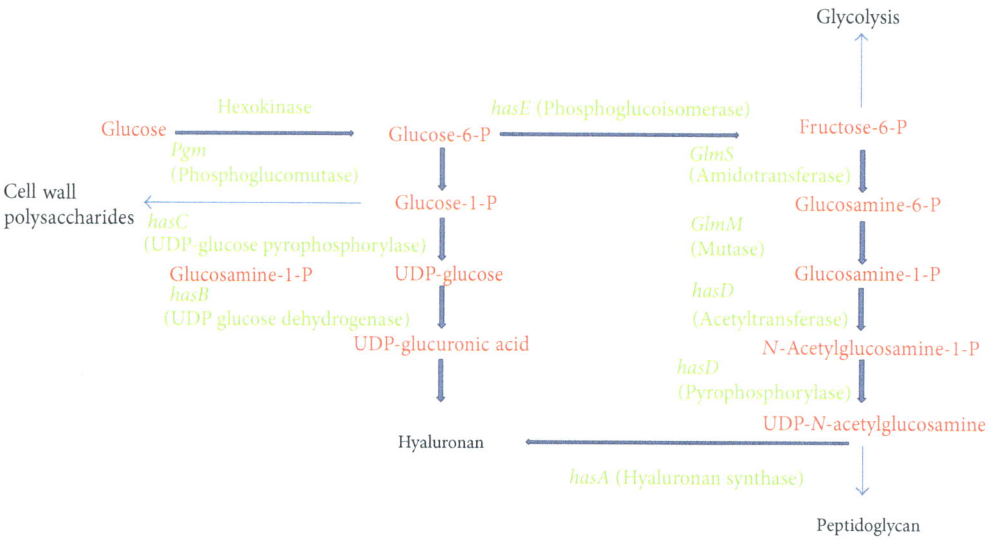

FIGURE 3: Schematic overview of the hyaluronan synthesis pathway in *S. zooepidemicus* adapted from [53] 2013 The American Society for Biochemistry and Molecular Biology.

higher than 10^5 Da [14]. Due to the high complexity of the regulation of hyaluronan synthesis in vertebrates a cohesive view on the relation between the regulatory components is not available yet.

Recently, the metabolic route of hyaluronan formation in *S. zooepidemicus* was elucidated. The streptococcal HA synthase is found in operons also encoding one or more enzymes involved in biosynthesis of the activated sugars [59]. The *has* operon in *S. zooepidemicus* encodes for five genes (Figure 3): hyaluronan synthase (*has*A), UDP-glucose dehydrogenase (*has*B), UDP-glucose pyrophosphorylase (*has*C), a *glm*U paralog encoding for a dual function enzyme acetyltransferase and pyrophosphorylase activity (*has*D), and a pgi paralog encoding for phosphoglucoisomerase (*has*E) [60]. *Has*B and *has*C are involved in UDP-GlcUA synthesis, while *has*D and *has*E are involved in the synthesis of UDP-GlcNAc. The hyaluronan biosynthesis is a costly process for bacteria in terms of carbon and energy resources and competes with cell growth because of the large pathway overlap with the cell wall biosynthesis (Figure 3). Other streptococci strains have *has* operons that contain besides hyaluronan synthase, only the *has*B and *has*C genes. This suggests that the superior hyaluronan synthesis observed for *S. zooepidemicus* relates to the availability of UDP-sugar precursors. Understanding the metabolic routes for hyaluronan synthesis had a significant role in the optimization of the microbial production of hyaluronan, allowing the control of the polymer chain length and increasing the product yield.

3. Production of Hyaluronan

Industrial manufacturing of hyaluronan is based on two main processes, the extraction from animal tissues and microbial fermentation using bacterial strains. Both technologies produce polydisperse high molecular weight hyaluronan ($M_w \geq 1 \times 10^6$ Da, polydispersity ranging from 1.2 to 2.3) for biomedical and cosmetic applications [12, 61, 62]. The first process, to be applied at industrial scale, was the extraction of hyaluronan from animal waste which is still an important technology for commercial products, but is hampered by several technical limitations. One drawback in the extraction process is the inevitable degradation of hyaluronan, caused by (a) the endogenous hyaluronidase activity in animal tissues, breaking down the polymer chain through enzymatic hydrolysis, and (b) the harsh conditions of extraction. Extraction protocols have been improved over the years, but still suffer from low yields, due to the intrinsic low concentration of hyaluronan in the tissue, and from high polydispersity of polymer products due to both the natural polydispersity of hyaluronan and to the uncontrolled degradation during extraction. As in any process for the production of therapeutic compounds from animal sources, there is a potential risk of contamination with proteins and viruses, but this can be minimized by using tissues from healthy animals and extensive purification. Nevertheless, concerns on viral (particularly avian) and protein (particularly bovine) contamination increased the interest in the biotechnological production of hyaluronan.

Production of hyaluronan by bacterial fermentation evolved to a mature technology in the last two decades of the 20th century. In the early developmental stages of bacterial fermentation using group A and C streptococci, optimization of culture media and cultivation conditions along with strain improvement were used to increase hyaluronan yield and quality. Hyaluronan yields reached 6-7 g L^{-1}, which is the upper technical limit of the process caused by mass transfer limitation due to the high viscosity of the fermentation broth. Bacterial fermentation produces hyaluronan with high molecular weight and purity, but risk of contamination with bacterial endotoxins, proteins, nucleic acids, and heavy metals exists. The identification of the genes involved in the biosynthesis of hyaluronan and of the sugar nucleotide

TABLE 4: A comparative view of technologies for manufacturing hyaluronan.

	Advantages	Disadvantages
Extraction from animal materials	(i) Well-established technology (ii) Available raw material at low costs (iii) Product with very high M_w up to 20 MDa (iv) Natural product	(i) Variation in product quality (ii) Risk of polymer degradation (iii) Risk of contamination with protein, nucleic acids, and viruses (iv) Extensive purification needed (v) Low yield
Bacterial production	(i) Mature technology (ii) High yield (iii) Product with high M_w (1–4 MDa)	(i) Use of genetically modified organism (GMO) (ii) Risk of contamination with bacterial endotoxins, proteins, nucleic acids, and heavy metals
Enzymatic synthesis	(i) Versatile technology (ii) Control of the M_w of the products. Products up to 0.55–2.5 MDa and also defined oligosaccharides can be obtained (iii) No risks of contamination (iv) Constant product quality	(i) Emerging technology in development stage (ii) Technological and economic viability must still be demonstrated

precursors in the 90s opened the way towards hyaluronan production using safe, nonpathogenic recombinant strains.

In the past decade a new technology emerged using isolated HA synthase to catalyze the polymerization of the UDP-sugar monomers. This novel enzymatic technology for hyaluronan synthesis is very versatile allowing producing both high molecular weight hyaluronan and hyaluronan oligosaccharides with defined chain length and low polydispersity. Production of monodisperse hyaluronan oligosaccharides on mg-scale using two single-action mutants of the *P. multocida* HA synthase was demonstrated by the group of DeAngelis [9], but large scale production was not achieved yet. A comparative analysis of established and emerging technologies for high molecular weight hyaluronan production is given in Table 4.

3.1. Extraction from Animal Tissues. Extraction of hyaluronan from animal tissues was initially used for laboratory investigations in order to identify and characterize the polymer and to elucidate its biological potential and biomedical application. Hyaluronan from almost all tissues of vertebrates including the vitreous body of the eye, umbilical cord, synovial fluid, pig skin, the pericardial fluid of the rabbit, and the cartilage of sharks was isolated and described [63]. Recently, the extraction of hyaluronan from fish eyes as an alternative resource outside animal husbandry was reported [64]. Nevertheless, the most accessible sources for large scale production high molecular weight hyaluronan are rooster combs (1.2×10^6 Da), the human umbilical cords (3.4×10^6 Da), the vitreous humor of cattle (7.7×10^4–1.7×10^6 Da), and the bovine synovial fluid (14×10^6 Da).

Despite the numerous methods for isolation and purification of hyaluronan reported in the early literature, it was only in 1979 that a method became available for the production of pharmaceutical grade hyaluronan by extraction from animal tissues. Balazs [5] developed an efficient procedure to isolate and purify hyaluronic acid from rooster combs and human umbilical cords that set the basis of the industrial production of hyaluronan from rooster combs for medical application.

Hyaluronan is soluble in water, but extraction of highly pure, high molecular weight hyaluronan from animal tissues is difficult, since in biological materials hyaluronan is usually present in a complex with other biopolymers including proteoglycans [65]. Methods to release hyaluronan from these complexes comprise the use of proteolytic enzymes (i.e., papain, pepsin, pronase, and trypsin), hyaluronan ion-pair precipitation (with e.g., cetylpyridinium chloride), precipitation with organic solvents, nonsolvent precipitation, detergents, and so forth [21]. Ultrafiltration and chromatography are used to remove the degradation products and other contaminants. Sterile filtration is used to remove all microbial cells prior to alcohol precipitation, drying, and conditioning of the end product.

Despite extensive purification, animal hyaluronan can be contaminated with proteins and nucleic acids. The quantity and nature of contaminants differ with the source, as was shown for hyaluronan isolates from human umbilical cord and bovine vitreous humor, containing elevated protein and nucleic acid levels compared to those from rooster comb and bacterial capsule isolates [12]. Diseases as bovine spongiform encephalopathy (BSE) made us aware of the risk of cross-species viral infection. Consequently, hyaluronan from animal sources has to be extensively purified to get rid of these contaminants and extraction and purification processes are continuously improved to fulfill the high quality standards for medical applications. To date, animal waste is still the most important source for the industrial manufacturing of hyaluronan for medical application and provides up to several tones of medical-grade hyaluronan preparations per year. Pharmacia (Sweden), Pfizer (USA), Genzyme (USA), and

Diosynth (The Netherlands) were among the first companies that produced hyaluronan from animal waste at industrial scale. Commercially available hyaluronan preparations from animal sources have a molecular weight ranging from several hundred thousand up to 2.5 million Da [66].

3.2. Bacterial Production. The development of hyaluronan production through bacterial fermentation started in the 60s, when it was acknowledged that animal derived hyaluronan sources can contain undesired proteins causing possible allergic inflammation responses [67]. Since the hyaluronan polymer produced in animals and bacteria is identical, bacterial hyaluronan is not immunogenic and therefore is an excellent source for medical grade hyaluronan. Extracting hyaluronan from microbial fermentation broth is a relatively simple process with high yields. An additional and important advantage of microbial hyaluronan production is that microbial cells can be physiologically and/or metabolically adapted to produce more hyaluronan of high molecular weight. Therefore, microbial hyaluronan production using either pathogenic streptococci or safe recombinant hosts, containing the necessary hyaluronan synthase, is nowadays more and more preferred.

3.2.1. Production of Hyaluronan with Streptococci Strains. The first reported hyaluronic acid isolation from group A hemolytic streptococci resulted in 60–140 mg L^{-1} hyaluronan [68]. Since then, many different attempts have been made to increase the amount of hyaluronan, including traditional techniques as optimizing the extraction process, adapting the culture media, and selecting strains with high hyaluronan productivity. In thirty years, the hyaluronan yields using batch fermentation increased from 300–400 mg L^{-1} [69] to 6-7 g L^{-1} [70], which is the practical limit due to mass transfer limitations [25]. Group C streptococci which are not human pathogens and have superior hyaluronan productivity are frequently used for hyaluronan synthesis instead of Group A streptococci or instead the animal pathogenic bacterium *P. multocida*. The most used strains are *S. equi* subsp. *equi* and *S. equi* subsp. *zooepidemicus*. From dairy food products *S. thermophilus* strains with high productivity of useful exopolysaccharides including hyaluronan were isolated [71], offering a safe hyaluronan producing organism. Recently, a recombinant *S. thermophilus* was constructed that was able to produce 1.2 g L^{-1} hyaluronan and the average molecular weight was comparable with the wild type strain [72].

Streptococci strains for hyaluronan production generally use glucose as carbon source. The yield of hyaluronan on glucose under aerobic fermentation conditions varies between 5 and 10% [25, 73, 74], which is significantly higher than typical yields for complex polysaccharides in lactic acid bacteria. Other carbon sources like starch [75], lactose, sucrose, and dextrin [76] can be used for hyaluronan productivity similar to glucose. The use of carbon sources like starch and lactose which are largely available at lower costs is important in view of the process economics.

Research on optimization of hyaluronan production through fermentation of streptococci strains addressed the improvement of the hyaluronan yield, the increase of the molecular weight, and the reduction of the polymer polydispersity, parameters which are essential for hyaluronan applications. The main question was if the molecular weight could be further regulated either through strain improvements, optimized process conditions, or through metabolic engineering. These topics will be discussed below.

Regulation within Bacterial Cells. Energy and Carbon Resources. Hyaluronan biosynthesis in streptococci requires a large amount of energy and competes with the bacterial cell growth for glucose as energy provider or as UDP-sugar precursor. Indeed, when unlimited glucose is available, the highest bacterial growth was observed at optimal cultivation conditions (pH 7 and 37°C) whilst the highest hyaluronan productivity and molecular weight were obtained at suboptimal growth conditions, since when cells are growing slowly, the carbon and energy resources are available for other processes [77]. Under glucose-limiting conditions first hyaluronan productivity and then molecular weight declines [25]. Decrease of the initial glucose concentrations from 60 to 20 g L^{-1} decreased the hyaluronan concentration and molecular weight from 4.20 g L^{-1} and 3.0 million Dalton, to 1.65 g L^{-1} and 2.65 million Dalton, respectively [77].

Aerobic culturing conditions increased the hyaluronan productivity by 50% as well as hyaluronan molecular weight whereas cell growth was unaffected [77–79]. Under aerobic fermentation conditions streptococci change their metabolism from producing lactate into producing acetate, formate, and ethanol. This generates four ATP molecules per hexose for the acetate production instead of two ATP molecules formed in lactate production [80]. It was assumed that the increase of hyaluronan production is due to the increased levels of ATP or to increased levels of NADH oxidase which removes the excess of NADH in the presence of oxygen. However, studies designed to improve ATP formation by increased acetate production through maltose metabolism [80] or increasing NADH oxidase levels by metabolic engineering [81] revealed that hyaluronan synthesis was not limited by energy resources.

Studies have shown that only a critical level of dissolved oxygen of 5% is needed to increase the hyaluronan yield during cell growth, above this value the yield is constant [79]. Interestingly, the expression of HA synthase in *S. zooepidemicus* is nine times higher under aerobic conditions than under anaerobic conditions [74], explaining the increase of hyaluronan synthesis. Furthermore, oxygen induces enzymes involved in the production of UDP-GlcNAc (*has*D) and acetoin recycling, which offers an additional ATP molecule and acetyl-CoA that can be used in hyaluronan production [82].

Based on these results, we can conclude that elevated HA synthase concentrations and UDP-GlcNAc availability seem to be the main reason for the increase in hyaluronan synthesis under aerobic conditions.

Influence of Process Conditions on Hyaluronan Yield and Molecular Weight. Streptococcal fermentation to produce hyaluronan is, like other fermentations, influenced by medium composition, pH, temperature, aeration, and agitation. Especially aerobic conditions and the initial glucose concentration had a large influence on hyaluronan yield and molecular weight, as was discussed above. Both agitation and fermentation modus have a rather complex effect on hyaluronan production and are, therefore, in greater detail described, below.

The effect of agitation on the yield and the molecular weight of hyaluronan in *Streptococcus* fermentation is complex. Increasing agitation increases the hyaluronan yield under both anaerobic and aerobic conditions, most probably due to an enhanced mass transfer induced by the reduced viscosity of the broth [77, 79, 83, 84]. However, hyaluronan molecular weight increases at moderate impeller speed due to mass transfer enhancement, but decreases at high impeller speed most probably due to degradation by the reactive oxygen species that are formed under aerobic conditions at high impeller speed [85]. Oxidative degradation of hyaluronan can be prevented by the addition of oxygen scavengers such as salicylic acid in the culture media resulting in an increase of the hyaluronan molecular weight [85, 86].

Batch versus Chemostat. In batch fermentation, the hyaluronan production is limited to 6-7 g L^{-1} due to high viscosity of the broth and mass transfer constraints. Further improvement of the hyaluronan yield is therefore not possible through high-cell-density fermentation or through a high-yield strain. A continuous culture is the best strategy to improve the volumetric productivity of hyaluronan synthesis for four reasons. First, cell growth of streptococci can be maintained at the exponential phase. Extension of the exponential phase could lead to an increased amount of hyaluronan, since it was shown that hyaluronan synthesis stops in the stationary phase [40, 87] and the excretion of the HA synthase and other cell wall proteins which takes place in the stationary phase is prevented [88]. Second, suboptimal growth conditions can be imposed by nutrient limitation in order to decrease the resource competition between cell growth and hyaluronan synthesis [77, 89]. Third, a turnaround phase is avoided by using a continuous culture [90]. Finally, the viscosity of the broth can be controlled by controlling the hyaluronan concentration, thus enhancing the mass transfer in the reactor. Chemostat cultivation of *S. zooepidemicus* by controlling the medium conditions has been reported [75, 90, 91] and this opens the way for further improvement of the hyaluronan production process.

Strain Improvement. Improved hyaluronan producing strains have been obtained by random mutagenesis and selection of mutant strains oriented to the reduction of the hyaluronan-depolymerising enzyme responsible for the decrease of molecular weight of hyaluronan during fermentation (hyaluronidase-negative strains) and to the reduction of streptolysin, the exotoxin responsible for the β-hemolysis (nonhemolytic strains). Fermentation of nonhemolytic and hyaluronidase-free mutants produced hyaluronan with a molecular weight varying from 3.5 to 5.9 million Da and yields around 6-7 g L^{-1} [70, 76, 92]. Random mutagenesis has also resulted in Streptococci strains that produce hyaluronan of extraordinary high molecular weight ranging from 6 to 9 million Da and with moderate yields of 100 to 410 mg L^{-1} [93].

Further improvement of strains requires genetic engineering of the producer microorganisms to engineer the hyaluronan metabolic pathway. Metabolic engineering in *S. zooepidemicus* and in recombinant hosts has demonstrated that the ratio between UDP-GlcNAc and UDP-GlcUA and the ratio between HA synthase and substrates are the most important factors to influence hyaluronan molecular weight. Chen et al. individually overexpressed each of the five genes in the *has* operon in *S. zooepidemicus*. They discovered that overexpression of *has*A, involved in hyaluronan biosynthesis, and *has*B and *has*C, involved in UDP-GlcUA biosynthesis, decreased molecular weight, while overexpression of *has*E, involved in UDP-GlcNAc biosynthesis, greatly enhanced molecular weight. Overexpression of *has*D had no effect, but molecular weight was further increased when combined with *has*E overexpression [60]. Upregulation of *has*D in *S. zooepidemicus* has been observed at aerobic conditions [82] but also in the presence of an empty plasmid [94], both accompanied by a production of higher M_w hyaluronan. Marcellin et al. [94] suggested that regulation of *has*D is at the translational or posttranslational level and that the plasmid burden is sufficient to remove the *has*D encoded step as a limiting step. In summary, these results indicate that, with a proper balance between the synthesis pathways to UDP-GlcNAc and UDP-GlcUA, the hyaluronan molecular weight can be further increased and controlled.

In addition to the ratio between the two precursors, the importance of the ratio between precursor UDP-GlcUA and HA synthase also influences hyaluronan molecular weight. This was demonstrated for the heterologous host *L. lactis*, two plasmids were constructed containing either *has*A or *has*B from *S. zooepidemicus* with the inducible expression promoters NICE and lacA. Hyaluronan molecular weight increased significantly at *has*A/*has*B ratios below 2, indicating that higher UDP-GlcUA availability per HA synthase enhances hyaluronan molecular weight [95].

To conclude, production of hyaluronan by fermentation of streptococci strains is a mature technology, affording high molecular weight, highly pure polymers suitable for medical, pharmaceutical and cosmetic applications. Several tons per year of bacterial hyaluronan are currently produced for medical and cosmetic applications. Companies that produce hyaluronan through streptococcal fermentation are, for instance, Q-Med (Uppsala, Sweden), Lifecore Biomedical (Chaska, USA), and Genzyme (Cambridge, USA).

3.2.2. Hyaluronan from Recombinant Nonpathogenic Microorganisms. To avoid the risk of exotoxin contamination in hyaluronan products from pathogenic streptococci strains, safe organisms have been genetically engineered into hyaluronan producers, by introducing HA synthase enzymes from either streptococci or *P. multocida*. Using this approach, hyaluronan producing strains of *Enterococcus faecalis* [7], *E. coli* [7, 48, 96, 97], *Bacillus subtilis* [98], *Agrobacterium* sp. [99], and *L. lactis* [95, 100–102] were obtained and exploited for hyaluronan production. Heterologous expression of HA synthase (*has*A gene in streptococci) was sufficient to induce production of hyaluronan. Since the hyaluronan biosynthesis has partially a common pathway with cell wall biosynthesis (Figure 3), production of hyaluronan by the host goes, however, on the expense of cell growth, due to the depletion of sugar precursors. Better hyaluronan producing heterologous strains with improved intracellular availability of sugar precursors were obtained by coexpression of the HA synthase *has*A gene derived from *S. equi* or *P. multocida* with *has*B homologue (UDP-glucose dehydrogenase) or *has*B + *has*C homologues (UDP-glucose dehydrogenase + UDP-glucose pyrophosphorylase) from *E. coli*. A recombinant *E. coli* strain obtained using this strategy produced 2 g L^{-1} hyaluronan, and the yield was increased to 3.8 g L^{-1} when the culture media was supplemented with glucosamine [99].

Using a similar strategy, namely the coexpression of the HA synthase *has*A gene from *P. multocida* with *has*B homologue (UDP-glucose dehydrogenase) from *E. coli*, into the food-grade microorganisms *Agrobacterium* sp. [99] and *L. lactis* [101], these strains were able to produce hyaluronan at levels up to 0.3 g L^{-1} for engineered *Agrobacterium* sp. and 0.65 g L^{-1} for the recombinant *L. lactis*. Although hyaluronan production yields were low, the hyaluronan production from these food-grade microorganisms has high potential for food and biomedical applications.

Without overexpression of *has*B or an analog encoding for UDP-glucose dehydrogenase, UDP-GlcUA levels are often limiting in heterologous hosts [7, 97, 98, 101], whereas UDP-GlcNAc seems to be adequately available in the hosts since it is a main component for bacterial cell wall synthesis. Attempts to overexpress both synthetic pathways to UDP-GlcNAc and UDP-GlcUA in heterologous hosts have been so far unsuccessful, but overexpression of *has*A, *has*B, and *has*C has led to considerable increases in hyaluronan synthesis varying between 200 and 800% depending on the host [96, 100].

A process for the production of ultrapure sodium hyaluronate by the fermentation of a novel, nonpathogenic recombinant strain of *B. subtilis* was developed by the biotech company Novozymes [103]. A major advantage of using *B. subtilis* as host microorganism is that it is cultivable at large scale, does not produce exo- and endotoxins, and does not produce hyaluronidase. To build up the recombinant *Bacillus* strain that produces hyaluronan, expression constructs utilizing the *has*A gene from *S. equisimilis* in combination with overexpression of one or more of the three native *B. subtilis* precursors genes—*tua*D (*has*B homologue, encoding UDP-glucose dehydrogenase), *gta*B (*has*C, encoding UDP-glucose pyrophosphorylase), and *gca*D (*has*D, encoding UDP-*N*-acetyl glucosamine pyrophosphorylase). The recombinant *B. subtilis* strain was able to produce up to 5 g L^{-1} of hyaluronan with a molecular weight of 1–1.2 million Da, when cultivated on a minimal medium based on sucrose at pH 7 and 37°C [94]. The hyaluronan is secreted into the medium and is not cell-associated, which simplifies the downstream processing significantly, making the use of water-based solvents possible for hyaluronan recovery. The *Bacillus* hyaluronan shows very low levels of contaminants (i.e., proteins, nucleic acids, metal ions, and exo- and endotoxins) and thus has a high safety profile, including no risks of viral contamination or of transmission of animal spongiform encephalopathy.

As an alternative to prokaryotic HA production, hyaluronan can be produced by infecting green algae cells of the genus Chlorella with a virus [45, 104], although the reported yields were low (0.5–1 g L^{-1}).

3.3. In Vitro Production of Hyaluronan Using Isolated HA Synthase. Synthesis of hyaluronan using isolated HA synthase becomes relevant when hyaluronan polymers of defined molecular weight and narrow polydispersity are needed. Isolated HA synthase is able to catalyze *in vitro* at well-defined conditions the same reaction as it catalyzes *in vivo*, namely, the synthesis of hyaluronan from the nucleotide sugars UDP-GlcNAc and UDP-GlcUA. Preparative enzymatic synthesis of hyaluronan using the crude membrane-bound HA synthase from *S. pyogenes* was demonstrated, although the yield was low, around 20% [105]. The hyaluronan yield was increased to 90% when the enzymatic hyaluronan synthesis was coupled with *in situ* enzymatic regeneration of the sugar nucleotides using UDP and relatively inexpensive substrates, Glc-1-P and GlcNAc-1-P in a one-pot reaction. The average molecular weight of the synthetic hyaluronan was around 5.5×10^5 Da, corresponding to a degree of polymerization of 1500.

High molecular weight monodisperse hyaluronan polymers with M_w up to 2.500 kDa (~12,000 sugar units) and polydispersity (M_w/M_n) of 1.01–1.20 were obtained by enzymatic polymerization using the recombinant *P. multocida* HA synthase, PmHAS, overexpressed in *E. coli* [106]. PmHAS uses two separate glycosyl transferase sites to add GlcNAc and GlcUA monosaccharides to the nascent polysaccharide chain. Hyaluronan synthesis with PmHAS was achieved either by *de novo* synthesis from the two UDP-sugars precursors (1) and by elongation of an hyaluronan-like acceptor oligosaccharide chain by alternating, repetitive addition of the UDP-sugars as follows:

$$n\,\text{UDP-GlcUA} + n\,\text{UDP-GlcNAc} + z[\text{GlcUA-GlcNAc}]_x \longrightarrow 2n\,\text{UDP} + [\text{GlcUA} + \text{GlcNAc}]_{x+n}. \quad (2)$$

The control of the chain length and polydispersity of the hyaluronan polymer is determined by the intrinsic enzymological properties of the recombinant PmHAS, (i) the rate limiting step of the *in vitro* polymerization appears to be the chain initiation, and (ii) *in vitro* enzymatic polymerization is a fast nonprocessive reaction. Therefore, the concentration of the hyaluronan acceptor controls the size and the polydispersity of the hyaluronan polymer in the presence of a finite amount of UDP-sugar monomers [106]. Using this synchronized, stoichiometrically-controlled enzymatic polymerization reaction, low molecular weight hyaluronan (~8 kDa) with narrow size distribution was synthesized. One important feature of the PmHAS is that chain elongation occurs at the nonreducing end of the growing chain and this makes the use of modified acceptors as substrates possible and consequently the synthesis of hyaluronan polymers with various end-moieties. The elongation of a hyaluronan tetrasaccharide labeled at the reducing end with the fluorophore 2-aminobenzoic acid using PmHAS and the quantitative formation of fluorescent hyaluronan oligomers and polymers was demonstrated [10, 107].

These studies have shown the high potential of the *in vitro* enzymatic synthesis of hyaluronan polymers and oligomers with low polydispersity, but for the large scale application further developments are needed. Technological bottlenecks that must be resolved are (a) production of robust enzymes for hyaluronan synthesis, (b) selective separation of the UDP byproduct from the reaction mixture, to prevent enzyme inactivation and to allow UDP recycling for the synthesis of UDP-sugar substrates, (c) development of efficient processes for the production of the hyaluronan-like template for elongation, and (d) the development of simple, low cost technologies for the synthesis of sugar nucleotide substrates starting from bulk carbohydrates. Since sugar nucleotides are expensive substrates their regeneration is a crucial step for the development of an economic process for the production of hyaluronan.

For *in vitro* production of hyaluronan the Class I HA synthases are less suitable due to the fact that they are integral membrane proteins. Research with these HA synthases is mostly performed on membrane fractions but this system is less suitable for larger scale production. An alternative could be the immobilization of the enzyme in the cell wall of for instance yeast. Production of various human oligosaccharides has shown to be feasible with yeast cells in which glycosyl transferases are expressed and anchored in the cell wall glucan [108].

A far more promising enzyme for *in vitro* production of hyaluronan is the *P. multocida* Class II HA synthase. This enzyme is not an integral membrane protein and the analysis of the subcellular location and enzyme activity has shown that a recombinant truncated version of the protein lacking a 216 carboxy terminal amino acids retained HA synthase activity but was located in the cytoplasm when expressed in *E. coli* [109]. Furthermore, it was shown that two distinct transferase activities are located on the *P. multocida* HA synthase each of which can be inactivated by mutations in the DXD amino acid motif present in the active site while retaining the other transferase activity [110].

UDP-Recycling and Sugar Nucleotide Synthesis (Enzymatic Cascade, Bacterial Production). The general metabolic routes for incorporation of sugar units into glycosaminoglycans are their prior conversion into sugar nucleotides such as UDP-GlcA and UDP-GlcNAc for hyaluronan synthesis. The biosynthetic reactions are known as LeLoir pathways and these pathways can be different in microorganisms [111]. For example, the pathways for the synthesis of UDP-GlcNAc in eukaryotes and prokaryotes are different. Production of sugar nucleotides needs also cofactor regeneration for preparative applications, because the cofactors are too expensive to be used as stoichiometric agents. Nowadays, several cofactors can be effectively regenerated using enzymes, for example, NAD^+ by glutamate dehydrogenase with α-ketoglutarate, or whole cell based methods, for example, *Corynebacterium ammoniagenes* that converts orotic acid into UTP [112–115].

4. Future Perspectives

Hyaluronan, with its exceptional properties and multiple and diverse applications, has proved to be an exceptional material for medical, pharmaceutical, and cosmetic applications. Tremendous developments have been achieved in the past decades in all hyaluronan-related fields, covering the whole chain from the production of hyaluronan polymers and oligomers, to the development of advanced materials for clinical applications. Latest accomplishments in hyaluronan production and in particular the elucidation of the biosynthetic pathways in hyaluronan producing microorganisms opens new premises for the further optimization of biotechnological process for hyaluronan production with safe hosts. Further understanding of the *in vivo* regulation of hyaluronan molecular weight will benefit the biotechnological production of defined hyaluronan products with narrow polydispersity. We believe that bacterial fermentation using nonpathogenic, safe heterologous hosts will become the main source of high molecular weight hyaluronan, and metabolic engineering will be used to improve and control molecular weight. Stimulated by the developments of production techniques and the growing knowledge on the biological function of hyaluronan, research to enhancing existing medical products and to validating new concepts in medical therapies will be set forth.

Abbreviations

BSE:	Bovine spongiform encephalopathy
HA synthase or HAS:	Hyaluronan synthase
M_w:	Weight average molecular mass
M_n:	Number average molecular mass
GlcAU:	Glucuronic acid
GT:	Glycosyl transferase
UDP:	Uridine diphosphate
GlcNac:	*N*-acetyl glucosamine
GMO:	Genetically modified organism.

References

[1] P. L. DeAngelis, "Glycosaminoglycan polysaccharide biosynthesis and production: today and tomorrow," *Applied Microbiology and Biotechnology*, vol. 94, no. 2, pp. 295–305, 2012.

[2] H. G. Garg and C. A. Hales, *Chemistry and Biology of Hyaluronan*, Elsevier, 2004.

[3] K. Meyer and J. W. Palmer, "The polysaccharde of the vitreous humor," *The Journal of Biological Chemistry*, vol. 107, no. 3, pp. 629–634, 1934.

[4] B. Weissmann and K. Meyer, "The structure of hyalobiuronic acid and of hyaluronic acid from umbilical cord," *Journal of the American Chemical Society*, vol. 76, no. 7, pp. 1753–1757, 1954.

[5] E. A. Balazs, "Ultrapure hyaluronic acid and the use thereof," U.S. Patent, US4141973, 1979.

[6] P. L. DeAngelis, J. Papaconstantinou, and P. H. Weigel, "Molecular cloning, identification, and sequence of the hyaluronan synthase gene from group A Streptococcus pyogenes," *The Journal of Biological Chemistry*, vol. 268, no. 26, pp. 19181–19184, 1993.

[7] P. L. DeAngelis, J. Papaconstantinou, and P. H. Weigel, "Isolation of a Streptococcus pyogenes gene locus that directs hyaluronan biosynthesis in acapsular mutants and in heterologous bacteria," *The Journal of Biological Chemistry*, vol. 268, no. 20, pp. 14568–14571, 1993.

[8] G. Blatter and J. C. Jacquinet, "The use of 2-deoxy-2-trichloroacetamido-D-glucopyranose derivatives in syntheses of hyaluronic acid-related tetra-, hexa-, and octa-saccharides having a methyl β-D-glucopyranosiduronic acid at the reducing end," *Carbohydrate Research*, vol. 288, pp. 109–125, 1996.

[9] P. L. DeAngelis, L. C. Oatman, and D. F. Gay, "Rapid chemoenzymatic synthesis of monodisperse hyaluronan oligosaccharides with immobilized enzyme reactors," *The Journal of Biological Chemistry*, vol. 278, no. 37, pp. 35199–35203, 2003.

[10] F. K. Kooy, *Enzymatic Production of Hyaluronan Oligo- and Polysaccharides*, Wageningen University, Wageningen, The Netherlands, 2010.

[11] B. Porsch, R. Laga, and C. Konak, "Batch and size-exclusion chromatographic characterization of ultra-high molar mass sodium hyaluronate containing low amounts of strongly scattering impurities by dual low angle light scattering/refractometric detection," *Journal of Liquid Chromatography and Related Technologies*, vol. 31, no. 20, pp. 3077–3093, 2008.

[12] A. Shiedlin, R. Bigelow, W. Christopher et al., "Evaluation of hyaluronan from different sources: streptococcus zooepidemicus, rooster comb, bovine vitreous, and human umbilical cord," *Biomacromolecules*, vol. 5, no. 6, pp. 2122–2127, 2004.

[13] R. Mendichi and L. Soltes, "Hyaluronan molecular weight and polydispersity in some commercial intra-articular injectable preparations and in synovial fluid," *Inflammation Research*, vol. 51, no. 3, pp. 115–116, 2002.

[14] R. Stern, "Devising a pathway for hyaluronan catabolism: are we there yet?" *Glycobiology*, vol. 13, no. 12, pp. 105R–115R, 2003.

[15] M. Erickson and R. Stern, "Chain gangs: new aspects of hyaluronan metabolism," *Biochemistry Research International*, vol. 2012, Article ID 893947, 9 pages, 2012.

[16] T. C. Laurent and J. R. E. Fraser, "Hyaluronan," *The FASEB Journal*, vol. 6, no. 7, pp. 2397–2404, 1992.

[17] P. L. DeAngelis, "Hyaluronan synthases: Fascinating glycosyltransferases from vertebrates, bacterial pathogens, and algal viruses," *Cellular and Molecular Life Sciences*, vol. 56, no. 7-8, pp. 670–682, 1999.

[18] T. E. Hardingham and H. Muir, "The specific interaction of hyaluronic acid with cartilage proteoglycans," *Biochimica et Biophysica Acta*, vol. 279, no. 2, pp. 401–405, 1972.

[19] V. C. Hascall, "Hyaluronan, a common thread," *Glycoconjugate Journal*, vol. 17, no. 7–9, pp. 607–616, 2000.

[20] N. Afratis, C. Gialeli, D. Nikitovic et al., "Glycosaminoglycans: key players in cancer cell biology and treatment," *FEBS Journal*, vol. 279, no. 7, pp. 1177–1197, 2012.

[21] C. Schiraldi, A. La Gatta, and M. De Rosa, "Biotechnological production and application of hyaluronan," in *Biopolymers*, M. Elnashar, Ed., pp. 387–412, InTech, Rijeka, Croatia, 2010.

[22] F. Freitas, V. D. Alves, and M. A. M. Reis, "Advances in bacterial exopolysaccharides: from production to biotechnological applications," *Trends in Biotechnology*, vol. 29, no. 8, pp. 388–398, 2011.

[23] B. D. Ulery, L. S. Nair, and C. T. Laurencin, "Biomedical applications of biodegradable polymers," *Journal of Polymer Science B*, vol. 49, no. 12, pp. 832–864, 2011.

[24] R. Stern, A. A. Asari, and K. N. Sugahara, "Hyaluronan fragments: an information-rich system," *European Journal of Cell Biology*, vol. 85, no. 8, pp. 699–715, 2006.

[25] B. F. Chong, L. M. Blank, R. Mclaughlin, and L. K. Nielsen, "Microbial hyaluronic acid production," *Applied Microbiology and Biotechnology*, vol. 66, no. 4, pp. 341–351, 2005.

[26] S. Ghatak, S. Misra, and B. P. Toole, "Hyaluronan oligosaccharides inhibit anchorage-independent growth of tumor cells by suppressing the phosphoinositide 3-kinase/Akt cell survival pathway," *The Journal of Biological Chemistry*, vol. 277, no. 41, pp. 38013–38020, 2002.

[27] J. Y. Lee and A. P. Spicer, "Hyaluronan: a multifunctional, megaDalton, stealth molecule," *Current Opinion in Cell Biology*, vol. 12, no. 5, pp. 581–586, 2000.

[28] P. L. DeAngelis and P. H. Weigel, "Immunochemical confirmation of the primary structure of streptococcal hyaluronan synthase and synthesis of high molecular weight product by the recombinant enzyme," *Biochemistry*, vol. 33, no. 31, pp. 9033–9039, 1994.

[29] P. Prehm, "Hyaluronate is synthesized at plasma membranes," *Biochemical Journal*, vol. 220, no. 2, pp. 597–600, 1984.

[30] A. A. E. Chavaroche, L. A. M. van den Broek, and G. Eggink, "Production methods for heparosan, a precursor of heparin and heparan sulfate," *Carbohydrate Polymers*, vol. 93, no. 1, pp. 38–47, 2013.

[31] N. Itano, "Hyaluronan biosynthesis: a multifaceted process," *Connective Tissue*, vol. 33, no. 3, pp. 221–226, 2001.

[32] C. A. de la Motte and J. A. Drazba, "Viewing hyaluronan: imaging contributes to imagining new roles for this amazing matrix polymer," *Journal of Histochemistry and Cytochemistry*, vol. 59, no. 3, pp. 252–257, 2011.

[33] N. Itano and K. Kimata, "Molecular cloning of human hyaluronan synthase," *Biochemical and Biophysical Research Communications*, vol. 222, no. 3, pp. 816–820, 1996.

[34] A. P. Spicer, M. L. Augustine, and J. A. McDonald, "Molecular cloning and characterization of a putative mouse hyaluronan synthase," *The Journal of Biological Chemistry*, vol. 271, no. 38, pp. 23400–23406, 1996.

[35] N. Itano, T. Sawai, M. Yoshida et al., "Three isoforms of mammalian hyaluronan synthases have distinct enzymatic properties," *The Journal of Biological Chemistry*, vol. 274, no. 35, pp. 25085–25092, 1999.

[36] P. Heldin, "Molecular mechanisms that regulate hyaluronan synthesis," *Gene Therapy and Molecular Biology*, vol. 3, pp. 465–474, 1999.

[37] A. Jacobson, J. Brinck, M. J. Briskin, A. P. Spicer, and P. Heldin, "Expression of human hyaluronan syntheses in response to external stimuli," *Biochemical Journal*, vol. 348, no. 1, pp. 29–35, 2000.

[38] B. A. Dougherty and I. Van de Rijn, "Molecular characterization of hasB from an operon required for hyaluronic acid synthesis in group A streptococci. Demonstration of UDP-glucose dehydrogenase activity," *The Journal of Biological Chemistry*, vol. 268, no. 10, pp. 7118–7124, 1993.

[39] B. A. Dougherty and I. Van de Rijn, "Molecular characterization of hasA from an operon required for hyaluronic acid synthesis in group A streptococci," *The Journal of Biological Chemistry*, vol. 269, no. 1, pp. 169–175, 1994.

[40] D. L. Crater, B. A. Dougherty, and I. Van de Rijn, "Molecular characterization of hasC from an operon required for

hyaluronic acid synthesis in group A streptococci. Demonstration of UDP-glucose pyrophosphorylase activity," *The Journal of Biological Chemistry*, vol. 270, no. 48, pp. 28676–28680, 1995.

[41] A. Markovitz, J. A. Cifonelli, and A. Dorfman, "The biosynthesis of hyaluronic acid by group A Streptococcus. VI. Biosynthesis from uridine nucleotides in cell-free extracts," *The Journal of Biological Chemistry*, vol. 234, pp. 2343–2350, 1959.

[42] K. Kumari and P. H. Weigel, "Molecular cloning, expression, and characterization of the authentic hyaluronan synthase from Group C Streptococcus equisimilis," *The Journal of Biological Chemistry*, vol. 272, no. 51, pp. 32539–32546, 1997.

[43] P. L. DeAngelis and A. M. Achyuthan, "Yeast-derived recombinant DG42 protein of Xenopus can synthesize hyaluronan in vitro," *The Journal of Biological Chemistry*, vol. 271, no. 39, pp. 23657–23660, 1996.

[44] P. H. Weigel, V. C. Hascall, and M. Tammi, "Hyaluronan synthases," *The Journal of Biological Chemistry*, vol. 272, no. 22, pp. 13997–14000, 1997.

[45] P. L. DeAngelis, W. Jing, M. V. Graves, D. E. Burbank, and J. L. Van Etten, "Hyaluronan synthase of chlorella virus PBCV-1," *Science*, vol. 278, no. 5344, pp. 1800–1803, 1997.

[46] P. M. Coutinho, E. Deleury, G. J. Davies, and B. Henrissat, "An evolving hierarchical family classification for glycosyltransferases," *Journal of Molecular Biology*, vol. 328, no. 2, pp. 307–317, 2003.

[47] A. Jong, C. H. Wu, H. M. Chen et al., "Identification and characterization of CPS1 as a hyaluronic acid synthase contributing to the pathogenesis of Cryptococcus neoformans infection," *Eukaryotic Cell*, vol. 6, no. 8, pp. 1486–1496, 2007.

[48] P. L. DeAngelis, W. Jing, R. R. Drake, and A. M. Achyuthan, "Identification and molecular cloning of a unique hyaluronan synthase from Pasteurella multocida," *The Journal of Biological Chemistry*, vol. 273, no. 14, pp. 8454–8458, 1998.

[49] P. H. Weigel and P. L. DeAngelis, "Hyaluronan synthases: a decade-plus of novel glycosyltransferases," *The Journal of Biological Chemistry*, vol. 282, no. 51, pp. 36777–36781, 2007.

[50] S. Bodevin-Authelet, M. Kusche-Gullberg, P. E. Pummill, P. L. DeAngelis, and U. Lindahl, "Biosynthesis of hyaluronan: direction of chain elongation," *The Journal of Biological Chemistry*, vol. 280, no. 10, pp. 8813–8818, 2005.

[51] P. E. Pummill, T. A. Kane, E. S. Kempner, and P. L. DeAngelis, "The functional molecular mass of the Pasteurella hyaluronan synthase is a monomer," *Biochimica et Biophysica Acta*, vol. 1770, no. 2, pp. 286–290, 2007.

[52] P. L. DeAngelis, "Microbial glycosaminoglycan glycosyltransferases," *Glycobiology*, vol. 12, no. 1, pp. 9R–16R, 2002.

[53] K. Rilla, H. Siiskonen, A. P. Spicer, J. M. T. Hyttinen, M. I. Tammi, and R. H. Tammi, "Plasma membrane residence of hyaluronan synthase is coupled to its enzymatic activity," *The Journal of Biological Chemistry*, vol. 280, no. 36, pp. 31890–31897, 2005.

[54] M. Nardini, M. Ori, D. Vigetti, R. Gornati, I. Nardi, and R. Perris, "Regulated gene expression of hyaluronan synthases during Xenopus laevis development," *Gene Expression Patterns*, vol. 4, no. 3, pp. 303–308, 2004.

[55] J. Y. L. Tien and A. P. Spicer, "Three vertebrate hyaluronan synthases are expressed during mouse development in distinct spatial and temporal patterns," *Developmental Dynamics*, vol. 233, no. 1, pp. 130–141, 2005.

[56] M. Suzuki, T. Asplund, H. Yamashita, C. H. Heldin, and P. Heldin, "Stimulation of hyaluronan biosynthesis by platelet-derived growth factor-BB and transforming growth factor-β1 involves activation of protein kinase C," *Biochemical Journal*, vol. 307, no. 3, pp. 817–821, 1995.

[57] H. Chao and A. P. Spicer, "Natural antisense mRNAs to hyaluronan synthase 2 inhibit hyaluronan biosynthesis and cell proliferation," *The Journal of Biological Chemistry*, vol. 280, no. 30, pp. 27513–27522, 2005.

[58] D. Vigetti, A. Genasetti, E. Karousou et al., "Modulation of hyaluronan synthase activity in cellular membrane fractions," *The Journal of Biological Chemistry*, vol. 284, no. 44, pp. 30684–30694, 2009.

[59] L. M. Blank, P. Hugenholtz, and L. K. Nielsen, "Evolution of the hyaluronic acid synthesis (has) operon in Streptococcus zooepidemicus and other pathogenic streptococci," *Journal of Molecular Evolution*, vol. 67, no. 1, pp. 13–22, 2008.

[60] W. Y. Chen, E. Marcellin, J. Hung, and L. K. Nielsen, "Hyaluronan molecular weight is controlled by UDP-N-acetylglucosamine concentration in Streptococcus zooepidemicus," *The Journal of Biological Chemistry*, vol. 284, no. 27, pp. 18007–18014, 2009.

[61] L. Soltes, R. Mendichi, D. Lath, M. Mach, and D. Bakoš, "Molecular characteristics of some commercial high-molecular-weight hyaluronans," *Biomedical Chromatography*, vol. 16, no. 7, pp. 459–462, 2002.

[62] P. S. Harmon, E. P. Maziarz, and X. M. Liu, "Detailed characterization of hyaluronan using aqueous size exclusion chromatography with triple detection and multiangle light scattering detection," *Journal of Biomedical Materials Research B*, vol. 100, no. 7, pp. 1955–1960, 2012.

[63] E. U. Ignatova and A. N. Gurov, "Principles of extraction and purification of hyaluronic acid," *Khimiko-Farmatsevticheskii Zhurnal*, vol. 24, no. 3, pp. 42–46, 1990.

[64] I. Amagai, Y. Tashiro, and H. Ogawa, "Improvement of the extraction procedure for hyaluronan from fish eyeball and the molecular characterization," *Fisheries Science*, vol. 75, no. 3, pp. 805–810, 2009.

[65] M. O'Regan, I. Martini, F. Crescenzi, C. De Luca, and M. Lansing, "Molecular mechanisms and genetics of hyaluronan biosynthesis," *International Journal of Biological Macromolecules*, vol. 16, no. 6, pp. 283–286, 1994.

[66] G. Kogan, L. Šoltés, R. Stern, and P. Gemeiner, "Hyaluronic acid: a natural biopolymer with a broad range of biomedical and industrial applications," *Biotechnology Letters*, vol. 29, no. 1, pp. 17–25, 2007.

[67] J. C. Thonard, S. A. Migliore, and R. Blustein, "Isolation of hyaluronic acid from broth cultures of Streptococci," *The Journal of Biological Chemistry*, vol. 239, pp. 726–728, 1964.

[68] F. E. Kendall, M. Heidelberger, and M. H. Dawson, "A serologically inactive polysaccharide elaborated by mocoid strains of group A hemolytic Streptococcus," *The Journal of Biological Chemistry*, vol. 118, no. 1, pp. 61–69, 1937.

[69] B. Holmstrom and J. Ricica, "Production of hyaluronic acid by a Streptococcal strain in batch culture," *Applied Environmental Microbiology*, vol. 15, no. 6, pp. 1409–1413, 1967.

[70] J. H. Kim, S. J. Yoo, D. K. Oh et al., "Selection of a Streptococcus equi mutant and optimization of culture conditions for the production of high molecular weight hyaluronic acid," *Enzyme and Microbial Technology*, vol. 19, no. 6, pp. 440–445, 1996.

[71] N. Izawa, T. Hanamizu, R. Iizuka et al., "Streptococcus thermophilus produces exopolysaccharides including hyaluronic acid," *Journal of Bioscience and Bioengineering*, vol. 107, no. 2, pp. 119–123, 2009.

[72] N. Izawa, M. Serata, T. Sone, T. Omasa, and H. Ohtake, "Hyaluronic acid production by recombinant Streptococcus thermophilus," *Journal of Bioscience and Bioengineering*, vol. 111, no. 6, pp. 665–670, 2011.

[73] H. J. Gao, G. C. Du, and J. Chen, "Analysis of metabolic fluxes for hyaluronic acid (HA) production by Streptococcus zooepidemicus," *World Journal of Microbiology and Biotechnology*, vol. 22, no. 4, pp. 399–408, 2006.

[74] X. J. Duan, H. X. Niu, W. S. Tan, and X. Zhang, "Mechanism analysis of effect of oxygen on molecular weight of hyaluronic acid produced by Streptococcus zooepidemicus," *Journal of Microbiology and Biotechnology*, vol. 19, no. 3, pp. 299–306, 2009.

[75] J. Zhang, X. Ding, L. Yang, and Z. Kong, "A serum-free medium for colony growth and hyaluronic acid production by Streptococcus zooepidemicus NJUST01," *Applied Microbiology and Biotechnology*, vol. 72, no. 1, pp. 168–172, 2006.

[76] J. H. Im, J. M. Song, J. H. Kang, and D. J. Kang, "Optimization of medium components for high-molecular-weight hyaluronic acid production by Streptococcus sp. ID9102 via a statistical approach," *Journal of Industrial Microbiology and Biotechnology*, vol. 36, no. 11, pp. 1337–1344, 2009.

[77] D. C. Armstrong, M. J. Cooney, and M. R. Johns, "Growth and amino acid requirements of hyaluronic-acid producing Streptococcus zooepidemicus," *Applied Microbiology and Biotechnology*, vol. 47, no. 3, pp. 309–312, 1997.

[78] M. J. Cooney, L. T. Goh, P. L. Lee, and M. R. Johns, "Structured model-based analysis and control of the hyaluronic acid fermentation by Streptococcus zooepidemicus: physiological implications of glucose and complex-nitrogen-limited growth," *Biotechnology Progress*, vol. 15, no. 5, pp. 898–910, 1999.

[79] W. C. Huang, S. J. Chen, and T. L. Chen, "The role of dissolved oxygen and function of agitation in hyaluronic acid fermentation," *Biochemical Engineering Journal*, vol. 32, no. 3, pp. 239–243, 2006.

[80] B. F. Chong and L. K. Nielsen, "Aerobic cultivation of Streptococcus zooepidemicus and the role of NADH oxidase," *Biochemical Engineering Journal*, vol. 16, no. 2, pp. 153–162, 2003.

[81] B. F. Chong and L. K. Nielsen, "Amplifying the cellular reduction potential of Streptococcus zooepidemicus," *Journal of Biotechnology*, vol. 100, no. 1, pp. 33–41, 2003.

[82] T. F. Wu, W. C. Huang, Y. C. Chen, Y. G. Tsay, and C. S. Chang, "Proteomic investigation of the impact of oxygen on the protein profiles of hyaluronic acid-producing Streptococcus zooepidemicus," *Proteomics*, vol. 9, no. 19, pp. 4507–4518, 2009.

[83] M. R. Johns, L. T. Goh, and A. Oeggerli, "Effect of pH, agitation and aeration on hyaluronic acid production by Streptococcus zooepidemicus," *Biotechnology Letters*, vol. 16, no. 5, pp. 507–512, 1994.

[84] S. Hasegawa, M. Nagatsuru, M. Shibutani, S. Yamamoto, and S. Hasebe, "Productivity of concentrated hyaluronic acid using a Maxblend (*R*) fermentor," *Journal of Bioscience and Bioengineering*, vol. 88, no. 1, pp. 68–71, 1999.

[85] X. Zhang, X. J. Duan, and W. S. Tan, "Mechanism for the effect of agitation on the molecular weight of hyaluronic acid produced by Streptococcus zooepidemicus," *Food Chemistry*, vol. 119, no. 4, pp. 1643–1646, 2010.

[86] M. Cazzola, F. O'regan, and V. Corsa, "Culture medium and process for the preparation of high molecular weight hyaluronic acid," Fidia Advanced Biopolymers S.R.L., 2003.

[87] I. Van De Rijn, "Streptococcal hyaluronic acid: Proposed mechanisms of degradation and loss of synthesis during stationary phase," *Journal of Bacteriology*, vol. 156, no. 3, pp. 1059–1065, 1983.

[88] A. Mausolf, J. Jungmann, H. Robenek, and P. Prehm, "Shedding of hyaluronate synthase from streptococci," *Biochemical Journal*, vol. 267, no. 1, pp. 191–196, 1990.

[89] D. C. Ellwood, C. G. T. Evans, G. M. Dunn et al., "Production of hyaluronic acid," Fermentech Medical Limited, 1996.

[90] W. C. Huang, S. J. Chen, and T. L. Chen, "Production of hyaluronic acid by repeated batch fermentation," *Biochemical Engineering Journal*, vol. 40, no. 3, pp. 460–464, 2008.

[91] L. M. Blank, R. L. McLaughlin, and L. K. Nielsen, "Stable production of hyaluronic acid in streptococcus zooepidemicus chemostats operated at high dilution rate," *Biotechnology and Bioengineering*, vol. 90, no. 6, pp. 685–693, 2005.

[92] H. Y. Han, S. H. Jang, E. C. Kim et al., "Microorganism producing hyaluronic acid and purification method of hyaluronic acid," p. 26, 2004.

[93] S. Stahl, "Methods and means for the production of hyaluronic acid," US6090596, 2000.

[94] E. Marcellin, W. Y. Chen, and L. K. Nielsen, "Understanding plasmid effect on hyaluronic acid molecular weight produced by Streptococcus equi subsp. zooepidemicus," *Metabolic Engineering*, vol. 12, no. 1, pp. 62–69, 2010.

[95] J. Z. Sheng, P. X. Ling, X. Q. Zhu et al., "Use of induction promoters to regulate hyaluronan synthase and UDP-glucose-6-dehydrogenase of Streptococcus zooepidemicus expression in Lactococcus lactis: a case study of the regulation mechanism of hyaluronic acid polymer," *Journal of Applied Microbiology*, vol. 107, no. 1, pp. 136–144, 2009.

[96] H. Yu and G. Stephanopoulos, "Metabolic engineering of Escherichia coli for biosynthesis of hyaluronic acid," *Metabolic Engineering*, vol. 10, no. 1, pp. 24–32, 2008.

[97] Z. Mao, H. D. Shin, and R. Chen, "A recombinant *E. coli* bioprocess for hyaluronan synthesis," *Applied Microbiology and Biotechnology*, vol. 84, no. 1, pp. 63–69, 2009.

[98] B. Widner, R. Behr, S. Von Dollen et al., "Hyaluronic acid production in Bacillus subtilis," *Applied and Environmental Microbiology*, vol. 71, no. 7, pp. 3747–3752, 2005.

[99] Z. Mao and R. R. Chen, "Recombinant synthesis of hyaluronan by Agrobacterium sp," *Biotechnology Progress*, vol. 23, no. 5, pp. 1038–1042, 2007.

[100] S. B. Prasad, G. Jayaraman, and K. B. Ramachandran, "Hyaluronic acid production is enhanced by the additional co-expression of UDP-glucose pyrophosphorylase in Lactococcus lactis," *Applied Microbiology and Biotechnology*, vol. 86, no. 1, pp. 273–283, 2010.

[101] L. J. Chien and C. K. Lee, "Hyaluronic acid production by recombinant Lactococcus lactis," *Applied Microbiology and Biotechnology*, vol. 77, no. 2, pp. 339–346, 2007.

[102] S. B. Prasad, K. B. Ramachandran, and G. Jayaraman, "Transcription analysis of hyaluronan biosynthesis genes in Streptococcus zooepidemicus and metabolically engineered Lactococcus lactis," *Applied Microbiology and Biotechnology*, vol. 94, no. 6, pp. 1593–1607, 2012.

[103] A. Sloma et al., "Recombinant expression of bacterial hyaluronan synthase operon genes in Bacillus and hyaluronic acid production," p. 218, 2003.

[104] M. V. Graves, D. E. Burbank, R. Roth, J. Heuser, P. L. Deangelis, and J. L. Van Etten, "Hyaluronan synthesis in virus PBCV-1-infected chlorella-like green algae," *Virology*, vol. 257, no. 1, pp. 15–23, 1999.

[105] C. De Luca, M. Lansing, I. Martini et al., "Enzymatic synthesis of hyaluronic acid with regeneration of sugar nucleotides," *Journal of the American Chemical Society*, vol. 117, no. 21, pp. 5869–5870, 1995.
[106] P. L. DeAngelis, "Monodisperse hyaluronan polymers: synthesis and potential applications," *Current Pharmaceutical Biotechnology*, vol. 9, no. 4, pp. 246–248, 2008.
[107] F. K. Kooy, M. Ma, H. H. Beeftink, G. Eggink, J. Tramper, and C. G. Boeriu, "Quantification and characterization of enzymatically produced hyaluronan with fluorophore-assisted carbohydrate electrophoresis," *Analytical Biochemistry*, vol. 384, no. 2, pp. 329–336, 2009.
[108] Y. I. Shimma, F. Saito, F. Oosawa, and Y. Jigami, "Construction of a library of human glycosyltransferases immobilized in the cell wall of Saccharomyces cerevisiae," *Applied and Environmental Microbiology*, vol. 72, no. 11, pp. 7003–7012, 2006.
[109] W. Jing and P. L. De Angelis, "Dissection of the two transferase activities of the Pasteurella multocida hyaluronan synthase: two active sites exist in one polypeptide," *Glycobiology*, vol. 10, no. 9, pp. 883–889, 2000.
[110] W. Jing and P. L. DeAngelis, "Analysis of the two active sites of the hyaluronan synthase and the chondroitin synthase of Pasteurella multocida," *Glycobiology*, vol. 13, no. 10, pp. 661–671, 2003.
[111] S. Milewski, I. Gabriel, and J. Olchowy, "Enzymes of UDP-GlcNAc biosynthesis in yeast," *Yeast*, vol. 23, no. 1, pp. 1–14, 2006.
[112] H. Zhao and W. A. Van Der Donk, "Regeneration of cofactors for use in biocatalysis," *Current Opinion in Biotechnology*, vol. 14, no. 6, pp. 583–589, 2003.
[113] A. Ruffing and R. R. Chen, "Metabolic engineering of microbes for oligosaccharide and polysaccharide synthesis," *Microbial Cell Factories*, vol. 5, p. 25, 2006.
[114] W. Liu and P. Wang, "Cofactor regeneration for sustainable enzymatic biosynthesis," *Biotechnology Advances*, vol. 25, no. 4, pp. 369–384, 2007.
[115] C. A. G. M. Weijers, M. C. R. Franssen, and G. M. Visser, "Glycosyltransferase-catalyzed synthesis of bioactive oligosaccharides," *Biotechnology Advances*, vol. 26, no. 5, pp. 436–456, 2008.

16

Identification and Quantification of Protein Glycosylation

Ziv Roth, Galit Yehezkel, and Isam Khalaila

Avram and Stella Goldstein-Goren Department of Biotechnology Engineering, Ben-Gurion University of the Negev, P.O. Box 653, 84105 Beer Sheva, Israel

Correspondence should be addressed to Isam Khalaila, isam@bgu.ac.il

Academic Editor: Margarita Stoytcheva

Glycosylation is one of the most abundant posttranslation modifications of proteins, and accumulating evidence indicate that the vast majority of proteins in eukaryotes are glycosylated. Glycosylation plays a role in protein folding, interaction, stability, and mobility, as well as in signal transduction. Thus, by regulating protein activity, glycosylation is involved in the normal functioning of the cell and in the development of diseases. Indeed, in the past few decades there has been a growing realization of the importance of protein glycosylation, as aberrant glycosylation has been implicated in metabolic, neurodegenerative, and neoplastic diseases. Thus, the identification and quantification of protein-borne oligosaccharides have become increasingly important both in the basic sciences of biochemistry and glycobiology and in the applicative sciences, particularly biomedicine and biotechnology. Here, we review the state-of-the-art methodologies for the identification and quantification of oligosaccharides, specifically *N*- and *O*-glycosylated proteins.

1. Protein Glycosylation

Carbohydrates are essential for cell metabolism and energy production and are building blocks of the extracellular matrix. Carbohydrates also play a role in a variety of biological processes, such as intra- and intercellular signaling and mediation of cell-cell interactions. In plants and animals alike, they may be found in various forms and structures—monosaccharides, oligosaccharides, or long and complex polysaccharides. Saccharides may be present in their free forms or they may be linked to other molecules, such as proteins and lipids. In proteoglycans the saccharide is the dominant component, whereas in glycoproteins, the saccharide constitutes only a small portion of the entire molecule.

Protein glycosylation, that is, the attachment of a saccharide moiety to a protein, is a modification that occurs either cotranslationally or posttranslationally. The two major types of glycosylation, *N*-linked and *O*-linked, are both involved in the maintenance of protein conformation and activity, in protein protection from proteolytic degradation, and in protein intracellular trafficking and secretion [1]. *N*-glycan moieties also play a key role in the folding, processing, and secretion of proteins from the endoplasmic reticulum (ER) and the Golgi apparatus [1]. In the past few decades, growing attention has been directed to protein glycosylation in the biomedical field, since aberrant glycosylations have been associated with various congenital [2], metabolic, neurodegenerative [3], and immune diseases [4] and with cancer [5]. Thus, the study of glycans and their localization and structural pattern will shed light on their role in protein regulation and function, both under normal conditions and in pathologies. In biotechnology, too, there is a need for an indepth understanding of glycosylation, since the efficient production of recombinant proteins, including glycoproteins [6], is becoming increasingly important. Thus, there is a great deal of interest in the use and development of tools for glycobiology research. This paper is focused on protein glycosylation and on state-of-the-art methods for the identification and quantification of glycoproteins.

1.1. N-Glycans. In proteins, the *N*-linked oligosaccharide is attached to the amide group of an asparagine residue within the consensus peptide sequence NXS/T, where X is any amino acid except proline. *N*-glycosylation occurs solely on proteins that shuttle via the secretory pathway [7]. The assembly of the *N*-glycan oligosaccharide begins in the cytosolic side of the ER and then it flips to the lumen side, where the assembly is completed, and attaches to the translated protein. The

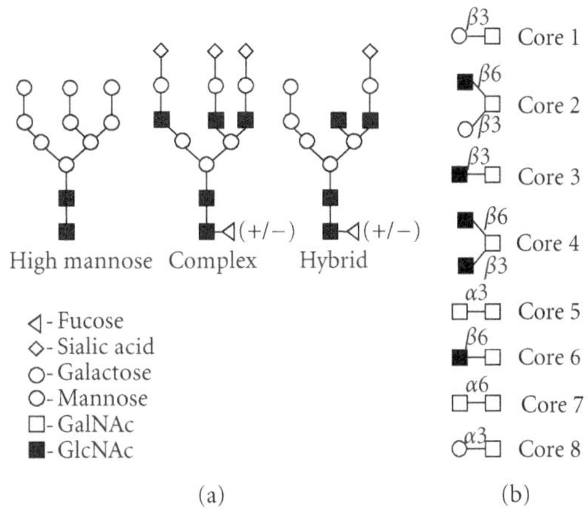

FIGURE 1: Basic core structures of *N*-glycans (a) and *O*-glycans (b).

attached oligosaccharide-protein complex has the form of $Glc_3Man_9GlcNAc_2$ in the ER, but following the correct folding of the protein, it is transferred to the Golgi apparatus as $Man_8GlcNAc_2$ [8]. In the Golgi apparatus, the glycan moiety can remain in the high-mannose form or it can be processed further—with some glycans being removed and a variety of other glycans added—to form a hybrid and complex structure. This glycan "tree" may be bi-, tri- or tetraantennary, and it may also be bisected by a GlcNAc attachment to the core mannose (Figure 1(a)).

1.2. O-Glycans. The most common *O*-glycosylations are *O*-acetylgalactosamine (*O*-GalNAc) and *O*-acetylglucosamine (*O*-GlcNAc). *O*-GalNAc is attached to the hydroxyl group of the protein serine or threonine residues through an α-linkage, while *O*-GlcNAc is attached through a β-linkage. Unlike *N*-linked glycosylation, no consensus sequence has been found for *O*-linked glycosylation [9], and it could occur at various intracellular sites. Structurally, *O*-glycosylation may involve monosaccharides and/or oligosaccharide complexes of several different monosaccharides [9]. *O*-GalNAcylation, which occurs in the Golgi and hence on secreted proteins, is a common *O*-glycosylation, with over 10% of human proteins being *O*-GalNAcylated. There are eight types of core *O*-GalNAc structures (Figure 1(b)), which can be further elongated. This is often a complex process, since the same protein can carry different *O*-glycan structures. The most abundant *O*-GalNAcylated proteins are mucins, the heavily glycosylated mucus proteins that are secreted mainly from the epithelium of the gastrointestinal and respiratory tracts and also from the salivary glands [9]. A number of congenital, developmental, and immune diseases have been attributed to mucin type *O*-glycan alterations [10].

Unlike *N*- and *O*-α-glycosylation, β-*O*-GlcNAcylation, which occurs on cytosolic and nuclear proteins, is a dynamic modification resembling protein phosphorylation [11]. Moreover, there is a dynamic crosstalk between *O*-GlcNAcylation and phosphorylation: the two modifications can reciprocally occur on the same Ser/Thr residue or occupy adjacent residues [12–14]. Since the discovery of *O*-GlcNAc [15], a myriad of *O*-GlcNAcylated cytosolic and nuclear proteins have been found in eukaryotes, including transcription and translation factors, metabolic and signaling components, cytoskeletal proteins, oncogenic products, and tumor suppressor proteins [12, 16–21]. Protein *O*-GlcNAcylation is known to be vital for cell life and to be linked to regulation of the cell cycle and cell growth, proliferation, and apoptosis.

2. Analysis of Protein Glycosylation

The analysis of protein glycans is complicated by their vast variety and the large number of potential glycosylation combinations: even a single protein can undergo a number of *N*- and *O*-glycosylations. Moreover, the same glycosylation site can be occupied by different glycans in different copies of a protein, a phenomenon termed microheterogeneity. Thus, reliable, detailed characterization of glycans often requires the use of several methods, as described below.

2.1. Glycoprotein Detection. For detection of glycosylated proteins, two types of methods are generally used: staining and affinity-based methods.

2.1.1. Staining Procedures. A basic, simple method to determine whether a protein is glycosylated is to resolve it on SDS-PAGE and to stain the gel for glycoproteins. Most gel-staining procedures are based on the periodic acid-Schiff (PAS) reaction, in which periodic acid oxidizes two vicinal diol groups to form an aldehyde, which reacts with the Schiff reagent to give a magenta color [22]. This chromogenic gel staining is most commonly performed with acid fuchsin [23], which can be detected fluorescently at 535 nm and is thus two- to fourfold more sensitive than the visible staining method [24]. Some commercially available fluorescent stains utilize periodate oxidation to attach a fluorescent hydrazide, with no further reduction being required. These fluorescent dyes are highly sensitive, detecting <18 ng of glycoproteins [25]. However, these periodate oxidation methods may be problematic, since nonspecific reactions can occur with proteins that possess endogenous oxidizable moieties, for example, aldehydes or ketones. In addition, the staining intensity is lower for proteins with fewer glycosylation sites and "poorer" glycan trees than for heavily glycosylated proteins.

Two other gel stains that should be mentioned are alcian blue and Stains-All [26, 27], which are used for detecting proteoglycans, glycosaminoglycans, and negatively charged glycoproteins. However, one should bear in mind that for these stains, too, non-specific reactions with phosphoproteins and other negatively charged proteins are also possible.

2.1.2. Affinity-Based Procedures. The following methods are more specific and facilitate the determination of the glycosylation type.

(a) Saccharide-Binding Proteins. Lectins are proteins (found in animals, plants, and microorganisms) that specifically bind mono- or oligosaccharides [28]. Over the years, numerous lectin-based applications for glycoprotein detection and characterization have been developed (Table 1). In the widely used lectin blot, a protein sample is resolved on SDS-PAGE and transferred onto a nitrocellulose or polyvinylidene fluoride (PVDF) membrane. The membrane is then incubated with a specific lectin and labeled with a group, such as digoxigenin (DIG) or biotin that will further bind to a secondary antibody or to avidin, respectively [29, 30]. Several lectins are commercially available from companies such as Roche Applied Science and Sigma. The antibody or avidin is conjugated to an enzyme that catalyzes a color-producing reaction (alkaline-phosphatase) or a more sensitive luminescence-producing reaction (horseradish peroxidase).

(b) Enzyme-Based Methods. Click chemistry, utilizing specific enzymes, may be used to label and isolate glycosylated proteins. For example, a protein that is *O*-β-GlcNAcylated or that carries *O*-α- and/or *N*-glycans with a terminal GlcNAc (either purified or in a protein mixture) is incubated with GALT (β-1,4-galactosyltransferase), which specifically adds an azidogalactose to GlcNAc. The modified *O*-GlcNAc is then allowed to react with a fluorescent alkyne, the sample is resolved on SDS-PAGE, and the modified protein is visualized by UV, using a gel imager [24]. Similarly, *O*-GlcNAcylated peptides can be labeled and further analyzed by mass spectrometry (MS).

(c) Antibody-Based Methods. *O*-β-GlcNAc may be analyzed in a method that is specific to this unique glycan moiety, since primary antibodies, for example, commercially available CTD 110.1 and RL2, have been developed against it. The cross-reaction of these antibodies with secondary antibodies conjugated to enzymes that catalyze a color- or luminescence-producing reaction can be exploited in procedures such as blotting to detect *O*-GlcNAcylated proteins [31–33].

2.2. Glycan Structure Analysis. After protein glycosylation has been confirmed, the glycan moiety structure can be elucidated by chromatography and mass spectrometry. The glycan could be analyzed when attached to the protein or following its release. We will focus on the latter approach, since it is more reliable.

2.2.1. Glycan Release from Proteins. Depending on its nature, the glycan moiety can be removed from the protein by enzymatic or chemical means.

(a) Enzymatic Cleavage of Glycans. Most of the commercially available glycan-cleaving enzymes are specific for *N*-glycans, with fewer being available for *O*-glycans. The large variety of *O*-glycans dictates that several enzymes might be required for the analysis of a single sample, and their utility is thus limited. Thus, chemical removal is more applicable for *O*-glycans.

The most efficient cleaving enzyme for *N*-glycan moiety would be a glycoamidase, particularly glycoamidase F (PNGase F), which cleaves the bond between GlcNAc and an Asn residue, thereby converting Asn to Asp. PNGase F cleaves most types of *N*-linked oligosaccharides, including complex structures (e.g., high mannose, hybrid, and multisialylated structure), up to tetraantennary trees and oligosaccharides containing sulfate-substituted residues [34]. However, for native proteins high concentrations of PNGase F may be required. If the oligosaccharide contains α-1,3 core fucosylation or an individual *N*-acetylglucosamine attached to asparagine, PNGase F will not cleave the *N*-glycan moiety, and in that case another glycoamidase, PNGase A, should be used [35]. However, the use of PNGase A may be problematic for glycan cleavage from intact protein or large peptides, probably due to its high molecular weight (80 kDa) that limits its accessibility to the cleavage site [34]. In addition, PNGase F and A activity is reduced when the *N*-glycan is close to the protein N- or C-terminus. Thus, for native glycoproteins or known glycan moieties, endoglycosidases that cleave the *N*-glycan tree between the two core GlcNAcs (GlcNAc1-4GlcNAc) are preferred. It should be remembered that endoglycosidases are rather specific: endo-H cleaves most hybrid-type oligosaccharides and high-mannose, core-fucose, and bisecting glycans; endo F_1 cleaves high-mannose and hybrid-type glycans [36]; endo F_2 cleaves high-mannose and biantennary glycans; endo F_3 cleaves bi- and tri-antennary glycans [34].

(b) Chemical Removal of Glycans. The two main chemical methods for removing *O*- and *N*-linked oligosaccharides are β-elimination and hydrazinolysis. In β-elimination, exposure of a glycoprotein to an alkaline environment (e.g., of sodium hydroxide) results in glycan release. The free glycan moiety can be further degraded into monosaccharides by additional β-elimination reactions, until complete degradation—termed a "peeling reaction." Yet, glycan degradation can be prevented by performing the reaction in the presence of sodium borohydride, which reduces the remaining aldehydes [37]. Although efficient, this method has a major disadvantage in that oligosaccharides possess only one group available for labeling, the reduced end aldehyde. This group is eliminated during the reaction, thus labeling following glycan release is not possible.

Table 1: Lectins and their specific affinity to glycans.

Lectin	Origin	Specific affinity
Mannose binding lectins		
Con A	*Canavalia ensiformis*	Branched α-mannosidic structures; high-mannose type, hybrid type and biantennary complex type *N*-glycans
GNA	*Galanthus nivalis*	α-1–3 and α-1–6-linked high-mannose structures
LCH	*Lens culinaris*	Fucosylated core region of bi- and triantennary complex type *N*-glycans
Galactose/*N*-acetylgalactosamine binding lectins		
RCA	*Ricinus communis*	Galβ1-4GlcNAcβ1-R
ECL	*Erythrina cristagalli lectin*	Galβ1-4GlcNAcβ1-R
PNA	*Arachis hypogaea*	Galβ1-3GalNAcα1-Ser/Thr (T-Antigen)
AIL	*Artocarpus integrifolia*	(Sia)Galβ1-3GalNAcα1-Ser/Thr (T-Antigen)
VVL	*Vicia villosa*	GalNAcα-Ser/Thr (Tn-Antigen)
Sialic acid/*N*-acetylglucosamine binding lectins		
WGA	*Triticum vulgaris*	GlcNAcβ1-4GlcNAcβ1-4GlcNAc, Neu5Ac (sialic acid)
PHA	*Phaseolus vulgaris*	*N*-Acetylglucosamine
SNA	*Sambucus nigra*	Neu5Acα2-6Gal(NAc)-R
MAL	*Maackia amurensis*	Neu5Ac/Gcα2-3Galβ1-4GlcNAcβ1-R
Fucose binding lectins		
UEA	*Ulex europaeus*	Fucα1-2Gal-R
AAL	*Aleuria aurantia*	Fucα1-2Galβ1-4(Fucα1-3/4)Galβ1-4GlcNAc; R_2-GlcNAcβ1-4(Fucα1-6)GlcNAc-R_1

In hydrazinolysis, anhydrous hydrazine is added to a salt-free, lyophilized glycoprotein to start the hydrolysis reaction. This method may be used for both *N*- and *O*-linked carbohydrates, which can be even released separately when mixed together, by altering the reaction conditions, specifically the temperature; *O*-linked oligosaccharides are released at 60°C, while *N*-linked oligosaccharides are released at 95°C [38]. The released glycan moiety remains intact, whereas the protein may be degraded.

2.2.2. Glycan Desalting and Purification. Following their release from the glycoprotein, free glycans are usually found in a solution that contains salts, detergents, proteins, peptides, amino acids, and so forth. Before further analysis, these contaminants must be removed. There are several methods for glycan desalting and purification. The graphitized carbon desalting method is based on the ability of oligosaccharides to bind to carbon beads, while unbound simple monosaccharides, salts, and detergents can be washed away with water [39]. The glycan moiety is eluted from the beads with acetonitrile (ACN), while strongly bound proteins and peptides remain attached. Polar oligosaccharides are usually eluted by 25% ACN, while charged oligosaccharides require the addition of 0.05% trifluoroacetic acid (TFA) [39]. In another desalting method, a drop of glycan sample is loaded onto a cation-exchange Nafion membrane, which is floated on water. Proteins, peptides, and salts within the sample will bind to the Nafion, and the purified glycan sample can be retrieved [40]. Similarly to the method with the Nafion membrane, the cation-exchange AG-50 resin can be used. Salts and some detergents may also be removed by dialysis [41].

2.2.3. Glycan Structure Analysis. Following purification, two approaches may be taken for glycan structure analysis: chromatography and MS.

(a) Glycan Analysis by Chromatography. In chromatography, fluorescent labeling of the glycan moiety usually improves the detection. The fluorescent molecule is conjugated to the glycan reducing end via reductive amination [42]. The most commonly used fluorophores are 2-aminobenzoic acid (2-AA), 2-aminobenzamide (2-AB), and 2-aminopyridine (2-AP). However, aminopyridine-based fluorophores can hinder specific glycan characterization, since their conjugation may cause desialylation [42], whereas the other two fluorophores, 2-AA and 2-AB, may cause only mild desialylation. The negative charge of 2-AA makes it less appropriate for chromatography and MS, but more suitable for electrophoresis [42]. The 2-AB fluorophore labels glycans nonselectively in a 1 : 1 ratio, allowing subpicomolar detection, while maintaining the right molar proportions [43].

The most widely used chromatographic methods for glycan separation are weak anion exchange (WAX) [44], gel filtration (on Bio-Gel P4) (Ashford,1987), high-performance anion-exchange chromatography with pulsed amperometric detection (HPAEC-PAD) [45], and normal-phase high-performance liquid chromatography (NP-HPLC) [43]. Although all methods are efficient, one should carefully choose the appropriate method in the light of the specific requirements and the specific properties of the particular glycans being analyzed. WAX and gel filtration chromatography require large amounts of oligosaccharides. HPAEC-PAD requires high-pH and high-salt buffers (which must be removed before further analysis of the separated glycans, such as in the case of exoglycosidase sequencing). NP-HPLC, having high resolution and reproducibility, is the most efficient method for glycan analysis. In this method, a partially hydrolyzed 2-AB-labeled dextran is used as an external standard, and the retention time of the unknown glycan (relative to the standard) is converted to glucose units (GU) [43]. This GU value is compared to a database of experimental values to obtain a preliminary structure assignment for the glycan [46, 47]. Yet, the NP-HPLC resolution may be too low for some glycans; thus, a complementary method, such as reverse-phase-HPLC (RP-HPLC), is required [48]. Since the dextran standard does not resolve well on RP-HPLC, the alternate standard is arabinose [48].

Further validation of the glycan sequence can be performed with an array of exoglycosidases, that is, the sequential application of specific exoglycosidases to cleave terminal monosaccharides from the nonreducing end of the glycoprotein [49]. This chromatographic method also allows relative glycan quantification: by calculating the peak area one can assess the percentage of a specific type of oligosaccharide out of the total glycan repertoire. Furthermore, relative quantification of the same oligosaccharide in several samples can be achieved [43].

(b) Glycan Analysis by MS. MS analysis is based on glycan ionization, fragmentation, and mass identification of the fragments. The two main MS methods for oligosaccharide analysis are electrospray ionization (ESI) and matrix-assisted laser desorption/ionization (MALDI). Before analysis, the glycan sample must be desalted to prevent ionization masking in ESI and MALDI analysis. In MALDI analysis, the analyte is ionized through the matrix, usually to a single sodiated ion, $[M+Na]^+$. Several matrices are applicable for glycans, with 2,5-dihydroxybenzoic acid (DHB) being the most widely used. Yet, in the light of the variety of glycans and their different polarities, 2,4,6-trihydroxyacetophenone (THAP) is a more suitable matrix [50]. In ESI and nanospray, the analyte in solution is converted to an aerosol by the electrospray. Here the ions are usually multiply charged, which may complicate the glycan analysis [51].

In MS, the ionization polarity mode should be considered. Although most proteomics and glycomics analyses are performed in a positive-ion mode, the glycan moieties composition is diverse, and if the particular moiety contains *N*-acetylation and acidic residues, such as sialic acid, ionization may be prevented. In that case, a negative-ion mode could be employed, but further MS/MS and fragmentation analysis (performed in a positive mode) is usually not applicable. The above notwithstanding, positive ionization can be improved by adopting one of the following two strategies. The glycan moiety can be desialylated by sialidase that cleaves the terminal sialic acid [50]. Although desialylation enables better ionization, some information is lost: in a pool of released oligosaccharides it is practically impossible to identify which oligosaccharide possessed the sialic acid. Alternatively, permethylation, in which all the glycan-free OH groups are methylated [52, 53], will stabilize oligosaccharides containing sialic acid residues. Moreover, permethylation masks highly polar groups and confers a slight hydrophobicity; thus, the oligosaccharides are more easily separated from contaminants (e.g., salts) that may interfere with the analysis, and they also become more uniformly and efficiently ionized [54]. Indeed, fragment ions obtained from permethylated derivatives are easily assignable to glycan sequences [55].

Glycan fragmentation is achieved by cleaving either the glycosidic bonds between monosaccharides or the bonds within a monosaccharide ring (termed cross-ring cleavage). The ions obtained from the reducing side are designated X in cross-ring cleavage and Y and Z in glycosidic cleavage. The nonreducing side ions are designated A in cross-ring cleavage and B and C in glycosidic cleavage (Figure 2) [41]. In collision-induced dissociation (CID) fragmentation, ions enter a collision cell filled with a gas (usually argon) and are subjected to high- or low-energy collisions. When performed with high-collision energy, CID results mostly in cross-ring cleavages, providing data on the monosaccharide linkages, while low-collision energy yields glycosidic cleavages, providing data on the monosaccharide sequence and branching [56]. Recently, high-energy C-trap dissociation (HCD) fragmentation was applied for glycan and glycoprotein analysis [57]. The resolution of this method in the low-mass spectrum is higher than that of CID, enabling the identification of the monosaccharide masses. Thus, all the fragment ions, obtained by a combination of techniques, are assigned, providing the full glycan sequence [57, 58]. For detailed structural characterization, multiple ion isolation and fragmentation cycles are needed. For that ion-trap analysis with tandem MS fragmentation is the method of choice as described by Prien et al. [59].

Although MALDI and ESI spectra can be used for the elucidation of glycan composition and structure, full glycan characterization by MS is rather complicated: glycan quantity limitations are common, and more importantly, the isobaric structure of many monosaccharides (i.e., having the same molecular weights) hinders their specific identification based solely on the number of carbon atoms. While glycan fragmentation may contribute some information on the glycan sequence, branching, and linkage, the combination of several methods is usually necessary to produce a highly reliable picture. For example, data acquired by MS and exoglycosidase array will give a near-complete characterization of the glycan sequence, branching, and linkage [60]. Similarly, oligosaccharide hydrolysis after permethylation, followed by gas-chromatography mass-spectrometry (GC-MS) analysis would identify unmethylated carbons of the monosaccharide. Moreover, structural characterization and branching can be best resolved by NMR analysis if enough material in sufficient purity is available. Thus, information on branching and inner oligosaccharide linkages could be obtained [53].

2.3. Glycan Site Mapping. Identification of a glycosylation site is important, since such knowledge could provide an indication of the function of that glycan. Moreover, assignment of a certain structure to specific site can also shed light on the protein's glycosylation profile and microheterogeneity and hence, plausibly, its activity. As mentioned above, removal of an entire *N*-glycan moiety by PNGase F or A results in

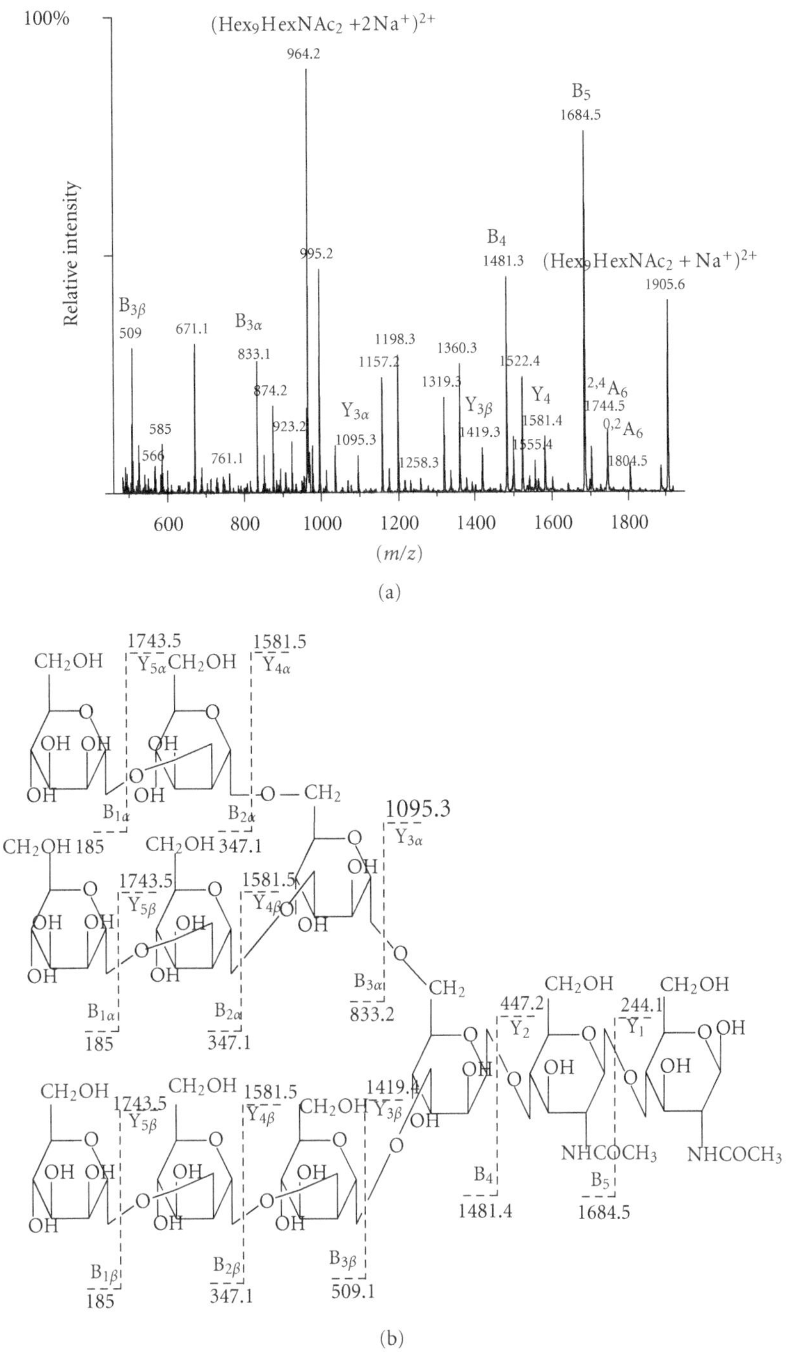

FIGURE 2: ESI-MS/MS for N-glycan structural analysis. MS/MS spectrum of the $Hex_9HexNAc_2$ N-glycan (a). Numbers above the parent ions peaks represent $[M+Na]^+$, while other peaks represent $[M+H]^+$ of the fragments. Expected fragment type and mass are indicated in the oligosaccharide structure (b).

the conversion of Asn to Asp and a shift of one mass unit for each N-glycosylation site. Thus, when a deglycosylated protein is further digested with trypsin, the peptides that are bound to the glycan moiety will be 1 Da heavier than the expected theoretical mass. By subjecting these peptides to MS/MS, each peptide that possesses Asp (instead of Asn) is identified as formerly attached to the glycan moiety [61]. A similar approach involves glycan removal with PNGase F in the presence of $H_2{}^{18}O$: the deglycosylated Asn would be labeled with ^{18}O, thus its mass, altered by 3 Da (1 Da for the Asn-Asp conversion and 2 Da contributed by ^{18}O), can be followed [62].

Glycan site mapping is not always straightforward, and some challenges may be encountered. One challenge is to identify a glycopeptide in a large pool with an absolute majority of nonglycosylated peptides. Moreover, the glycan

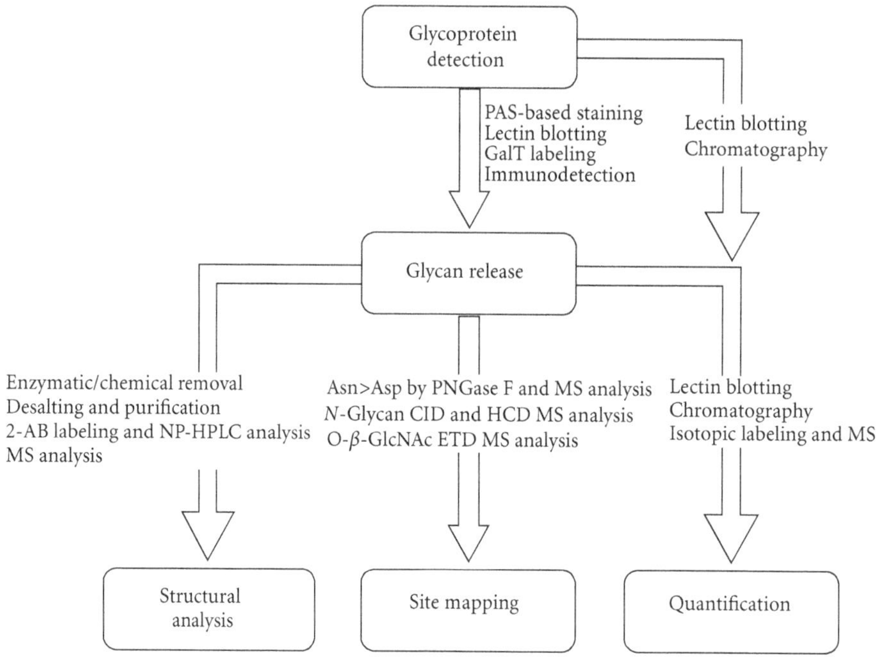

Figure 3: Schematic representation of glycoprotein identification, glycan structural elucidation, and protein glycosylation site mapping and quantification.

moiety tends to suppress ionization compared to unmodified peptides. Therefore, in such a case glycopeptide enrichment should be employed. A vast variety of commercial resin-bound lectins are available for column chromatography [63]. Another challenge involves the analysis of modified peptides, specifically glycopeptides; in this case, the commonly used CID and HCD fragmentation methods result in intensive fragmentation of the glycan moiety, while the peptide remains intact. Two recently developed fragmentation methods, electron capture dissociation (ECD) and electron transfer dissociation (ETD), may solve this problem. Both these methods, based on electronic excitation energy, result in peptide fragmentation that leaves the modification intact and attached to the amino acid. Thus, enriched glycopeptides can be subjected to CID-ETD fragmentation, in which the CID fragments the glycan moiety and the ETD sequences the glycopeptide with the attached glycan moiety [58].

A study on *N*-glycan site mapping by HCD was recently published [57]: a linear ion-trap orbitrap hybrid mass spectrometer (LTQ-orbitrap) was used to sequence a glycopeptide through repeated fragmentation (MS^n). In the first round of fragmentation (MS^2), CID was used for analyzing glycan sequencing and branching, and HCD (analyzed in the orbitrap) was used—due to its ability to form glycan oxonium ions—to detect glycopeptides. HCD yields a unique ion, designated Y1, which represents the mass of the peptide with Asn-GlcNAc. The Y1 intensity in the HCD spectra is high, and the ion can thus be subjected to further MS^3 for peptide sequencing [57]. The newly released GlypID 2.0 software, utilizing the accurate precursor mass and both the CID and HCD spectra, enables the glycosylation site characterization [64].

In contrast, localization of the dynamic mono-glycosylation *O*-*β*-GlcNAc is elusive. Due to its extreme labile nature in the gas phase and due to ion suppression, *O*-*β*-GlcNAc detection and site mapping of an intact *O*-*β*-GlcNAcylated peptide are virtually impossible by standard MS methods relying on CID; ETD is the only applicable type of fragmentation. ETD-produced C and Z fragment ions that contain Ser or Thr residues would carry an additional mass of 203 Da, indicating the attachment of *O*-*β*-GlcNAc [65]. Alternatively, *O*-*β*-GlcNAcylated peptides may be identified following *O*-GlcNAc removal. Use of conventional *β*-elimination to remove the *O*-*β*-GlcNAc results in the conversion of glycosylserine to 2-aminopropenoic acid (2-ap) and glycothreonine to 2-amino-2-butenoic acid (2-ab) [66], which decreases the glycopeptide mass by *m*/*z* 222. The obtained CID fragment ion represents a loss of *m*/*z* 69 (2-ap) instead of *m*/*z* 87 for serine, while the mass of threonine and its derivative 2-ab remains *m*/*z* 101. This method enables rapid identification of *O*-GlcNAc-modified peptides in a complex mixture, as well as its site mapping at the low picomole level.

2.4. Glycoprotein Quantification. The substoichiometric relation between glycosylation and the protein constitutes a further challenge in glycoprotein quantification. Yet, lectin purification, followed by a labeling procedure, can be used for relative quantification of a specific glycoprotein. The simplest method is lectin blotting (see the previous), followed by densitometry of the lectin cross-reactive band. In addition, lectin chromatography can be used for comparative quantification of a specific glycoprotein in several samples; the relative amounts of a particular glycoprotein, purified by a specific lectin column, in different samples, are determined by densitometry of SDS-PAGE protein bands, stained with either Coomassie blue or another specific stain.

Chromatography by NP-HPLC has previously been used for relative quantification of 2-AB-labeled oligosaccharides in one or more samples [43]; calculation of the area under the chromatographic peaks indicates the amount of the specific glycan. In addition, chromatography and MS are often coupled to provide a reproducible and reliable glycan characterization and quantification. MS is also applicable for the relative quantitation of glycopeptides by state-of-the-art methods based on stable isotope labeling, for example, isotope-coded affinity tagging (ICAT) [67] or isobaric tags for relative and absolute quantitation (iTRAQ) [68]. In these methods, lectin-purified glycoproteins from different samples are each labeled with either light or heavy isotopes during trypsin digestion, mixed together, and analyzed by MS. The relative abundance of the light- and heavy-isotopic peaks of the same peptide from the two different samples indicates their relative quantities. In case of microheterogeneity, that is, the existence of several glycoforms (Figure 1), quantification by a modified permethylation procedure, using isotopic labeling, is applicable [54]. Following glycan release, each free glycan sample is labeled with either ^{12}C- or ^{13}C-methyl iodide; the two samples are mixed in equal proportions, and the labeled glycans are analyzed by MS. This method has a high-dynamic range, adequate linearity, and high reproducibility [54].

3. Concluding Remarks

Glycosylation is fundamental for protein function and for cell physiology. Analysis of the structure and localization of protein glycans is necessary in life sciences and biotechnology. The most reliable analytical tools currently available are chromatography and mass spectrometry. To begin with, characterization of glycans by lectin chromatography is the primary method to be used. For structure elucidation, exoglycosidase array followed by a combination of NP-HPLC and MS is the method of choice. MS analysis, by either ETD or isotopic labeling, is suitable for glycosylation site mapping. Isotopic labeling is further appropriate for quantification of glycans (Figure 3).

Acknowledgments

The authors would like to thank Ms. Inez Mureinik for styling the manuscript. I. Khalaila is the incumbent of Ma'of Fellowship from the Kahanoff Foundation.

References

[1] A. Varki, "Biological roles of oligosaccharides: all of the theories are correct," *Glycobiology*, vol. 3, no. 2, pp. 97–130, 1993.

[2] E. A. Eklund and H. H. Freeze, "The congenital disorders of glycosylation: a multifaceted group of syndromes," *NeuroRx*, vol. 3, no. 2, pp. 254–263, 2006.

[3] T. Lefebvre, C. Guinez, V. Dehennaut, O. Beseme-Dekeyser, W. Morelle, and J. C. Michalski, "Does O-GlcNAc play a role in neurodegenerative diseases?" *Expert Review of Proteomics*, vol. 2, no. 2, pp. 265–275, 2005.

[4] P. M. Rudd, T. Elliott, P. Cresswell, I. A. Wilson, and R. A. Dwek, "Glycosylation and the immune system," *Science*, vol. 291, no. 5512, pp. 2370–2376, 2001.

[5] M. E. Taylor and K. Drickamer, *Introduction to Glycobiology*, Oxford University Press, New York, NY, USA, 2nd edition, 2006.

[6] S. Dube, J. W. Fisher, and J. S. Powell, "Glycosylation at specific sites of erythropoietin is essential for biosynthesis, secretion, and biological function," *Journal of Biological Chemistry*, vol. 263, no. 33, pp. 17516–17521, 1988.

[7] A. Helenius and M. Aebi, "Intracellular functions of *N*-linked glycans," *Science*, vol. 291, no. 5512, pp. 2364–2369, 2001.

[8] L. Ellgaard, M. Molinari, and A. Helenius, "Setting the standards: quality control in the secretory pathway," *Science*, vol. 286, no. 5446, pp. 1882–1888, 1999.

[9] A. Varki, *Essentials of Glycobiology*, Cold Spring Harbor Laboratory Press, New York, NY, USA, 2nd edition, 2009.

[10] L. A. Tabak, "The role of mucin-type O-glycans in eukaryotic development," *Seminars in Cell and Developmental Biology*, vol. 21, no. 6, pp. 616–621, 2010.

[11] G. W. Hart, "Dynamic O-linked glycosylation of nuclear and cytoskeletal proteins," *Annual Review of Biochemistry*, vol. 66, pp. 315–335, 1997.

[12] T. Y. Chou, C. V. Dang, and G. W. Hart, "Glycosylation of the c-Myc transactivation domain," *Proceedings of the National Academy of Sciences of the United States of America*, vol. 92, no. 10, pp. 4417–4421, 1995.

[13] W. H. Yang, J. E. Kim, H. W. Nam et al., "Modification of p53 with O-linked *N*-acetylglucosamine regulates p53 activity and stability," *Nature Cell Biology*, vol. 8, no. 10, pp. 1074–1083, 2006.

[14] G. W. Hart, M. P. Housley, and C. Slawson, "Cycling of O-linked β-*N*-acetylglucosamine on nucleocytoplasmic proteins," *Nature*, vol. 446, no. 7139, pp. 1017–1022, 2007.

[15] C. R. Torres and G. W. Hart, "Topography and polypeptide distribution of terminal *N*-acetylglucosamine residues on the surfaces of intact lymphocytes. Evidence for O-linked GlcNAc," *Journal of Biological Chemistry*, vol. 259, no. 5, pp. 3308–3317, 1984.

[16] C. F. Chou, A. J. Smith, and M. B. Omary, "Characterization and dynamics of O-linked glycosylation of human cytokeratin 8 and 18," *Journal of Biological Chemistry*, vol. 267, no. 6, pp. 3901–3906, 1992.

[17] W. G. Kelly, M. E. Dahmus, and G. W. Hart, "RNA polymerase II is a glycoprotein. Modification of the COOH-terminal domain by O-GlcNAc," *Journal of Biological Chemistry*, vol. 268, no. 14, pp. 10416–10424, 1993.

[18] I. Han, E. S. Oh, and J. E. Kudlow, "Responsiveness of the state of O-linked *N*-acetylglucosamine modification of nuclear pore protein p62 to the extracellular glucose concentration," *Biochemical Journal*, vol. 350, no. 1, pp. 109–114, 2000.

[19] L. Wells, K. Vosseller, and G. W. Hart, "Glycosylation of nucleocytoplasmic proteins: signal transduction and O-GlcNAc," *Science*, vol. 291, no. 5512, pp. 2376–2378, 2001.

[20] D. C. Love and J. A. Hanover, "The hexosamine signaling pathway: deciphering the "O-GlcNAc code"," *Science's STKE*, vol. 2005, no. 312, p. re13, 2005.

[21] N. E. Zachara and G. W. Hart, "Cell signaling, the essential role of O-GlcNAc!," *Biochimica et Biophysica Acta - Molecular and Cell Biology of Lipids*, vol. 1761, no. 5-6, pp. 599–617, 2006.

[22] J. T. Clarke, "Simplified "Disc" (Polyaceylamide Gel) electrophoresis," *Annals of the New York Academy of Sciences*, vol. 121, no. 2, pp. 428–436, 1964.

[23] R. M. Zacharius, T. E. Zell, J. H. Morrison, and J. J. Woodlock, "Glycoprotein staining following electrophoresis on acrylamide gels," *Analytical Biochemistry*, vol. 30, no. 1, pp. 148–152, 1969.

[24] T. H. Steinberg, "Chapter 31 protein Gel staining methods. An introduction and overview," *Methods in Enzymology*, vol. 463, no. C, pp. 541–563, 2009.

[25] C. Hart, B. Schulenberg, T. H. Steinberg, W. Y. Leung, and W. F. Patton, "Detection of glycoproteins in polyacrylamide gels and on electroblots using Pro-Q Emerald 488 dye, a fluorescent periodate Schiff-base stain," *Electrophoresis*, vol. 24, no. 4, pp. 588–598, 2003.

[26] A. H. Wardi and G. A. Michos, "Alcian blue staining of glycoproteins in acrylamide disc electrophoresis," *Analytical Biochemistry*, vol. 49, no. 2, pp. 607–609, 1972.

[27] M. R. Green, J. V. Pastewka, and A. C. Peacock, "Differential staining of phosphoproteins on polyacrylamide gels with a cationic carbocyanine dye," *Analytical Biochemistry*, vol. 56, no. 1, pp. 43–51, 1973.

[28] H. Lis and N. Sharon, "Lectins as molecules and as tools," *Annual Review of Biochemistry*, vol. 55, pp. 35–67, 1986.

[29] T. Sata, C. Zuber, and J. Roth, "Lectin-digoxigenin conjugates: a new hapten system for glycoconjugate cytochemistry," *Histochemistry*, vol. 94, no. 1, pp. 1–11, 1990.

[30] P. Gravel, O. Golaz, C. Walzer, D. F. Hochstrasser, H. Turler, and L. P. Balant, "Analysis of glycoproteins separated by two-dimensional gel electrophoresis using lectin blotting revealed by chemiluminescence," *Analytical Biochemistry*, vol. 221, no. 1, pp. 66–71, 1994.

[31] F. I. Comer, K. Vosseller, L. Wells, M. A. Accavitti, and G. W. Hart, "Characterization of a mouse monoclonal antibody specific for *O*-linked *N*-acetylglucosamine," *Analytical Biochemistry*, vol. 293, no. 2, pp. 169–177, 2001.

[32] G. D. Holt, C. M. Snow, and A. Senior, "Nuclear pore complex glycoproteins contain cytoplasmically disposed *O*-linked *N*-acetylglucosamine," *Journal of Cell Biology*, vol. 104, no. 5, pp. 1157–1164, 1987.

[33] C. M. Snow, A. Senior, and L. Gerace, "Monoclonal antibodies identify a group of nuclear pore complex glycoproteins," *Journal of Cell Biology*, vol. 104, no. 5, pp. 1143–1156, 1987.

[34] R. A. O'Neill, "Enzymatic release of oligosaccharides from glycoproteins for chromatographic and electrophoretic analysis," *Journal of Chromatography A*, vol. 720, no. 1-2, pp. 201–215, 1996.

[35] V. Tretter, F. Altmann, and L. Marz, "Peptide-N4-(*N*-acetyl-β-glucosaminyl)asparagine amidase F cannot release glycans with fucose attached $\alpha 1 \rightarrow 3$ to the asparagine-linked *N*-acetylglucosamine residue," *European Journal of Biochemistry*, vol. 199, no. 3, pp. 647–652, 1991.

[36] R. B. Trimble and A. L. Tarentino, "Identification of distinct endoglycosidase (endo) activities in *Flavobacterium meningosepticum*: endo F1, endo F2, and endo F3. Endo F1 and endo H hydrolyze only high mannose and hybrid glycans," *Journal of Biological Chemistry*, vol. 266, no. 3, pp. 1646–1651, 1991.

[37] D. M. Carlson, "Structures and immunochemical properties of oligosaccharides isolated from pig submaxillary mucins," *Journal of Biological Chemistry*, vol. 243, no. 3, pp. 616–626, 1968.

[38] T. Patel, J. Bruce, A. Merry et al., "Use of hydrazine to release in intact and unreduced form both *N*- and *O*-linked oligosaccharides from glycoproteins," *Biochemistry*, vol. 32, no. 2, pp. 679–693, 1993.

[39] N. H. Packer, M. A. Lawson, D. R. Jardine, and J. W. Redmond, "A general approach to desalting oligosaccharides released from glycoproteins," *Glycoconjugate Journal*, vol. 15, no. 8, pp. 737–747, 1998.

[40] K. O. Bornsen, M. D. Mohr, and H. M. Widmer, "Ion-exchange and purification of carbohydrates on a Nafion membrane as a new sample pretreatment for matrix-assisted laser-desorption ionization mass-spectrometry," *Rapid Communications in Mass Spectrometry*, vol. 9, no. 11, pp. 1031–1034, 1995.

[41] Y. Mechref and M. V. Novotny, "Structural investigations of glycoconjugates at high sensitivity," *Chemical Reviews*, vol. 102, no. 2, pp. 321–369, 2002.

[42] J. C. Bigge, T. P. Patel, J. A. Bruce, P. N. Goulding, S. M. Charles, and R. B. Parekh, "Nonselective and efficient fluorescent labeling of glycans using 2-amino benzamide and anthranilic acid," *Analytical Biochemistry*, vol. 230, no. 2, pp. 229–238, 1995.

[43] G. R. Guile, P. M. Rudd, D. R. Wing, S. B. Prime, and R. A. Dwek, "A rapid high-resolution high-performance liquid chromatographic method for separating glycan mixtures and analyzing oligosaccharide profiles," *Analytical Biochemistry*, vol. 240, no. 2, pp. 210–226, 1996.

[44] G. R. Guile, S. Y. C. Wong, and R. A. Dwek, "Analytical and preparative separation of anionic oligosaccharides by weak anion-exchange high-performance liquid chromatography on an inert polymer column," *Analytical Biochemistry*, vol. 222, no. 1, pp. 231–235, 1994.

[45] R. R. Townsend and M. R. Hardy, "Analysis of glycoprotein oligosaccharides using high-pH anion exchange chromatography," *Glycobiology*, vol. 1, no. 2, pp. 139–147, 1991.

[46] M. P. Campbell, L. Royle, C. M. Radcliffe, R. A. Dwek, and P. M. Rudd, "GlycoBase and autoGU: tools for HPLC-based glycan analysis," *Bioinformatics*, vol. 24, no. 9, pp. 1214–1216, 2008.

[47] L. Royle, T. S. Mattu, E. Hart et al., "An analytical and structural database provides a strategy for sequencing *O*-glycans from microgram quantities of glycoproteins," *Analytical Biochemistry*, vol. 304, no. 1, pp. 70–90, 2002.

[48] G. R. Guile, D. J. Harvey, N. O'Donnell et al., "Identification of highly fucosylated *N*-linked oligosaccharides from the human parotid gland," *European Journal of Biochemistry*, vol. 258, no. 2, pp. 623–656, 1998.

[49] C. J. Edge, T. W. Rademacher, M. R. Wormald et al., "Fast sequencing of oligosaccharides: the reagent-array analysis method," *Proceedings of the National Academy of Sciences of the United States of America*, vol. 89, no. 14, pp. 6338–6342, 1992.

[50] D. Šagi, P. Kienz, J. Denecke, T. Marquardt, and J. Peter-Katalinić, "Glycoproteomics of N-glycosylation by in-gel deglycosylation and matrix-assisted laser desorption/ionisation-time of flight mass spectrometry mapping: application to congenital disorders of glycosylation," *Proteomics*, vol. 5, no. 10, pp. 2689–2701, 2005.

[51] Y. Mechref and M. V. Novotny, "Structural investigations of glycoconjugates at high sensitivity," *Chemical Reviews*, vol. 102, no. 2, pp. 321–369, 2002.

[52] I. Ciucanu and F. Kerek, "A simple and rapid method for the permethylation of carbohydrates," *Carbohydrate Research*, vol. 131, no. 2, pp. 209–217, 1984.

[53] J. Jang-Lee, S. J. North, M. Sutton-Smith et al., "Glycomic profiling of cells and tissues by mass spectrometry: fingerprinting and sequencing methodologies," *Methods in Enzymology*, vol. 415, pp. 59–86, 2006.

[54] G. Alvarez-Manilla, N. L. Warren, T. Abney et al., "Tools for glycomics: relative quantitation of glycans by isotopic

permethylation using 13CH3I," *Glycobiology*, vol. 17, no. 7, pp. 677–687, 2007.
[55] A. Dell, "F.A.B.-Mass Spectrometry of Carbohydrates," *Advances in Carbohydrate Chemistry and Biochemistry*, vol. 45, no. C, pp. 19–72, 1987.
[56] D. J. Harvey, R. H. Bateman, and M. R. Green, "High-energy collision-induced fragmentation of complex oligosaccharides ionized by matrix-assisted laser desorption/ionization mass spectrometry," *Journal of Mass Spectrometry*, vol. 32, no. 2, pp. 167–187, 1997.
[57] Z. M. Segu and Y. Mechref, "Characterizing protein glycosylation sites through higher-energy C-trap dissociation," *Rapid Communications in Mass Spectrometry*, vol. 24, no. 9, pp. 1217–1225, 2010.
[58] N. E. Scott, B. L. Parker, A. M. Connolly et al., "Simultaneous glycan-peptide characterization using hydrophilic interaction chromatography and parallel fragmentation by CID, higher energy collisional dissociation, and electron transfer dissociation MS applied to the *N*-linked glycoproteome of Campylobacter jejuni," *Molecular and Cellular Proteomics*, vol. 10, no. 2, 2011.
[59] J. M. Prien, D. J. Ashline, A. J. Lapadula, H. Zhang, and V. N. Reinhold, "The high mannose glycans from bovine ribonuclease B isomer characterization by ion trap MS," *Journal of the American Society for Mass Spectrometry*, vol. 20, no. 4, pp. 539–556, 2009.
[60] B. Küster, S. F. Wheeler, A. P. Hunter, R. A. Dwek, and D. J. Harvey, "Sequencing of *N*-linked oligosaccharides directly from protein gels: in- gel deglycosylation followed by matrix-assisted laser desorption/ionization mass spectrometry and normal-phase high-performance liquid chromatography," *Analytical Biochemistry*, vol. 250, no. 1, pp. 82–101, 1997.
[61] Z. Roth, S. Parnes, S. Wiel et al., "*N*-glycan moieties of the crustacean egg yolk protein and their glycosylation sites," *Glycoconjugate Journal*, vol. 27, no. 1, pp. 159–169, 2010.
[62] B. Küster and M. Mann, "O-18-labeling of *N*-glycosylation sites to improve the identification of gel-separated glycoproteins using peptide mass mapping and database searching," *Analytical Chemistry*, vol. 71, no. 7, pp. 1431–1440, 1999.
[63] H. Hwang, J. Zhaang, K. A. Chung et al., "Glycoproteomics in neurodegenerative diseases," *Mass Spectrometry Reviews*, vol. 29, no. 1, pp. 79–125, 2010.
[64] A. M. Mayampurath, Y. Wu, Z. M. Segu, Y. Mechref, and H. Tang, "Improving confidence in detection and characterization of protein n-glycosylation sites and microheterogeneity," *Rapid Communications in Mass Spectrometry*, vol. 25, no. 14, pp. 2007–2019, 2011.
[65] S. A. Whelan and G. W. Hart, "Proteomic approaches to analyze the dynamic relationships between nucleocytoplasmic protein glycosylation and phosphorylation," *Circulation Research*, vol. 93, no. 11, pp. 1047–1058, 2003.
[66] K. D. Greis, B. K. Hayes, F. I. Comer et al., "Selective detection and site-analysis of O-GlcNAc-modified glycopeptides by β-elimination and tandem electrospray mass spectrometry," *Analytical Biochemistry*, vol. 234, no. 1, pp. 38–49, 1996.
[67] S. P. Gygi, B. Rist, S. A. Gerber, F. Turecek, M. H. Gelb, and R. Aebersold, "Quantitative analysis of complex protein mixtures using isotope-coded affinity tags," *Nature Biotechnology*, vol. 17, no. 10, pp. 994–999, 1999.
[68] P. L. Ross, Y. N. Huang, J. N. Marchese et al., "Multiplexed protein quantitation in Saccharomyces cerevisiae using amine-reactive isobaric tagging reagents," *Molecular and Cellular Proteomics*, vol. 3, no. 12, pp. 1154–1169, 2004.

17

The Effect of Tribomechanical Micronization and Activation on Rheological, Thermophysical, and Some Physical Properties of Tapioca Starch

Zoran Herceg,[1] Verica Batur,[2] Anet Režek Jambrak,[1] Tomislava Vukušić,[1] Ines Gmajnički,[3] and Igor Špoljarić[3]

[1] *Faculty of Food Technology and Biotechnology, University of Zagreb, Pierottijeva 6, Zagreb, Croatia*
[2] *High School of Food Technology, Gjure Prejca 2, Zagreb, Croatia*
[3] *Forensic Science Centre "Ivan Vučetić", Ilica 110, Zagreb, Croatia*

Correspondence should be addressed to Zoran Herceg; zherceg@pbf.hr

Academic Editor: R. J. Linhardt

The aim of this research was to investigate the effect of tribomechanical treatments on rheological, thermophysical, and some physical properties of tapioca starch. Samples of tapioca starch were treated using laboratory equipment for tribomechanical micronization and activation (TMA equipment). Before and after the TMA treatment, analysis of the particle size and particle size distribution was carried out, in addition to scanning electron micrography in tapioca starch. Scanning electron micrography showed that tribomechanical processing of tapioca starch resulted in breaking accumulations of starch granules in the form of granules. Pasting parameters have shown that maximal viscosities of model starch suspension have been decreasing after tribomechanical treatment. On the basis of gelatinization curves, it can be concluded that there are changes in the gelatinization point after treatment, and there is decrease in enthalpy of gelatinization for model suspension. After tribomechanical treatment, changes in physical properties of starch suspensions were determined, as well as specific swelling capacity, solubility index, and turbidity of tapioca starch suspensions.

1. Introduction

Starch is present in food, whether it comes as a raw material or added as ingredient in order to achieve or improve certain properties. It has significant applications in the food industry, and is used as a tool for: thickening, stabilization of colloidal systems, moisture retention, flavor binding, gelling, film formation, improvement of product quality and other. Starch isolation from natural sources (seeds, tubers and fruit crops) gets a native starch which retains the original structure and properties. Due to its capability of improving water retention, it is used to improve textural properties, and due to low energy value-in the production of low-caloric foodstuffs [1–3].

Starch is a useful polymer and a cheap, natural material, the reason being the ease with which its physicochemical properties can be altered through chemical or enzyme modification and/or physical treatment [1, 4, 5]. There is great interest in novel methods to modify the structure of starch in the crystalline regions [6]. One of novel methods to modify the structure of starch is the process of tribomechanical micronization and activation (TMA process).

The process of tribomechanical micronization and activation (TMA), as well as the appropriate equipment, was patented in the year 1998 at the International Bureau of the WPO PCT Receiving Office in Geneva Switzerland, under number PCT/1B 99/00757 [7]. The TMA equipment is made up of housing and two rotor discs placed against each other. Each disc is furnished with 3–7 concentric circles—rings of teeth projecting from the surface of the disc—with specially constructed hard metal elements. The discs rotate in opposite directions at the same angular rate. The starting material enters the equipment through the central part of the rotor system by ventilator air streaming. Therefore, the particles are accelerated and, because of repeated changes in the direction of motion, collide, causing friction over short time intervals (less than 0.001 s).

Application of tribomechanical micronization in food processing presents a novel nonthermal method, with low energy consumption and short treatment time as compared to traditional technological processes, and presents novel method for food processing that can be used for new product development as improvement of technological process which is the purpose of this paper. The investigation with tribomechanical activation (TMA) has been primarily focused on inorganic materials [8], but recently, investigations were directed to organic materials such as powdered whey proteins [8–10]. Previous investigations have shown that TMA treatment, besides breaking up developed agglomerates, may significantly change structural characteristics of starches and its reactivity [11]. Because of such effects on the physical and structural properties of proteins and corn starch, it is to be expected that the procedure of TMA treatment can cause similar changes on tapioca starch.

Application of tapioca starch is used in numerous industrial and food applications, including thickening and gelling. In many applications, the properties of native starch are not optimal, and therefore starch is used with different additives and ingredients or is modified in order to change and improve its functions [12–15]. The use of native tapioca starches may be highly inconvenient due to retrogradation and syneresis phenomena as well as low mechanical and thermal stability observed in such material. Starch is modified in a different manner in order to change its physicochemical properties as well as to obtain desired functional properties. Tapioca starch has no taste or smell; so, adding color, flavor, eggs, milk, cream and sweeteners can be recast in age at variety of products especially sweets.

The purpose of this research was to establish a possible significant influence of TMA treatment on rheological, thermophysical, and some physical properties of tapioca starch. The examinations of the influence of TMA treatment on the morphological characteristic of the tapioca starch were also performed.

2. Materials and Methods

2.1. Sample Preparation. Powdered tapioca starch sample (commercial name: Cream Gel 70001) has been used. Its composition has been declared by manufacturer (Cargill Benelux BV, SAS van Gent, The Netherland) as follows: water 13.11%, starch 86.89%, and pH-5.5. Suspensions have been prepared by stirring the appropriated amount of tapioca starch powder and distilled water as stated in Table 1.

2.2. Tribomechanical Micronization. Powder tapioca starch was laboratory treated in the equipment for TMA (Figure 1). The laboratory TMA device is capable of treating samples within a wide range of rotor speeds (from 10,000 rpm up to of 22,000 rpm). The TMA optimized treatment was carried out at rotor speed (20,000 rpm) and at ambient temperature. Tapioca starch was treated using the TMA equipment under atmospheric pressure. Overheating of the materials was prevented by water cooling of the equipment through its housing. Intake of the material for treatment was continuous at 1 kg/min [7].

TABLE 1: Composition of tapioca starch suspensions: 1A—untreated; 1B—TMA treated.

Sample	Preparation	Tapioca starch (g)	Water (g)
10% (w/w) suspensions[c]	1A	11.37	98.62
	1B	11.34	98.65
2% (w/w) suspensions[b]	1A	2.27	97.72
	1B	2.26	97.73
1% (w/w) suspensions[a]	1A	1.14	98.86
	1B	1.13	98.87

[a] Used for turbidity measurements.
[b] Used for swelling power measurements.
[c] Used for rheological measurements.

FIGURE 1: Laboratory equipment for tribomechanical micronization and activation (TMA equipment) [7].

2.3. Particle Size Distribution. The particle size distribution of the powdered tapioca starch before and after TMA treatment was measured using a laser light scattering instrument, the Malvern Mastersizer 2000 (with a lens of 100 mm focal length). For analysis of the results obtained, MIE theory was applied [16]. Each analysis was made in triplicate, and the mean values were taken.

2.4. Scanning Electron Micrography of Tapioca Starch. For this investigation, scanning electron microscope Philips XL30 ESEM Tungsten, Philips, Eindhoven, The Netherlands (Edax detector, type PV 9760/68 ME, resolution 134.30 eV; BSE detector, Philips PW 6848/00), and software EDAX Genesis v. 5.21 were used for the characterization of air-dried, lyophilized, and spray-dried specimens. Photographs were taken at an accelerating voltage of 25.0 kV under recording time of 5 seconds. Diameter of observed area was 10 mm.

2.5. IR Spectroscopy. FTIR-Paragon 1000, Perkin Elmer, USA, Fouriers transformations,ART PIKE-s krystal (Instrument I-114/2-1/1), and software Spectrum, range of wavelength 550 cm^{-1}–2 4400 cm^{-1}, with the resolution 4 cm^{-1}.

2.6. Solubility Index and Swelling Power Determination of Tapioca Starch Model Systems. The swelling power and solubility index of tapioca starch model systems before and after TMA treatment were determined in triplicate according to the method of Leach et al. [17]. Aqueous suspensions of 2% starch (w/w) were heated in a water bath at constant temperatures and shaked, for 30 min. Each suspension was cooled and centrifuged at 3000 rpm for 15 min; the decanted was weighed, and the supernatant was placed in a vacuum stove at 120°C for 4 h. Solubility index, SI (%), has been calculated according to the equation:

$$SI = (W_s - W_0) \times 100, \tag{1}$$

W_s—percentage of the dry matter in supernatant (%);

W_0—percentage of the dry matter in the starting suspension (%).

Swelling power, SP (g of hydrated molecules/g starch dry matter), is calculated according to

$$SP = \frac{W_G}{W_{GDM}}, \tag{2}$$

W_G—gel mass (g);

W_{GDM}—mass of gel dry matter (g).

2.7. Turbidity (Light Transmittance %). Turbidity (light transmittance %) of tapioca starch pastes was measured as described by Perera and Hoover [18]. An aqueous suspension of starch from each tapioca variety (1 g/100 g solids) was heated in a water bath at 90°C for 1 h with constant stirring. The starch pastes were cooled for 1 h at 30°C, and the light transmittance was measured at 640 nm. The samples were stored for 5 days at 4°C in a refrigerator, and transmittance was determined every 24 h with a Shimadzu UV-1601 spectrophotometer (Shimadzu Corporation, Kyoto, Japan).

2.8. Determination of Rheological Properties of Tapioca Starch Model Systems. Rheological properties of tapioca starch suspensions were determined according to the method of Hagenimana and Ding [19]. Tapioca starch suspension of 10% (w/w) in 100 mL volume was directly placed into a stainless steel measuring bowl of a Brabender Micro-Visco-AmyloGraph (Duisburg, Germany). System operating at 250 rpm was then heated from 30 to 93°C, with a heating rate of 7.5°C. Samples were held for 1 min at 52°C and 5 min at maximal temperature of 93°C. The parameters recorded were maximum viscosity, breakdown viscosity, setback viscosity, and beginning of gelatinization (°C).

2.9. Thermal Properties of Tapioca Starch. Gelatinization properties of tapioca starch suspensions were analyzed using a differential scanning calorimeter DSC 822e (Mettler Toledo) equipped with STARe software. An empty pan was used as a reference. Prepared and TMA-treated corn starch suspensions were weighed into a standard aluminum pan. The pans were sealed and equilibrated for 24 h at room temperature before heat treatment in the DSC. The starch slurry was gelatinized in the DSC using a heat rate of 10°C/min from 25 to 95°C. After the heat treatment, the samples were cooled to 25°C and removed from DSC. The changes in enthalpy (ΔH in kJ kg^{-1} of dry starch), onset temperature (T_o), peak temperature (T_P), and conclusion temperature (T_c) for gelatinization were obtained from the exothermal DSC curves. All the measurements were performed in triplicate.

2.10. Statistical Analyses. The whole study was repeated and each value represents the mean of three measurements from three independent tribomechanical treatments. The effect of tribomechanical treatment on tested parameters was determined by analysis of variance, using statistical analyses with SPSS for Windows version 13.0 (SPSS Inc., Chicago, IL). Analysis of variance (one-way ANOVA), significant level used was 5% ($\alpha = 0.05$). The values not statistically different are accompanied by the letter (a) and the values statistically different by the letter (b).

3. Results and Discussion

3.1. Particle Size Distribution, Scanning Electron Micrography, and IR Spectroscopy of Tapioca Starch. The procedure of TMA treatment significantly changes the physical characteristics of the treated material [11]. Tribomechanical treatment causes micronization of various solid materials those results in nanoparticles and changes in structure and electrical potential of molecules [8]. Powdered tapioca starch during the TMA treatment is exposed to the strong mechanical forces of striking hammers and ventilation paddles placed on rotating discs which produce turbulent movement. In this way, tapioca starch particles collided and rubbed against each order in short time intervals (less than 0.001 s). The strong mechanical forces which were producing turbulent cause collision between particles of tapioca starch and were rubbing against each order had caused rupturing and mechanical damages of starch granules and changes in their size and shape. TMA treatment significantly changes particle size distribution of tapioca starch. The changes of particle size distribution and specific surface area tapioca starch during the tribomechanical activation can be seen in Figure 2. Scanning electron micrography showed that tribomechanical processing of tapioca starch resulted in breaking accumulations of starch granules in the form of granules and surface defects granules (Figure 3). The content of amylose (%) in starch is decrease after tribomechanical micronization [20, 21]. A significant reduction in amylose content after tribomechanical micronization was observed. These results are in order with the results of the solubility index (%) and swelling power (SP). Decrease in the content of amylose in granules of starch causes increase in swelling capacity and solubility of starch granules (Table 2). It is speculated that linear structure and random arrangement of amylose make amylose more susceptible to degradation [22].

IR spectroscopy showed that tribomechanical micronization and activation causes changes of tapioca starch granular

TABLE 2: Values of solubility index (%) and swelling power for native and modified tapioca starch suspensions after TMA treatment (1A (untreated) and 1B (TMA treated)).

Sample	Determination conditions	Swelling power (g of hydrated molecules/g starch dry matter)	Solubility index (%)
1A	20°C/5 min	2.89 ± 0.12^a	0.22 ± 0.05^a
	20°C/15 min	3.16 ± 0.08^a	0.29 ± 0.09^a
	70°C/5 min	4.17 ± 0.07^a	0.43 ± 0.11^a
	70°C/15 min	42.03 ± 0.11^a	0.69 ± 0.06^a
1B	20°C/5 min	3.31 ± 0.14^a	0.93 ± 0.09^b
	20°C/15 min	3.42 ± 0.04^a	1.27 ± 0.01^b
	70°C/5 min	26.31 ± 0.01^b	4.29 ± 0.06^b
	70°C/15 min	52.16 ± 0.02^b	6.95 ± 0.08^b

The values not statistically different are accompanied by the letter (a) and the values statistically different by the letter (b).

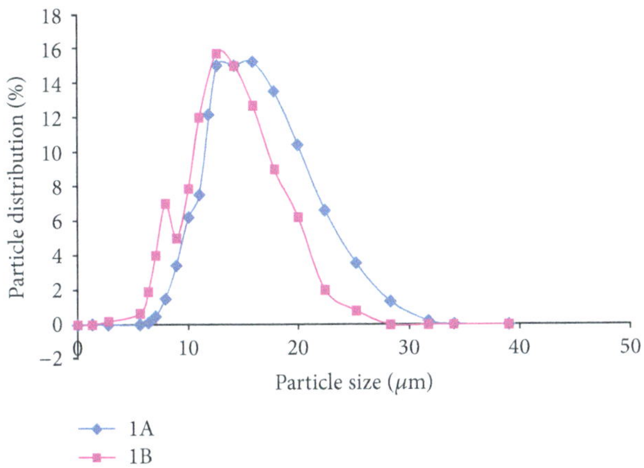

FIGURE 2: Granule size distribution of powered tapioca starch before and after tribomechanical treatment: 1A—untreated tapioca starch; 1B—tribomechanically treated tapioca starch.

structure (Figure 4). The strong mechanical forces during the TMA treatment cause shear forces that are capable of breaking the chains of polymers and damaging granules [23]. After TMA treatment was expected splitting of long polymer chains of amylase and the separation of the carbohydrate molecules less of the molar mass (disaccharides, threesaccharides and so forth), FT-IR showed that in this case did not happen. Thus, there was no separation of a large number of short carbohydrate chains, and FT-IR did not detect them. It can be concluded that the changes that are visible on Figure 4, have consequence of tearing of long chains of amylase into smaller oligosaccharide chains wich are still extremely high molar masses. Huang et al. have explored the influence of mechanical activation on tapioca starch and have come up with similar results. In this study, infrared spectroscopy showed no new functional groups formed during the process of mechanical activation [24].

3.2. Physical Properties of Tapioca Starch Model Systems. The procedure of TMA treatment significantly changes the physical characteristics of the treated material [25]. Table 2 shows the changes in the swelling power and solubility of tapioca starch before and after TMA treatment. The solubility index and swelling power of the tapioca starch were analyzed in order to obtain information about the structural differences and molecular arrangement of the granules. The solubility index (SI) and swelling power (SP) of the starch gel prepared with suspension of TMA-treated tapioca starch had higher values when compared with untreated suspensions. It is caused by disruption of starch granules and molecule by TMA treatment. Swelling capacity and solubility index show us the size of the interaction of starch chains within the amorphous and crystalline areas of starch granules. Their interaction is determined by the ratio of amylose and amylopectin, phosphorus content, and their conformation and degree of branching [26]. The processing of starch leads to structural reorganization of the starch granules resulting in the rejection of the starch molecules, thereby facilitating the binding of water between the amorphous regions of granules and the consequent increase in the swelling capacity and solubility [27]. The increase in solubility comes as a result of the depolymerization and structural weakening of the starch granule [28]. Increases in solubility and swelling power when applying TMA treatment are also higher for ṣamples at higher temperature (70°C) than at 20°C. Based on the results of measuring the absorbance at 640 nm on spectrophotometer for 5 days, significant differences after TMA treatment (Figure 5) in turbidity had been observed. Turbidity starch suspensions depend on the concentration of suspended particles in the volume of the measured absorbance. Perera and Hoover [18] explored the turbidity of native and modified starches and determined the existence of interaction between starch granules after completion of the modification. The demage on starch granules after mechanical modifications (tribomechanical) are mostly oriented to the surface of starch granules form conglomerates, which are rapidly precipitated from untreated starch granules themselves. The previous explanation is confirmed by the results of determining the turbidity of model starch suspensions tribomechanically treated tapioca starch, where after 5 days (up to 96 hours) of storage, a decrease in turbidity had been identified compared with the untreated samples (Figure 5). Singh et al. [26] also reported that the increase in turbidity affects the separation of amylose starch granules during swelling.

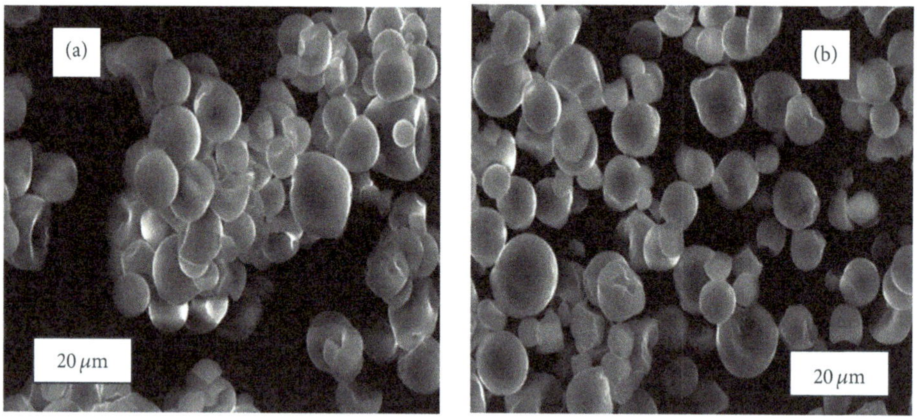

(a) (b)

FIGURE 3: Scanning electron micrography of tapioca starch: (a) untreated; (b) TMA treated.

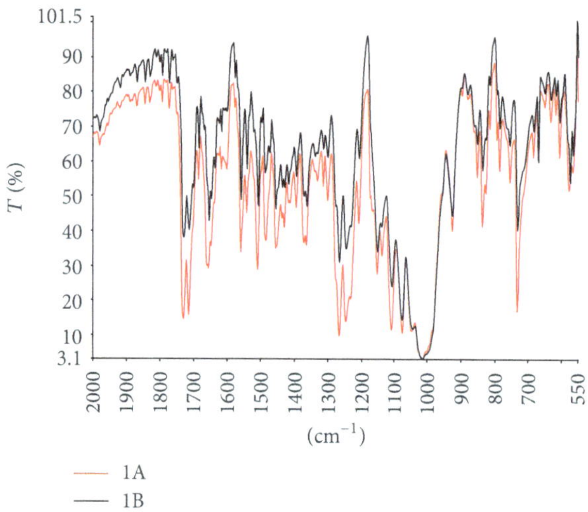

FIGURE 4: FTIR of tapioca starch: 1A—untreated tapioca starch; 1B—tribomechanically treated tapioca starch.

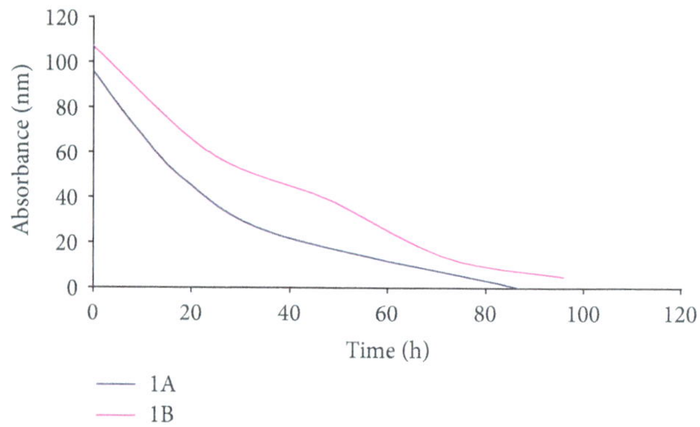

FIGURE 5: Turbidity of tapioca starch pastes before and after TMA treatment: 1A—untreated; 1B—TMA treated.

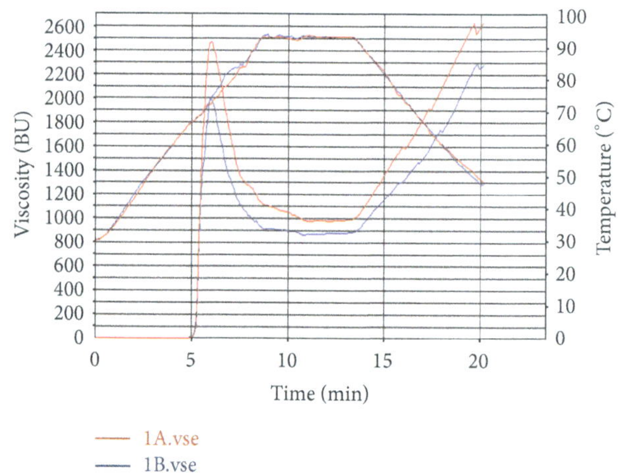

FIGURE 6: Viscosity of tapioca starch suspensions: 1A—untreated tapioca starch; 1B—tribomechanically treated tapioca starch.

3.3. Rheological and Thermophysical Properties of Tapioca Starch Model Systems. The recorded pasting properties of untreated and tribomechanically treated tapioca starch were pasting temperature, peak viscosity, trough viscosity (minimum viscosity at 95°C), final viscosity (viscosity at 50°C), breakdown viscosity (peak-trough viscosity), and setback viscosity (final-trough viscosity), as can be seen in Table 3. The representative curves for untreated and TMA-treated tapioca corn starch are shown in Figure 6. TMA treatment has affected a decrease of starch suspension viscosity (Figure 6). Pasting parameters have shown that maximum viscosity of tapioca starch model suspensions (2474 BU) was decreasing after tribomechanical treatment (1990 BU). Breakdown viscosity (1475 BU) have also decreased (1105 BU) after TMA treatment but there is increase of setback viscosity (2091 BU) as compared with native tapioca starch (1904 BU). There is no change of gelatinization temperatures of tapioca starch after tribomechanical activation.

The enthalpy of gelatinization (ΔH_{gel}), which represents the amount of thermal energy involved in the gelatinization process, was reduced after tribomechanical micronization of tapioca starch. TMA treatment brought about the weakening

TABLE 3: Pasting properties of tapioca starch suspensions TMA treatment (1A (untreated) and 1B (TMA treated)).

Sample	Maximum viscosity (BU)	Breakdown viscosity (BU)	Setback viscosity (BU)	Beginning of gelatinization (°C)
1A	2474 ± 3.97^a	1475 ± 5.77^a	1490 ± 5.36^a	67.2 ± 0.07^a
1B	1990 ± 4.61^b	1105 ± 3.79^b	2091 ± 6.59^b	67.2 ± 0.03^a

The values not statistically different are accompanied by the letter (a) and the values statistically different by the letter (b).

TABLE 4: DSC parameters of native and modified tapioca starch suspensions after TMA treatment (1A (untreated) and 1B (TMA treated)).

Sample	Onset temperature T_o (°C)	Peak temperature T_p (°C)	Conclusion temperature T_c (°C)	ΔT (°C) $T_c - T_o$	Entalpy of gelatinization ΔH (kJ/kg)
1A	65.93 ± 0.03^a	70.48 ± 0.01^a	81.55 ± 0.02^a	15.62 ± 0.03^a	50.626 ± 0.05^a
1B	65.10 ± 0.01^a	70.12 ± 0.01^a	80.65 ± 0.01^a	15.55 ± 0.02^a	49.738 ± 0.02^a

The values not statistically different are accompanied by the letter (a), the values statistically different by the letter (b).

of starch granules, and, consequently, less energy was needed to gelatinize the starch granules. This explains the lower ΔH model suspensions of treated tapioca starch ($\Delta H_{gel} = 49.738$ kJ/kg) as compared with model suspensions of untreated tapioca starch (50.626 kJ/kg). Model suspension of TMA-treated tapioca starch showed lower transition temperatures (T_o, T_P, and T_c) as compared to untreated (Table 4). The decreases in the T_o, T_p, and T_c of tapioca starches upon tribomechanical activation might be due to the weakening of the starch granules, which led to the early rupture of the amylopectin double helices [28]. TMA treatment of starch distorts the crystalline region in starch granules prior to a reversible hydration of the amorphous phase, which results in the destruction of the granular structure [11].

4. Conclusion

This study shows that tribomechanical micronization and activation, which presents a novel method for food processing, gives opportunity to modify and to improve rheological, thermophysical, and some physical properties of tapioca starch. Tribomechanical processing of tapioca starch resulted in breaking accumulations of starch granules in the form of granules. Pasting parameters have shown that maximal viscosities of model starch suspension were decreasing after tribomechanical treatment. Statistically significant increase in solubility and swelling power was observed. TMA treatment caused disrupting of starch granules by mechanical forces and made the granule more permeable to water during the heating step.

The enthalpy of gelatinization was reduced after tribomechanical micronization of tapioca starch. The decreases in the T_o, T_P, and T_c of tapioca starches upon tribomechanical activation might be due to the weakening of the starch granules, which led to the early rupture of the amylopectin double helices and shear thinning of tapioca starch and increased its cold-water solubility. Such behavior is consistent with mechanical activation increasing the amorphous regions of the starch granules while weakening and decreasing the crystalline regions of the starch. Comparing TMA with other methods minimal food processing, tribomechanical activation has the advantages with its simple process, minimal environmental problems and convenient operation.

References

[1] A. Golachowski, "Application of starch and its preparationin food industry," *Zeszyty Naukowe Akademii Rolniczej we Wrocławiu*, vol. 328, pp. 117–124, 1998 (Polish).

[2] P. Tomasik, "Modified starches and their application," *Polish Journal of Food and Nutrition Sciences*, vol. 54, no. 4, pp. 16–18, 2000 (Polish).

[3] T. Funami, Y. Kataoka, T. Omoto, Y. Goto, I. Asai, and K. Nishinari, "Effects of non-ionic polysaccharides on the gelatinization and retrogradation behavior of wheat starch," *Food Hydrocolloids*, vol. 19, no. 1, pp. 1–13, 2005.

[4] L. M. Che, D. Li, L. J. Wang, X. D. Chen, and Z. H. Mao, "Micronization and hydrophobic modification of cassava starch," *International Journal of Food Properties*, vol. 10, no. 3, pp. 527–536, 2007.

[5] L. M. Che, D. Li, L. J. Wang, N. Özkan, X. D. Chen, and Z. H. Mao, "Effect of high-pressure homogenization on the structure of cassava starch," *International Journal of Food Properties*, vol. 10, no. 4, pp. 911–922, 2007.

[6] Y. Liang, B. S. Zhang, L. S. Yang, and D. W. Gao, "Chemicalreactionativity of tapioca starch with non-cristallized granule state," *Journal of Zhengzhou Institute of Technology*, vol. 25, pp. 9–13, 2004.

[7] T. Lelas, "Vorrichtung zum Micronisieren vom Materialien und neuartige Verwendungsmoglichkeiten derartig mikronisierter Materialien," Patent PCT/ 1B99/00757, Geneve, Switzerland, 1998.

[8] V. Lelas, "Novel food processing technologies," *Dairy*, vol. 56, pp. 311–330, 2006.

[9] Z. Herceg, V. Lelas, and M. Skreblin, "Influence of tribomechanical micronisation on the rheological properties of whey proteins," *Food Technology and Biotechnology*, vol. 40, no. 2, pp. 145–156, 2002.

[10] Z. Herceg, V. Lelas, and G. Krešić, "Influence of tribomechanical micronization on the physical and functional properties of whey proteins," *International Journal of Dairy Technology*, vol. 58, no. 4, pp. 225–232, 2005.

[11] Z. Herceg, V. Batur, A. R. Jambrak, M. Badanjak, S. R. Brnčić, and V. Lalas, "Modification of rheological, thermophysical,

textural and some physical properties of corn starch by tribomechanical treatment," *Carbohydrate Polymers*, vol. 80, no. 4, pp. 1072–1077, 2010.

[12] F. E. Ortega-Ojeda and A. C. Elliason, "Gelatinisation and retrogradation behaviour of some starch mixtures," *Starch*, vol. 53, pp. 520–529, 2001.

[13] J. Babić, D. Šubarić, N. Nedić Tiban, and M. Kopjar, "The effect of lactose and whey powder on the gelatinization and retrogradation of tapioca starch," in *Proceedings of the 4th International Symposium on Food Rheology and Structure (ISFRS '06)*, P. Fischer and E. J. Windhab, Eds., pp. 707–708, Zürich, Švicarska, 2006.

[14] J. Babić, D. Šubarić, Đ. Ačkar, V. Piližota, M. Kopjar, and N. Nedić Tiban, "Effect of pectin and carrageenan on thermophysical and rheological properties of tapioca starch," *Czech Journal of Food Sciences*, vol. 24, no. 6, pp. 275–282, 2006.

[15] D. Šubarić, J. Babić, Đ. Ačkar et al., "Effect of galactomannan hydrocolloids on gelatinization and retrogradation of tapioca and corn starch," *Croatian Journal of Food Science and Technology*, vol. 3, no. 1, pp. 26–31, 2011.

[16] C. F. Bohren and D. R. Huffman, *Absorption and Scattering of Light By Small Particles*, Wiley, New York, NY, USA, 1983.

[17] H. W. Leach, L. D. McCowen, and T. J. Schoch, "Structure of the starch granule. Swelling and solubility patterns of various starches," *Cereal Chemistry*, vol. 36, pp. 534–544, 1959.

[18] C. Perera and R. Hoover, "Influence of hydroxypropylation on retrogradation properties of native, defatted and heat-moisture treated potato starches," *Food Chemistry*, vol. 64, no. 3, pp. 361–375, 1999.

[19] A. Hagenimana and X. Ding, "A comparative study on pasting and hydration properties of native rice starches and their mixtures," *Cereal Chemistry*, vol. 82, no. 1, pp. 70–76, 2005.

[20] B. Vidić, T. Vukušić, I. L. Herceg et al., "Influence od high intensity ultrasound on physical and textural properties of wheat starch," *Croatian Journal of Food Technology, Biotechnology*, vol. 4, pp. 134–140, 2011.

[21] Z. Herceg, S. Mededovic Thagard, V. Batur, and A. Režek Jambrak, "Effect of tribomechanical micronisation and activation treatment on textural and thermophysical properties of wheat and potato starch gels," *Food technology and Biotechnology*. In press.

[22] K. S. Sandhu, M. Kaur, N. Singh, and S. T. Lim, "A comparison of native and oxidized normal and waxy corn starches: physicochemical, thermal, morphological and pasting properties," *Food Science and Technology*, vol. 41, no. 6, pp. 1000–1010, 2008.

[23] V. V. Boldyrev, "Mechanochemistry and sonochemistry," *Ultrasonics*, vol. 2, no. 2, pp. S143–S145, 1995.

[24] Z. Q. Huang, J. P. Lu, X. H. Li, and Z. F. Tong, "Effect of mechanical activation on physico-chemical properties and structure of cassava starch," *Carbohydrate Polymers*, vol. 68, no. 1, pp. 128–135, 2007.

[25] G. Krešić, V. Lelas, A. Režek Jambrak, Z. Herceg, and S. Rimac Brnčić, "Influence of novel food processing technologies on the rheological and thermo-physical properties of whey proteins," *Journal of Food Engineering*, vol. 87, pp. 64–73, 2008.

[26] N. Singh, K. S. Sandhu, and M. Kaur, "Characterization of starches separated from Indian chickpea (*Cicer arietinum* L.) cultivars," *Journal of Food Engineering*, vol. 63, no. 4, pp. 441–449, 2004.

[27] J. Singh, L. Kaur, and O. J. McCarthy, "Factors influencing the physico-chemical, morphological, thermal and rheological properties of some chemically modified starches for food applications-a review," *Food Hydrocolloids*, vol. 21, no. 1, pp. 1–22, 2007.

[28] J. E. Hodge and E. M. Osman, "Carbohydrates," in *Food Chemistry*, O. R. Fennema, Ed., Marcel Dekker, New York, NY, USA, 1996.

ZnO Nanoparticles-Chitosan Composite as Antibacterial Finish for Textiles

Asmaa Farouk,[1,2] Shaaban Moussa,[3] Mathias Ulbricht,[4] and Torsten Textor[1,5]

[1] *Deutsches Textilforschungszentrum Nord-West gGmbH (DTNW), 47798 Krefeld, Germany*
[2] *National Research Centre, Textile Division, Textile Chemistry and Technology, Department of Preparation and Finishing of Cellulosic Fibers, Tahrir Street, Dokki, P.O. Box 12622, Giza, Egypt*
[3] *Genetic Engineering and Biotechnology Research Institute, Minoufiya University, P.O. Box 79/22857, El-Sadat City, Egypt*
[4] *Universität Duisburg-Essen, Technical Chemistry II, Universitätsstrasse 7, 45141 Essen, Germany*
[5] *Center for Nanointegration Duisburg-Essen CeNIDE, 47048 Duisburg, Germany*

Correspondence should be addressed to Asmaa Farouk, asmaafarouk2003@yahoo.com

Academic Editor: J. F. Vliegenthart

The antibacterial performance of sol-gel-derived inorganic-organic hybrid polymers filled with ZnO nanoparticles-chitosan against a gram-negative bacterium *Escherichia coli* and a gram-positive *Micrococcus luteus* has been investigated. Three different molecular weights (MW) of chitosan (CTS) $1.36 \cdot 10^5$, $2.2 \cdot 10^5$, and $3.0 \cdot 10^5$ Da with equal degree of deacetylation (DD, 85%) (coded as S 85-60, He 85-250, and He 85-500) with equal degree of deacetylation (DD, 85%) were examined. ZnO was prepared by the base hydrolysis of zinc acetate in isopropanol using lithium hydroxide ($LiOH \cdot H_2O$) to hydrolyze the precursor. Sol-gel-based inorganic-organic hybrid polymers were modified with these oxides and were applied to cellulosic cotton (100%) and cotton/polyester (65/35%) fabrics. Inorganic-organic hybrids polymers were based on 3-glycidyloxypropyltrimethoxysilane (GPTMS). Bacteriological tests were performed in nutrient agar media on solid agar plates and in liquid broth systems using ZnO nanoparticles with average particle size of (40 nm). Our study showed the enhanced antibacterial activity of ZnO nanoparticles chitosan (different MW) of against a gram-negative bacterium *Escherichia coli* DSMZ 498 and a gram-positive *Micrococcus luteus* ATCC 9341 in repeated experiments. The antibacterial activity of textile treated with ZnO nanoparticles chitosan increases with decreasing the molecular weight of chitosan.

1. Introduction

In the recent years, the environmental pollution and the population explosion have forced researchers to find new health-connected products for the human using. For textile industry, there are growing needs to develop finishes for textile materials which can offer improved protection to the user from UV radiation, bacteria, or fungi [1].

There is a great demand for antibacterial finishes on textile, since textile considered an ideal environment for bacterial growth. A huge number of chemicals are used in textile processes. But many of these chemicals are toxic to humans and do not easily degrade in the environment. The textile industry always looks for eco-friendly processes that replace toxic textile chemicals. These facts have facilitated the use of chitosan as a textile chemical. Chitosan, an important and commercially available biopolymer, has many chemical attributes to make it an interesting candidate for textile application. Chitosan has been found to be an interesting biopolymer for immobilization of desired biomolecules because of excellent film-forming ability, high permeability, and mechanical strength, nontoxicity, biocompatibility, easy availability, low cost, and containing reactive amino groups [2].

Chitosan is the deacetylated derivative of chitin (Figure 1), which is the second most abundant polysaccharide found on earth next to cellulose, it is obtained commercially from shrimp or crab shell chitin (a N-acetylglucosamine polymer) by alkaline deacetylation in various grades [3, 4]. It has been reported that chitosan gives considerable antibacterial activity against a broad spectrum with bacteria [5].

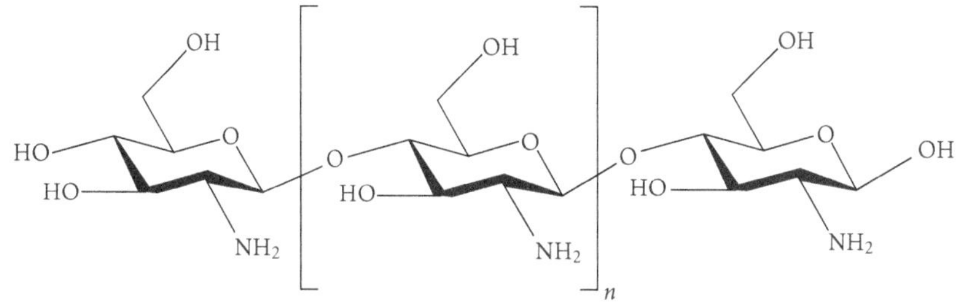

Figure 1: Chemical structure of chitosan.

However, the lack of strong chemical bonding with textile substrate became the major drawback for chitosan antibacterial application because of the problem in durability against repeated laundering [6].

As antibacterial agent, there are many studies that investigate the antibacterial action of ZnO nanoparticles. For example, some of the authors reported that ZnO particles have antibacterial activity against both Gram-positive and Gram-negative bacteria [7]. Yamamoto et al. reported that both the concentration and the surface area of the particles play an important part in the antibacterial activity evaluation, but the crystalline structure and the particle shape have lower effect. Yamamoto also reported that the antibacterial activity increased as the particle size decreased [8, 9]. Adams et al., Brayner et al., and Jeng and Swanson reported that ZnO has a better effect on the microorganisms than other metal oxides such as SiO_2, MgO, TiO_2, and CaO [10–12]. The observations reported by the researchers were explained by a number of mechanisms, these mechanisms include the production of active oxygen species due to the existence of the nanoparticles [13], damage of membrane cell wall because of the binding of the particles on the bacteria surface due to the electrostatic forces [14], penetration through the cell membrane [15], and interaction between the active oxygen species and the cell [16].

The present work was carried out with an objective to study the effect of ZnO nanoparticles sol-chitosan (different MW) composite on the antibacterial properties together with the physical and performance properties of treated fabrics. Figure 2 shows the proposed scheme for ZnO nanoparticle sol/chitosan composite.

2. Experimental

2.1. Materials and Chemicals. The coating process was carried out using 100% cotton (250 g/m^2) and 65/35% cotton/polyester (162 g/m^2) fabrics. Chitosan with average molecular weight of $1.36 \cdot 10^5$, $2.7 \cdot 10^5$, and $3.0 \cdot 10^5$ Da with a degree of deacetylation of 85% and with code name S85-60, 85-250, and 85-500, respectively (Heppe GmbH. Germany); acetic acid, nonionic detergent (Marlipal) from Sasol AG, Germany. 2,3,5-triphenyltetrazolium chloride (TTC) (Merck); Standard I-nutrient broth medium (SI) and Standard I-nutrient agar (SI-agar) (Merk). Zinc acetate dihydrate (ZnAc), lithium hydroxide monohydrate (98%), and 2-propanol were obtained from Merck, (3-Glycidyloxypropyl) trimethoxysilane (GPTMS, 98%) from ABCR. For catalysing the crosslinking reaction of epoxy group of the GPTMS 1-methylimidazole (97%, Fluka) was used.

2.2. Microorganism. *Escherichia coli (E. coli)* DSMZ498 and *Micrococcus luteus* ATCC 9341 were incubated overnight at 37°C in nutrient broth. The culture obtained was diluted with autoclaved nutrient broth to obtain cell suspensions which was adjusted to a final working concentration of 1×10^5 colony forming units (CFU) per mL.

2.3. Preparation of a GPTMS Sol. 10 mL GPTMS are dissolved in 100 mL isopropanol before hydrolysation using 0.01 M hydrochloric acid. The resulting sol is stirred for at least 3 h to form the basis sol (concentration of GPTMS sol 9,1 vol.%).

2.4. Preparation of Zinc Oxide Nanoparticles. The preparation procedure employed has been described in [17]. A two neck round bottom distillation flask was used to suspend 2.8 g of ($ZnAc \cdot 2H_2O$) in 100 mL of isopropanol by reflux heating for three hours. 0.75 g lithium hydroxide was dissolved in 100 mL isopropanol at room temperature by magnetic stirring. The ZnAc suspension was cooled to °C before the lithium hydroxide solution was added dropwise under vigorous stirring. The mixture was treated in an ultrasonic bath (SONOREX TK 52H) at room temperature for about 2 h.

2.5. Characterization of ZnO Nanoparticles. The size of the ZnO nanoparticles was measured by dynamic light scattering (DLS), using Zetasizer, Nano-S, produced by Malvern.

2.6. Preparation of Zinc Oxide Nanoparticles/Chitosan Composite. 0.5% (w/v) of chitosan was dissolved in 1% (v/v) aqueous acetic acid solution by stirring for 2-3 h. The finishing solutions were prepared: zinc oxide nanoparticle (200 mL/45 nm) solution, GPTMS sol (110 mL), and solutions of 100 mL/0.5% chitosan were mixed together and sonicated for 2 h. The pH of the solution containing chitosan was adjusted to be 5.5. To the finishing solution 50 μL of nonionic detergent was added to improve the wettability. 1-methylimidazol (0.5 mL/10 mL GPTMS) was added as

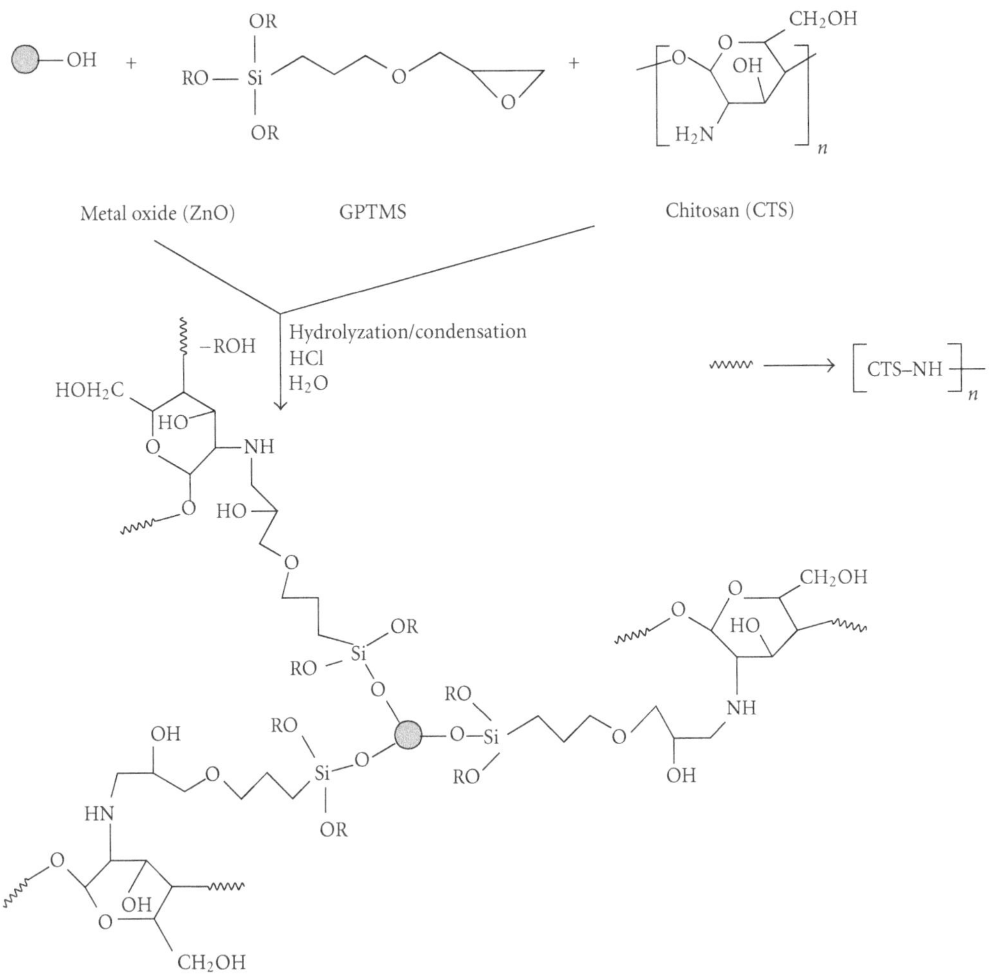

Figure 2: Proposed scheme for the reaction of ZnO sol/chitosan composite.

a catalyst for the cross-linking reaction of the epoxy group of the GPTMS.

2.7. Coating Process. The fabrics were immersed in ZnO nanoparticles solution containing chitosan of different MW for 5–10 min at room temperature. Coatings were carried out using a lab padder (Werner Mathis AG, Switzerland) to a liquor uptake of ~100%. After padding samples were dried in a Labcoater LTE-S (Werner Mathis AG) at 130°C for 30 minutes. Washing was done using a laboratory washing machine at 40°C.

2.8. Testing and Analysis of Zinc Oxide Nanoparticles/Chitosan Composite-Coated Fabrics. Different techniques were used to characterize the finished samples.

(i) *Nitrogen content* of the treated samples was determined according to Kjeldahl method [18].

(ii) *Crease recovery angles (CRA)* of the treated and untreated samples were measured according to DIN 53890 standard test method [19].

(iii) *Tensile strength* (TS) and elongation at break (E%) were measured according to DIN standard test method [20].

(iv) *The wettability* test according to TEGAWA was applied [21].

(v) Stiffness of the fabric was tested using Shirley stiffness tester according to DIN 53362.

(vi) Changes in the degree of whiteness of the treated fabrics were evaluated with a Datacolor spectrophotometer 3880 (cocos Manual Version 2, 3). The degree of whiteness is given according to Berger (light source D65/10).

(vii) *Antibacterial activity* assays were carried out according to the zone of inhibition test, this is the testing method to evaluate the antibacterial activity of textiles treated with antibacterial finish (ZnO nanoparticles/chitosan composite) by the existence of halos, and this is applicable when antibacterial finishing chemicals can easily diffuse into plate agar culture medium.

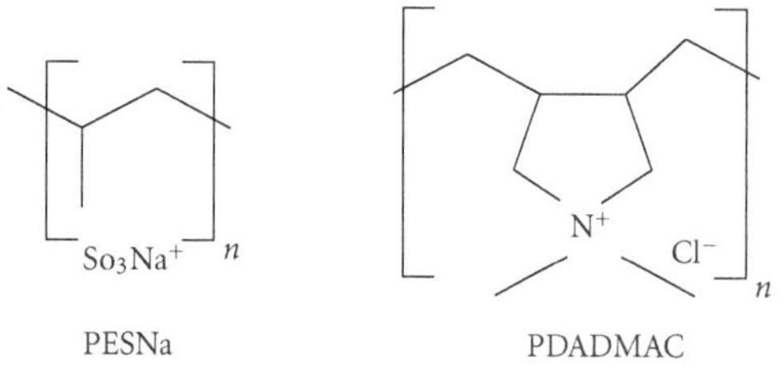

Figure 3: Structure of cationic and anionic polyelectrolyte.

(viii) AATTC Test Method 100–2004 and tetrazolium/formazan test (TTC) which is described in details in [22].

(ix) Laundering was performed in a laboratory washing machine (Linitest, ATLAS) according to 6330 : 2000 for 5 washing cycles using standard ECE (European Colour fastness Establishment) detergent without fluorescent whitening agents (FWA's).

(x) *Polyelectrolyte titration method*: polyelectrolyte titration was carried out using polyelectrolyte titration system, particle charge detector (PCD 03 PH), was from Mütek Analytic GmbH, Herrsching, Germany. One gram climatized treated cottons were immersed in 100 mL acetate buffer solution (pH = 4.7). To this solution 100 μL Marlipal was added. The fabric samples were left under shaking for 2-3 h. Then the samples were washed several times with distilled water and dried at ambient conditions. 50 mL anionic polyelectrolyte (0.001 N polyethlenesulfonic acid sodium salt; PES-Na) (Figure 1) and 100 μL Marlipal were added with stirring, and the solutions with the fabrics were filtered out. 10 mL from the filtrate was titrated against cationic polyelectrolyte; polydiallyl-di-methyl-ammonium-chloride (PDADMAC 0.001 N) (Figure 3). The volume consumed during the titration was called (V_1). For calibration, 10 mL (PESNa) was titrated against (PDADMAC), and the volume determined was called (V_2). The amount of titer consumption at the end point of titration can be used for the calculation of accessible charges according to the following equation:

$$\text{MilliVolt} \cdot \text{equivalent } (\text{mV} \cdot \text{eq.}) = \frac{[(V_2 - V_1) \times 0.001]}{0.2}, \quad (1)$$

that is, V_1 = consumption of (PDADMAC) which is equivalent to the unreacted (PESNa) (eq/mL), V_2 = consumption of (PDADMAC) equivalent to 10 mL (PESNa) (eq/mL), for the correction of the volume which equivalent to the 1 g fabric sample, we multiply the ($V_2 - V_1$) by factor 5 [4].

3. Results and Discussion

3.1. SEM Investigation. SEM investigations were carried out to examine changes in the topography after treatment process compared to the blank sample. The corresponding SEM micrographs are shown in Figure 4. As shown from the figure, after GPTMS-ZnO (30 nm) nanoparticles/chitosan composite the surface of the treated fabric became smoother compared with the untreated fabrics or fabric treated with only CTS due to the homogenious distribution of ZnO-CTS composite within the coating layer.

3.2. Antibacterial Activity Measurement of ZnO Nanoparticles Solution Containing Chitosan of Different MW

3.2.1. Qualitative Evaluation. In this study, the cellulosic fabrics were selected for the evaluation of the relative antibacterial activity of ZnO sol and ZnO sol/chitosan composite against a gram-negative bacterium *E. coli,* and a gram-positive *Micrococcus lutues* was studied qualitatively in aqueous LB broth by halo method (zone of inhibition). As shown in Figure 5, the antibacterial activity of finished fabric treated with ZnO sol/chitosan increased with chitosan MW decreased, this phenomenon was explained as the decrease of the MW of chitosan improves its solubility and improves the movement of polymer chains in the solution by decreasing the viscosity.

In Figure 5, the antibacterial activity against *E. coli* and *M. luteus* shows a clear zone of inhibition within and around the disc impregnated with (1) ZnO sol, (2) treated with ZnO sol/chitosan (higher MW) composite, (3) treated with ZnO sol/chitosan (medium MW), and (4) treated with ZnO sol/chitosan (lower MW).

3.2.2. Quantitative Evaluation. (The AATTC test method 100-2004) was first developed in 1961 by the AATCC Committee RA31, revised several times during the years, and reaffirmed in 1986 and 2004. The purpose of the AATCC Test Method is to "provide a quantitative procedure for the evaluation of the degree of antibacterial activity" of finishes on textile materials [23]. The method is based on inoculating test and control textile materials with chosen test organisms at $1-2\times10^5$ (colony-forming units) CFU/mL, usually in Nutrient broth. After incubation at 37°C for 18–24 hours, the microorganisms are eluted from the test samples by vigorous shaking in known amounts of solution, chosen to neutralize the antimicrobial effect. The number of microorganisms in the liquid is then determined via agar plating, and the percentage reduction by the treated samples is then calculated [23]. After that the number of bacteria was counted, and the reduction of bacteria, R, was calculated as follows:

$$R\% = \frac{B - P}{B} \times 100, \quad (2)$$

where: R = reduction in percentage, B = CFU/test sample in blank, P = CFU/test sample for treated samples.

The antibacterial activity of the coated fabrics was estimated for the *E. coli* and *M. lutues* according to the AATTC 100-2004 standard method. Figure 6 shows *E. coli*

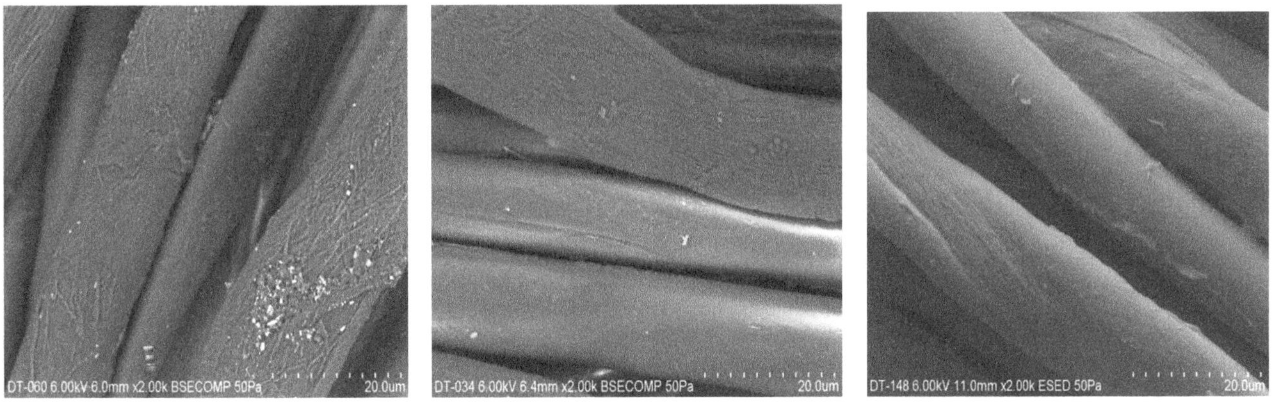

(a) (b) (c)

FIGURE 4: SEM photograph of the (a) untreated cotton/polester (65/35%) fabric, (b) chitosan-treated fabric, and (c) ZnO nanoparticle sol/chitosan composite.

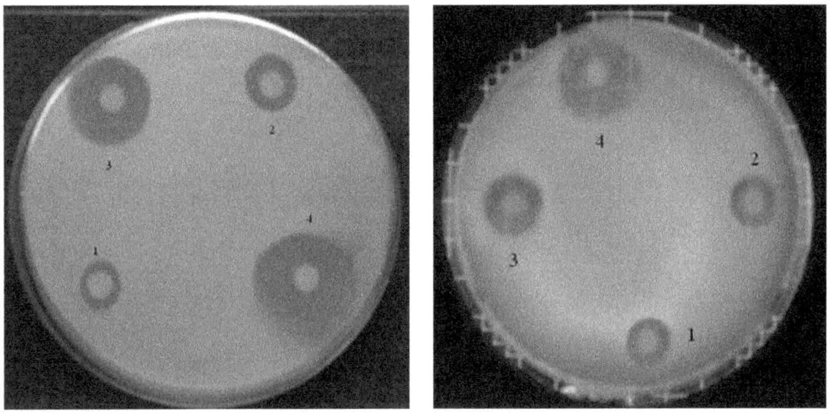

(a) (b)

FIGURE 5: The qualitative disc diffusion test of cotton treated with ZnO sol/chitosan composite for the growth inhibition of (a) *E. coli* and (b) *M. luteus*. Both photographs show the clear zone of inhibition around the sample impregnated with ZnO nanoparticles compared with ZnO sol/chitosan (with different MW) composite.

and *M. lutues* bacterial growth with the time for GPTMS-ZnO nanoparticles/chitosan (0.5%) (low MW) composite-treated fabrics, (a) *E. coli* bacterial growth, and (b) *M. Lutues* bacterial growth.

From Figure 6, it was observed that, in case of both *E. coli* and *M. lutues*, there is nearly 100% reduction of the number of colonies after 3 h in case of cotton 100% treated fabrics against *E. coli* and *M. lutues* bacteria. In case of cotton 100% fabrics treated with GPTMS-CTS and GPTMS-ZnO the complete reduction of the number of colonies was observed after 4 h and 5 h, respectively.

From Figure 6 we can note that there is nearly 100% reduction of the number of colonies after 3 h in case of cotton 100% treated fabric and cotton/polyester (65/35%) treated fabric against *E. coli* and *M. lutues* bacteria.

It has been reported that chitosan confers considerable antibacterial activity against a broad spectrum of bacteria [24], and the interaction between positively charged chitosan and negatively charged bacterial cell wall leads to the leakage of intracellular constituents.

3.2.3. Quantitative and Qualitative Evaluation. The antibacterial activity of the treated fabric samples was evaluated against *E. coli* and *Micrococcus luteus* using the TTC test method. Details about this method are described by Fouda et al. [22]. This test serves as an indicating system for the determination of the viability of microorganisms, since that absorbance of formazane is directly proportional to the amount of living bacteria as shown in Figure 7.

As we observed from the formazan absorbance value in Figure 7, cotton/polyester fabric treated with ZnO nanoparticles sol/chitosan (low MW) composite has the lower absorbance value of formazan, indicating the higher antibacterial activity, compared to the formazan absorbance value of cotton/polyester treated with GPTMS-ZnO sol or cotton/polyester treated with GPTMS-CTS, we can see also that the antibacterial activity of cotton/polyester treated with GPTMS-CTS composite is more higher than that treated with GPTMS-ZnO sol. A presence of ZnO in the coating yields a reduction of more than 70% within three hours; a presence of CTS in the coating improves this value to more

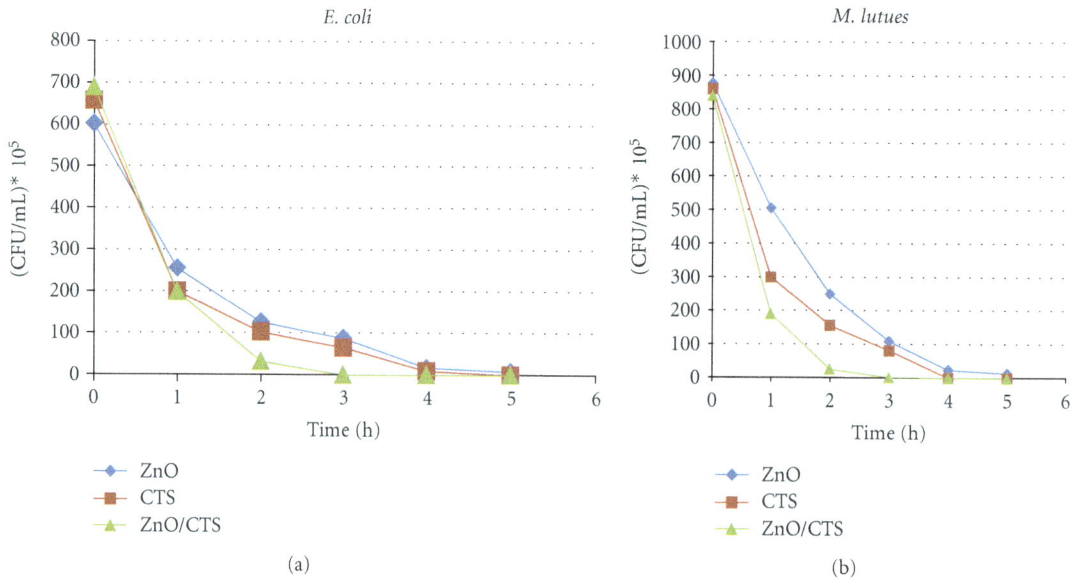

FIGURE 6: *E. coli* and *M. lutues* bacterial growth for ZnO nanoparticles sol/chitosan (low MW) composite-treated fabrics, (a) cotton 100% treated fabric, and (b) cotton/polyester (65/35%) treated fabric.

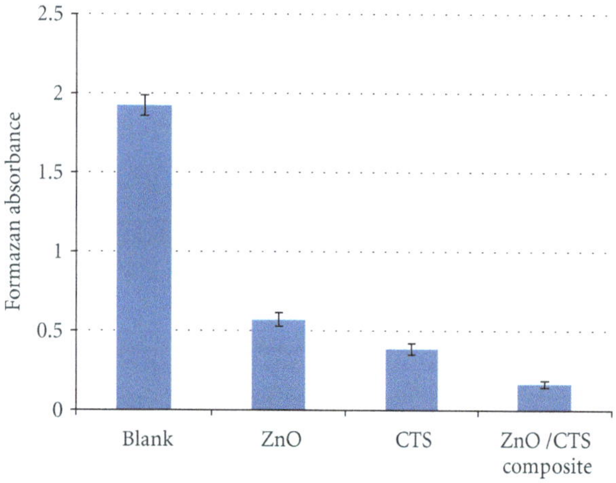

FIGURE 7: Absorbance of formazan for different cotton/polyester fabric treatment.

(a) (b) (c)

FIGURE 8: Dropping test for the identification of amino groups immobilized on treated cotton surface. (a) Cotton blank, (b) cotton treated with ZnO sol/chitosan composite, and (c) cotton treated with ZnO sol/chitosan (low MW) composite after washing.

than 80%, and the presence of ZnO/CTS composite in the coating yields a reduction of approximately 92%.

The antibacterial activity is in this order:

> GPTMS-ZnO-CTS composite > GPTMS-CTS composite > GPTMS-ZnO sol.

3.3. Characterization of the Amino Groups by Dropping Colour. After the application of GPTMS-ZnO-CTS composite to the textile surface using the pad-dry-cure method, ninhydrin test as colouring test was done to determine the present degree of the $-NH_2$ group of the Chitosan molecule on textile surface, since $-NH_2$ group serves as basis for the test of Chitosan. This test is used mainly in order to discover the immobilized amino groups which are at the surface of finished fabrics which is considered most important factor for the antibacterial activity of chitosan.

Figure 8 is a dropping test for the identification of immobilized amino groups on treated cotton surface before and after washing. The figure shows that this test was positive for all finished chitosan-containing fabrics since the brownish colour of ninhydrin converts to violet as a result of the reaction with the chitosan amino groups (ninhydrin reacts with the free amino group of chitosan and develops a violet colour). The respective blank test showed no effect as shown.

3.4. Polyelectrolyte Titration and Nitrogen Content of the Treated Fabrics. Polyelectrolyte titration was carried out in order to determine the amount of charged amino groups immobilized on cotton and cotton/polyester surfaces. Polyelectrolyte titration measurements on the treated fabrics may be used to assess the availability of ionic charges on the fabrics which give information on the (statistical) sites

Table 1: Polyelectrolyte titration and nitrogen content results.

Treated cotton with	Amount of nitrogen (mmol/g fibre)	Amount of cationic charges (meq/g fibre)
0.5% chitosan (lower MW)	0.32	0.040
0.5% chitosan (lower MW)/GPTMS	0.53	0.057
GPTMS-ZnO (10%, 30 nm)	—	0.032
GPTMS-ZnO (10%, 30 nm)/0.5% chitosan composite	0.55	0.065

Table 2: Effect of ZnO sol—chitosan on some performance properties of cotton and cotton blend fabric samples.

Substrate		Tensile strength (daN)	Elongation (%)	Air permeability ($l/dm^2 \cdot min.$)	Degree of whiteness (Berger)	Bending Stiffness ($cN\,cm^2$)	CRA (W+F)°
Untreated	Cotton	102.1 ± 0.75	21.9 ± 1.06	250.5	66.0	12.19 ± .025	143°
	CO/PET	87.4 ± 1.02	31.7 ± 1.04	334.0	70.9	3.23 ± .029	175°
Treated with ZnO in GPTMS-sol (10%)	Cotton	96.5 ± 1.25	22.9 ± 1.15	258.8	56.1	13.64 ± .014	150°
	CO/PET	86.6 ± 1.35	33.3 ± 1.45	350.7	68.3	5.47 ± .017	181°
Treated with chitosan/GPTMS	Cotton	107.7 ± 1.47	23.8 ± 0.95	245.3	53.7	15.88 ± .014	180°
	CO/PET	94.2 ± 1.025	33.7 ± 1.02	328.4	64.9	7.38 ± .022	196°
Treated with chitosan of low MW	Cotton	104.8 ± 1.04	21.9 ± 1.14	247.0	54.8	15.12 ± .024	178°
	CO/PET	91.3 ± 1.14	31.9 ± 0.85	330.5	65.6	6.65 ± .041	187°
Treated with ZnO sol/chitosan	Cotton	107.2 ± 1.03	23.5 ± 1.12	243.2	53.1	16.93 ± .017	181°
	CO/PET	92.7 ± 0.98	33.0 ± 1.07	326.1	64.3	8.81 ± .026	197°
Treated with GPTMS	Cotton	99.8 ± 1.01	21.5 ± 1.22	257.6	56.9	13.11 ± .028	151°
	CO/PET	86.3 ± 1.41	31.4 ± 1.31	350.0	69.4	4.83 ± .025	181°

of crosslinker and on possible chain segment mobility of biopolymer on fibre [22].

As it was observed from Table 1 that the amount of nitrogen and the amount of cationic charges of only chitosan treated fabric are weak compared to fabric treated with 0.5% chitosan (lower MW)/GPTMS which were explained by effect of GPTMS which works as crosslinking agent for the chitosan into the cotton fabric surface. Also the amount of cationic charges increased in case of cotton treated with GPTMSZnO (10%, 30 nm)/0.5% chitosan composite due to the presence of cationic charge from both chitosan as well as ZnO nanoparticles surface.

In case of nitrogen content measurements all the amount of nitrogen can be detected, but in case of polyelectrolyte titration the amount of cationic charges which was detected is lower than the expected value. This is because of that not all nitrogen charges are available for the polyelectrolyte titration, since the cationic charges which were detected are only the charges on the surface.

The mechanical data of the treated and untreated samples and the influence of the ZnO sol/chitosan composite treatment for the degree of whiteness, the air permeability, stiffness, tensile strength, and crease recovery angle were investigated and the corresponding data are given in Table 2.

(i) As it was expected, the tensile strength of fabric samples treated with chitosan, chitosan/GPTMS, and ZnO sol/chitosan slightly increased. The increase may be attributed to the penetration of chitosan molecule which made some sort of sizing to the treated fabrics and enhance the tensile strength.

(ii) The CRA enhancement of the chitosan finished fabrics suggests that chitosan undergoes cross-linking with GPTMS on the fabric to form network hybrid matrix.

(iii) Also the whiteness of the fabrics is slightly decreased, but in a tolerable degree.

(iv) Air permeability is an important factor in the performance of textile materials used to give an indication to the breathability of the coated fabrics, and within the investigation summarized in Table 2 it can be seen that there is no reduction of the air permeability values in case of fabrics treated with GPTMS and ZnO in GPTMS-sol but at least slight increased, but in case of fabrics treated with chitosan a slight decrease could be due to closing of open interstices caused by mechanical deposition of the finish [25].

(v) Bending stiffness of the fabrics increasing could be due to the higher uptake of the inorganic-organic hybrids polymers containing chitosan molecules on cellulosic fabric surface which made the structure more rigid.

4. Conclusion

The present work is a complete part of previously work done on studying of the antibacterial activity of ZnO nanoparticle sol. In order to investigate the effect of the MW on the antimicrobial activity of chitosan, cellulosic fabrics were treated with three chitosans with different MWs and similar DDA. The antimicrobial activity of the treated fabrics was tested in terms of the reduction rate using two strains of bacteria. In addition, the performance properties of the treated fabrics were evaluated. From the results we found that ZnO sol/chitosan (low MW) has better antibacterial activity since the decrease of molecular weight of chitosan increases the antibacterial activity, and this phenomenon was explained as the decrease of chitosan molecular weight in the composite improves the movement of the chains in the solution by decreasing the viscosity.

Acknowledgment

Very special thanks are due to the Egyptian Missions Department for giving A. Farouk the opportunity and the financial support to pursue her Ph.D. degree in Germany. Further financial support by the Ministerium für Innovation, Wissenschaft, Forschung und Technologie des Landes Nordrhein-Westfalen (Department of Innovation, Science, Research and Technology of the state of Nordrhein-Westfalen) is thankfully acknowledged.

References

[1] V. R. G. Dev, J. Venugopal, S. Sudha, G. Deepika, and S. Ramakrishna, "Dyeing and antimicrobial characteristics of chitosan treated wool fabrics with henna dye," *Carbohydrate Polymers*, vol. 75, no. 4, pp. 646–650, 2009.

[2] R. Khan, A. Kaushik, P. R. Solanki, A. A. Ansari, M. K. Pandey, and B. D. Malhotra, "Zinc oxide nanoparticles-chitosan composite film for cholesterol biosensor," *Analytica Chimica Acta*, vol. 616, no. 2, pp. 207–213, 2008.

[3] D. Knittel and E. Schollmeyer, "Chitosan und seine derivate für die textilveredlung," *Melliand Textil Berichte*, vol. 1, pp. 58–61, 2002.

[4] M. M. G. Fouda, R. Wittke, D. Knittel, and E. Schollmeyer, "Use of chitosan/polyamine biopolymers based cotton as a model system to prepare antimicrobial wound dressing," *International Journal of Diabetes Mellitus*, vol. 1, no. 1, pp. 61–64, 2009.

[5] C. H. Su, C. S. Sun, S. W. Juan, C. H. Hu, W. T. Ke, and M. T. Sheu, "Fungal mycelia as the source of chitin and polysaccharides and their applications as skin substitutes," *Biomaterials*, vol. 18, no. 17, pp. 1169–1174, 1997.

[6] W. Ye, J. H. Xin, P. Li, K. L. D. Lee, and T. L. Kwong, "Durable antibacterial finish on cotton fabric by using chitosan-based polymeric core-shell particles," *Journal of Applied Polymer Science*, vol. 102, no. 2, pp. 1787–1793, 2006.

[7] J. Sawai, "Quantitative evaluation of antibacterial activities of metallic oxide powders (ZnO, MgO and CaO) by conductimetric assay," *Journal of Microbiological Methods*, vol. 54, no. 2, pp. 177–182, 2003.

[8] O. Yamamoto, M. Hotta, J. Sawai, T. Sasamoto, and H. Kojima, "Influence of powder characteristic of ZnO on antibacterial activity—effect of specific surface area," *Journal of the Ceramic Society of Japan*, vol. 106, no. 10, pp. 1007–1011, 1998.

[9] O. Yamamoto, "Influence of particle size on the antibacterial activity of zinc oxide," *International Journal of Inorganic Materials*, vol. 3, no. 7, pp. 643–646, 2001.

[10] L. K. Adams, D. Y. Lyon, and P. J. J. Alvarez, "Comparative ecotoxicity of nanoscale TiO_2, SiO_2, and ZnO water suspensions," *Water Research*, vol. 40, no. 19, pp. 3527–3532, 2006.

[11] R. Brayner, R. Ferrari-Iliou, N. Brivois, S. Djediat, M. F. Benedetti, and F. Fiévet, "Toxicological impact studies based on *Escherichia coli* bacteria in ultrafine ZnO nanoparticles colloidal medium," *Nano Letters*, vol. 6, no. 4, pp. 866–870, 2006.

[12] H. A. Jeng and J. Swanson, "Toxicity of metal oxide nanoparticles in mammalian cells," *Journal of Environmental Science and Health Part A*, vol. 41, no. 12, pp. 2699–2711, 2006.

[13] S. Makhluf, R. Dror, Y. Nitzan, Y. Abramovich, R. Jelinek, and A. Gedanken, "Microwave-assisted synthesis of nanocrystalline MgO and its use as a bacteriocide," *Advanced Functional Materials*, vol. 15, no. 10, pp. 1708–1715, 2005.

[14] P. K. Stoimenov, R. L. Klinger, G. L. Marchin, and K. J. Klabunde, "Metal oxide nanoparticles as bactericidal agents," *Langmuir*, vol. 18, no. 17, pp. 6679–6686, 2002.

[15] H. Xu, H. Wang, Y. Zhang et al., "Hydrothermal synthesis of zinc oxide powders with controllable morphology," *Ceramics International*, vol. 30, no. 1, pp. 93–97, 2004.

[16] L. Zhang, Y. Jiang, Y. Ding, M. Povey, and D. York, "Investigation into the antibacterial behaviour of suspensions of ZnO nanoparticles (ZnO nanofluids)," *Journal of Nanoparticle Research*, vol. 9, no. 3, pp. 479–489, 2010.

[17] A. Farouk, T. Textor, E. Schollmeyer, A. Tarbuk, and A. M. Grancacic, "Sol-gel-derived inorganic-organic hybrid polymers filled with zno nanoparticles as an ultraviolet protection finish for textiles," *Autex Research Journal*, vol. 10, no. 3, pp. 58–63, 2010.

[18] A. I. Vogel, *Text-Book of Quantitative Inorganic Analysis (Elemental Practical Inorganic Chemistry) Part 3, Quantitative Inorganic Analysis*, Longman, 1975.

[19] German DIN Book, "Methods of Analysis," 53890, January 1972.

[20] German DIN Book, "Methods of Analysis," 255–262, 1979.

[21] S. Breichter, P. Wurster, and G. Schmidt, *Melliand Text Berichte*, vol. 8, pp. 581–583, 1987.

[22] M. Fouda, D. Knittel, R. Zimehl, C. Hipler, and E. Schollmeyer, "Functionalization of fibre surfaces by thin layers of chitosan and the antimicrobial activity," *Advances in Chitin Science*, vol. 8, p. 418, 2005.

[23] AATCC Committee, Antibacterial Finishes on Textile Materials: Assessment of 2004, AATCC, 2004.

[24] C. H. Su, C. S. Sun, S. W. Juan, C. H. Hu, W. T. Ke, and M. T. Sheu, "Fungal mycelia as the source of chitin and polysaccharides and their applications as skin substitutes," *Biomaterials*, vol. 18, no. 17, pp. 1169–1174, 1997.

[25] A. Karolia and S. Mendapara, "Imparting antimicrobial and fragrance finish on cotton using chitosan with silicon softener," *Indian Journal of Fibre and Textile Research*, vol. 32, no. 1, pp. 99–104, 2007.

Conformational Search on the Lewis X Structure by Molecular Dynamic: Study of Tri- and Pentasaccharide

N. Khebichat, K. Nekkaz, and S. Ghalem

Laboratory of Natural Products and Bio Actives, Department of Chemistry, Faculty of Sciences, Aboubakr Belkaid University, Tlemcen 13000, Algeria

Correspondence should be addressed to S. Ghalem, s_ghalem2002@yahoo.fr

Academic Editor: Yuriy A. Knirel

Carbohydrates play vital roles in many biological processes, such as recognition, adhesion, and signalling between cells. The Lewis X determinant is a trisaccharide fragment implicated as a specific differentiation antigen, tumor antigen, and key component of the ligand for the endothelial leukocyte adhesion molecule, so it is necessary or essential to determine and to know their conformational and structural properties. In this work, conformational analysis was performed using molecular dynamics (MD) simulation with the AMBER10 program package in order to study the dynamic behavior of of the Lewis X trisaccharide (β-D-Gal-(1,4)-[α-L-Fuc-(1,3)]-β-D-GlcNAc-OMe) and the Lewis X pentasaccharide (β-D-Gal-(1,4)-[α-L-Fuc-(1,3)]-β-D-GlcNAc-(1,3)-β-D-Gal-(1,4)-β-D-Glu-OMe) in explicit water model at 300 K for 10 ns using the GLYCAM 06 force field.

1. Introduction

In nature, carbohydrates form an important family of biomolecules, as simple or complex carbohydrates, either alone or covalently linked to proteins or lipids [1]. They play vital roles in many biological processes, such as recognition, adhesion, and communication between cells. Carbohydrates that are covalently linked to a nonsugar moiety (proteins, peptides, or lipids) are the most prominent cell-surface-exposed structures. The highly diverse structural variability of carbohydrates makes them good candidates for cell receptors and recognition molecules [2].

Determination of the three-dimensional structure of oligosaccharides and understanding the molecular basis of their recognition by receptors represent the main challenges of structural glycobiology. Elucidation of the three-dimensional structures and the dynamical properties of oligosaccharides is a prerequisite for a better understanding of the biochemistry of recognition processes and for the rational design of carbohydrate-derived drugs [1].

The human histo-blood group ABH (O) systems were the first major human alloantigens to be identified. The carbohydrate nature of the A, B, H, and Lewis antigens was established in 1950s [3, 4]. Figure 1 shows a schematic representation of current histo-blood group antigens although additional rare types also exist [5]. The oligosaccharidic epitope of ABH (O) and Lewis histo-blood groups have been the subject of many structural investigations. The energy maps of all of the disaccharides have been established by molecular mechanics methods [6, 7].

The determination of conformational preferences of oligosaccharides is best approached by describing their preferred conformations on potential energy surfaces as a function of the glycosidic linkage φ, ψ torsional angles. In the early days of conformational analysis of oligosaccharides, the calculations of either rigid or, later, relaxed conformational maps predominated. Today, complex simulation techniques such as molecular dynamics are increasingly being used. Several examples can be found in the literature of "state-of-the-art" conformational search strategies in order to simplify the complexity of the multiple-minima problem [8].

In this work, we carried out extensive, explicit solvent MD simulation studies for tri- and pentasaccharide Lewis X for 10 ns. Each monosaccharide unit in the Lewis X molecules adopts the typical $^4C_1$ chair conformation, with no significant deviation from the classical pyranose ring shape.

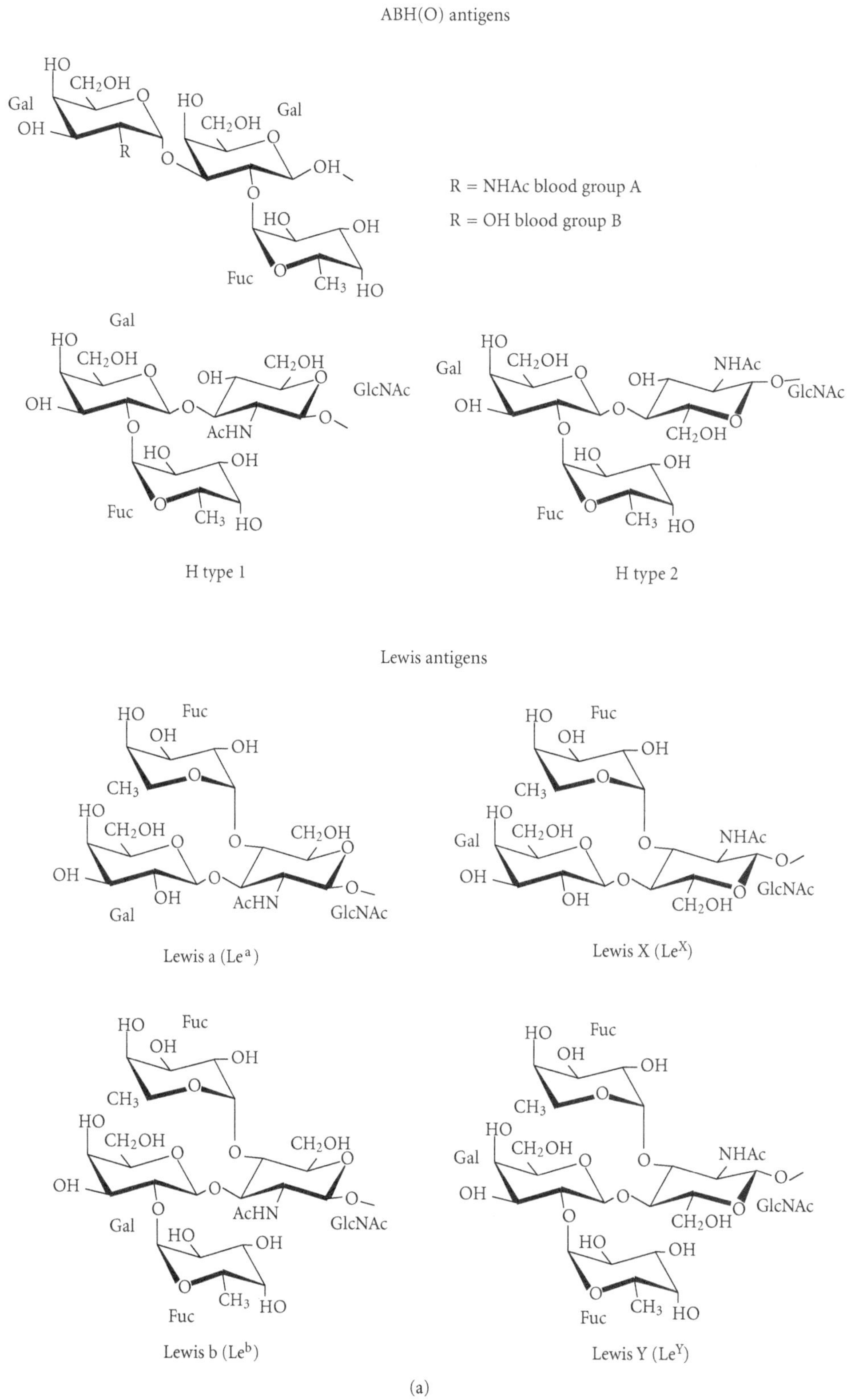

Figure 1: Continued.

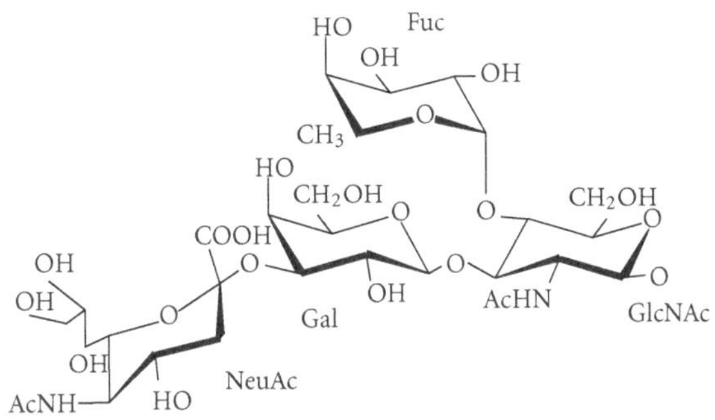

Sialyl Lewis a (SLe^a)

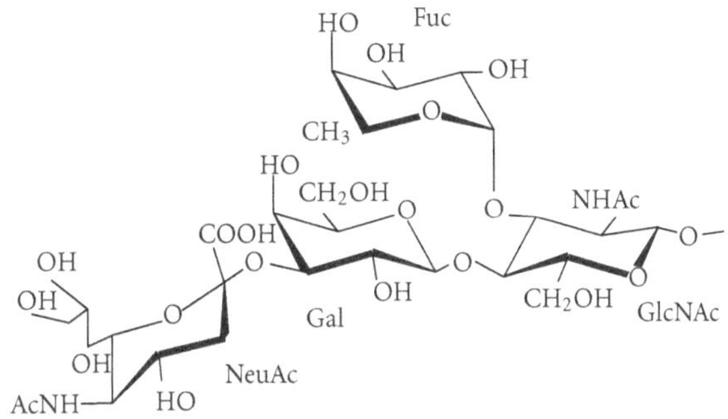

Sialyl Lewis X (SLe^X)

(b)

FIGURE 1: Schematic representation of histo-blood group oligosaccharides.

The objective of this study was to explore the conformational behaviour of those oligosaccharides in explicit water model at 300 K for 10 ns using the GLYCAM 06 force field.

2. Theoretical Approach

The atomic labelling and the torsion angles of interest are given in Figure 2. The torsion angles φ, ψ are defined $\varphi = \vartheta$ (05-Cl-0′2-C′2), $\psi = \vartheta$ (Cl-0′2-C′2-C′3), and $\omega = \vartheta$ (O5-C5-C6-O6). The conformations with specific torsion angles ω are referred to as gauche-trans (gt; $\omega = 60°$), gauche-gauche (gg; $\omega = -60°$) and trans-gauche (tg; $\omega = 180°$).

The MD simulations were carried out using AMBER10 [9] at a temperature of 300 K. This temperature was chosen because our primary objective was to study the dynamics of the oligosaccharides at physiologically relevant conditions. The starting Lewis X pentasaccharide and trisaccharide monomer coordinates and parameters were constructed with the "Glycam Biomolecule Builder" available online from the website of Woods group [10] (see Schemes 1 and 2).

Since these coordinates are not optimized, a first minimization by molecular mechanics (1,000 steps of steepest descent followed by 1,000 steps of conjugate gradient minimization with a gradient tolerance of 0.05 kcal mol^{-1}) was made with AMBER package version 10 [9]. Force field GLYCAM 06 was used for oligosaccharides (it used to simulate the interatomic interactions responsible for oligosaccharide and glycoprotein dynamics and is currently suitable for all biologically relevant N- and O-linkages in oligo- and polysaccharides and glycoproteins [11]). The simulations were carried out using an explicit solvent environment with AMBER package version 10. The solvated system for the oligosaccharides molecules was prepared using an explicit water box of 8 Å in the xleap module of AMBER10. The number of water molecules was 569 for the trisaccharide and 1072 for the pentasaccharide. All simulations were run with the SANDER module of AMBER with SHAKE algorithm [12] (tolerance = 0.0005 Å) to constrain covalent bonds involving hydrogens, using periodic boundary conditions, a 2 fs time step, a temperature coupling [13], a 8 Å cutoff was applied to the Lennard-Jones interaction, and constant pressure of 1 atm. The nonbonded list was updated every

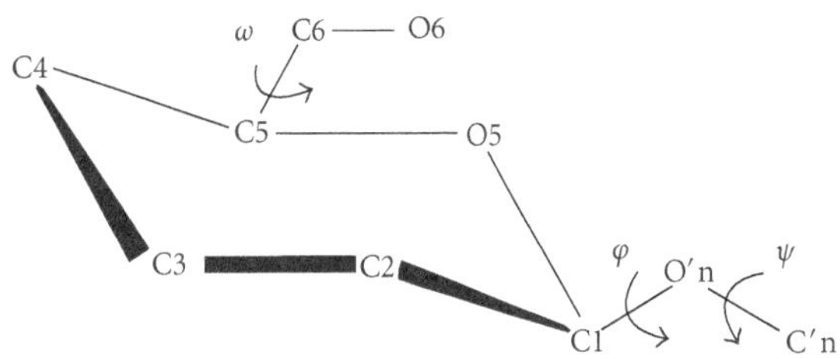

FIGURE 2: Atomic labeling and torsion angles of interest.

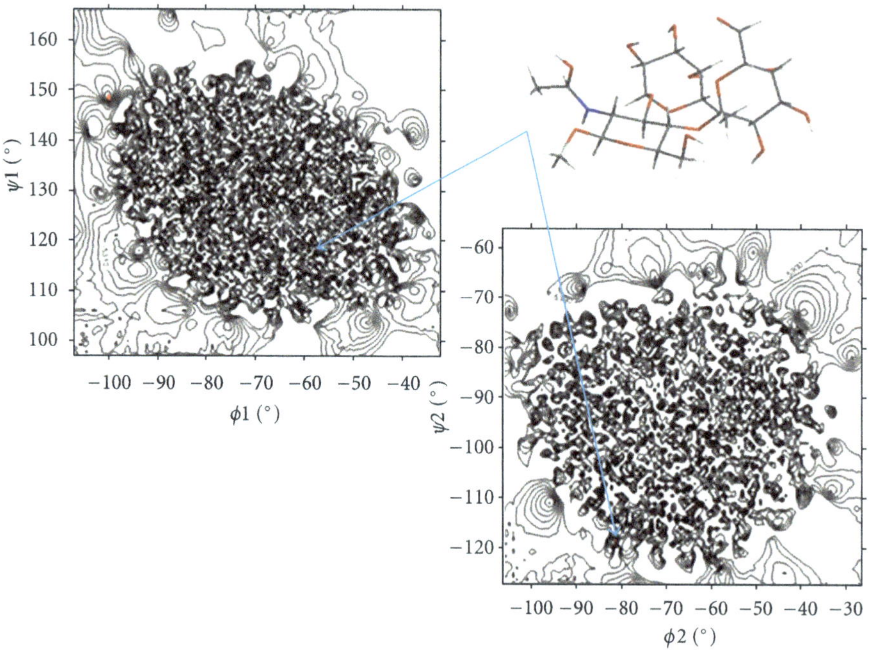

FIGURE 3: Relaxed conformational energy maps for the trisaccharide Lewis X calculated using the GLYCAM 06 force field.

10 steps. This preliminary study was aimed at finding the most energetically favorable structure which was found at 4856.9 ps for the trisaccharide and at 2411.8 ps for the pentasaccharide. Those conformations were found with a "home-made" Perl script that merely checked all energies values, finding the lowest one and giving the associated structure [2].

The PTRAJ module of AMBER10 was used to analyze and process the trajectory and coordinate files from SANDER, where the analyses might include carrying out superimposition, clustering analysis of hydrogen bonds, calculating fluctuations in bonds, angles, or dihedrals, correlation functions, and so forth, and the root mean-square deviations (RMSDs).

Finally, isoenergy contour maps were generated with ORIGIN and SURFER 9 (Golden Software, Inc., Golden Co) [14]. The positions of the minima in these maps were subsequently calculated by minimizing the energy without restraints of those grid points indicating minima on the maps.

3. Results

The calculated adiabatic conformational energy surfaces (φ, ψ) for the tri- and pentasaccharide Lewis X are presented as potential energy contour maps in Figures 3 and 4, respectively. The contours are separated by 1 kcal/mol and range from 1 to 10 kcal/mol. The energies are given relative to the lowest minimum. Geometries and relative potential energies of the predicted minima are listed in Tables 1 and 2, and the lowest energy conformers for each compound are shown in Figures 3 and 4.

TABLE 1: Potential Energy, torsional angles for the conformers with low energies of the trisaccharide Lewis X.

	PE (kcal/mol)	φ_1 (°)	ψ_1(°)	φ_2 (°)	ψ_2 (°)
Conformer 1	−5446.95	−61.84	124.7	−61.54	−104.33
Conformer 2	−5442.50	−84.91	155.63	−69.95	−92.95
Conformer 3	−5440.08	−80.60	139.25	−55.46	−98.51
Conformer 4	−5437.12	−56.58	130.28	−59.61	−80.28

3.1. System Root-Mean-Square Displacement. RMSD is an important parameter for describing the system stability. Figures 5 and 6 show the function curves between mass weighted RMSD fit and the time of the tri- and pentasaccharide Lewis X during dynamics simulation process, respectively.

3.2. Computation of Phi/Psi for Modeled Oligosaccharide. In the present study, structure prediction of oligosaccharides has been made through the torsion angles phi (ϕ)/psi (ψ). The combination of ϕ and ψ angles fully determine the conformation of an oligosaccharide. Figures 3 and 4 show the distribution of sterically allowed and energy minimized conformations in ϕ/ψ glycosidic space obtained from our explicit solvent MD simulations for 10 ns in the case of the tri- and pentasaccharide Lewis X, respectively. These figures also show the best conformers (which have the low energies). These conformers and their corresponding ϕ/ψ angular values are reported in Tables 1 and 2 for the tri- and pentasaccharide Lewis X, respectively.

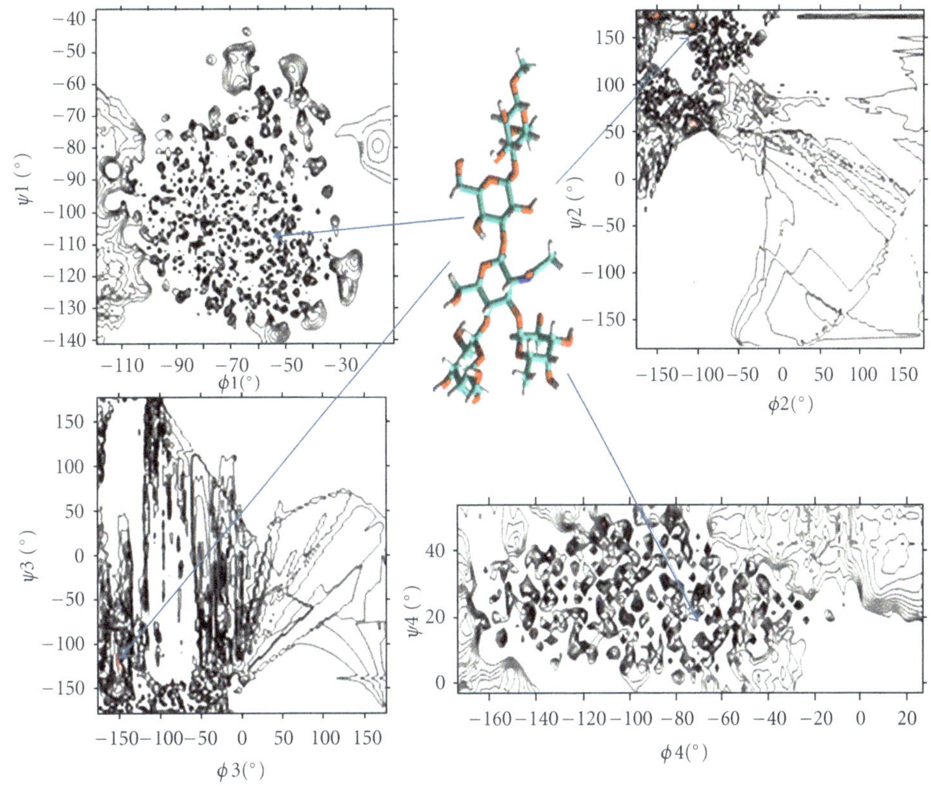

Figure 4: Relaxed conformational energy maps for the pentasaccharide Lewis X calculated using the GLYCAM 06 force field.

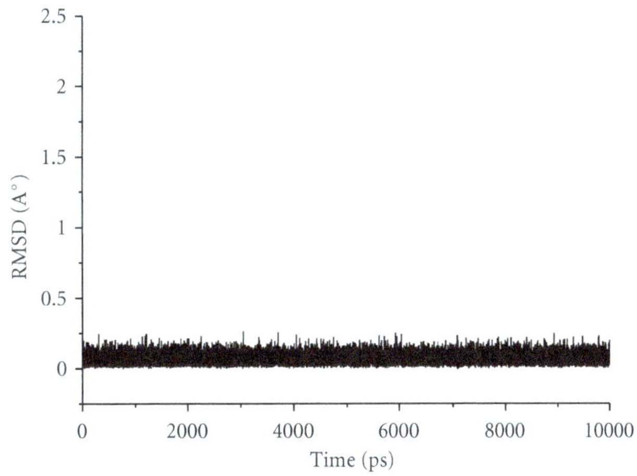

Figure 5: RMSD of the trisaccharide Lewis X as a function of time during MD, compared to the lowest-energy conformation.

Table 2: Potential energy, torsional angles for the conformers with low energies of the pentasaccharide Lewis X.

	PE (kcal/mol)	φ_1(°)	ψ_1 (°)	φ_2 (°)	ψ_2 (°)	φ_3 (°)	ψ_4 (°)	φ_4 (°)	ψ_4 (°)
Conformer 1	−10191.15	−78.30	−97.88	−124.4	64.97	−108.27	−173.37	−57.54	15.18
Conformer 2	−10162.71	−65.65	−106.09	−73.57	136.95	−52.97	−155.29	−71.26	22.90
Conformer 3	−10162.25	−56.49	−112.26	−78.58	144.67	−55.95	−157.92	−69.85	23.71
Conformer 4	−10152.28	−71.67	−108.12	−75.73	152.13	−73.72	−163.60	−74.94	18.04

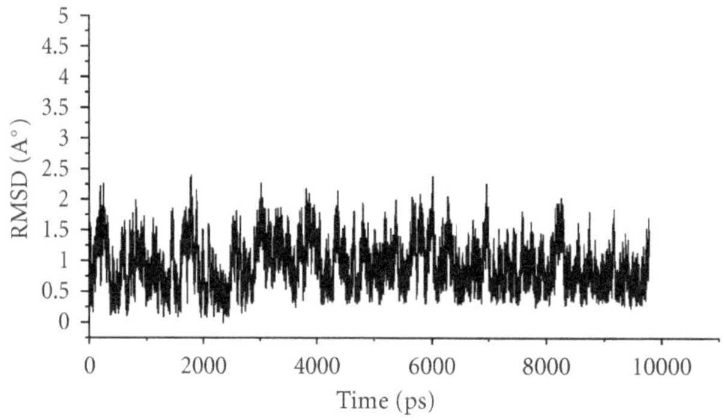

FIGURE 6: RMSD of the pentasaccharide Lewis X as a function of time during MD, compared to the lowest-energy conformation.

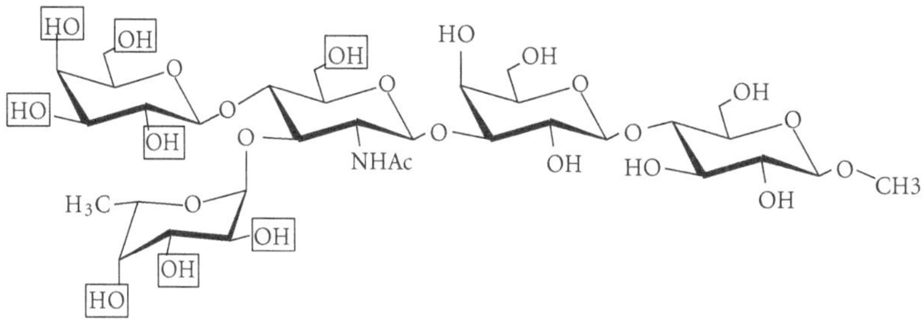

Lewis X pentasaccharide

SCHEME 1: Lewis X trisaccharide.

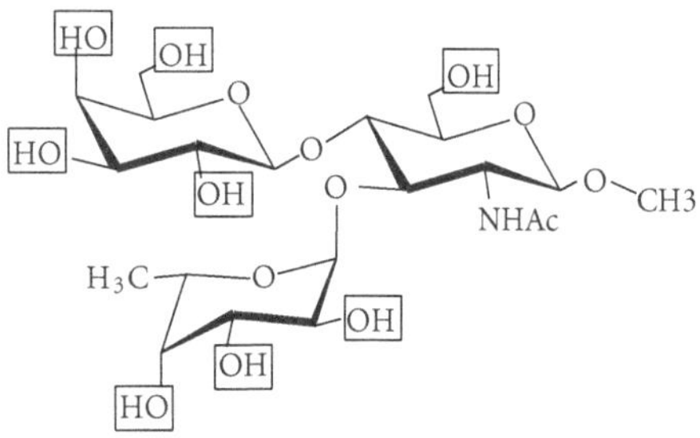

Lewis X trisaccharide

SCHEME 2: Lewis X pentasaccharide.

4. Discussion

The monitoring of kinetic, global, and potential energies along the trajectory, as well as the pressure and temperature, indicate that the global periodic system is stable and does not present simulation artefacts. Kinetic energy increased slowly, whereas potential energy decreased gradually during the simulations. The total energy of a system to be simulated should be conserved, whereas the fluctuation of a temperature is caused by the interconvention of the kinetic and potential energy components [15].

The RMS variations as a function of time are displayed on Figures 5 and 6 for the tri- and penta-saccharide Lewis X respectively, and indicate that both systems are equilibrated after 500 ps. The visual examination of the two 10 ns trajectories shows that both oligosaccharides are stable. An RMSD analysis was carried out to quantify the structural differences. The RMSD between the PDB starting structure of the simulations (obtained from Glycam biomolecule builder), and the low-energy conformers structure considering only the heavy atoms is 0.03 Å and 0.20 Å for the tri- and penta-saccharide respectively. This shows that both oligosaccharides are stable.

During the simulation of those oligosaccharides, we notice that for the pentasaccharide the most stable conformer is obtained at 2411.8 ps and that for the trisaccharide was obtained at 4856.9 ps. This means that the pentasaccharide is less flexible than the trisaccharide due to the steric effect.

5. Conclusion

A complete understanding of the role of carbohydrates in biological systems is to a large extent dependent on the information available about the equilibrium mixture and about the preferred conformation of the carbohydrate molecules in solution. The conformational analysis offers a tool which can determine all possible conformations which influence the solution behavior of carbohydrates. This work represents a theoretical study of tow oligosaccharides tri- and pentasaccharide Lewis X using molecular dynamic simulations in explicit water with the AMBER10 program package. RMSD measures of the accuracy of carbohydrates around their average conformations play an important indicator of many biological processes such as macromolecular recognition and complex formations [16].

Acknowledgments

The authors are grateful to Dr. R. Borsali, Dr. K. Mazeau, Dr. A. Rivet, and Dr. A. Nurisso (CERMAV) France, and they thank particularly Mr. Guendouzi A for his help.

References

[1] A. Imberty and S. Perez, "Structure, conformation, and dynamics of bioactive oligosaccharides: theoretical approaches and experimental validations," *Chemical Reviews*, vol. 100, no. 12, pp. 4567–4588, 2000.

[2] Y. Luo, F. Barbault, C. Gourmala et al., "Cellular interaction through LewisX cluster: theoretical studies," *Journal of Molecular Modeling*, vol. 14, no. 10, pp. 901–910, 2008.

[3] E. A. Kabat, "Immunochemical studies on the carbohydrate moiety of water-soluble blood group A, B, H, Le^a, and Le^b substances and their precursor I antigens," in *Carbohydrates in Solution*, H. Isbell, Ed., vol. 117, pp. 334–355, American Chemical Society, Washington, DC, USA, 1973.

[4] W. M. Watkins, "Biochemistry and genetics of the ABO, Lewis, and P blood group systems," in *Advances in Human Genetics*, H. Harris and K. Hirsch-Horn, Eds., vol. 10, pp. 1–136, Plenum Press, New York, NY, USA, 10th edition, 1980.

[5] S. I. Hakomori, "Antigen structure and genetic basis of histo-blood groups A, B and O: their changes associated with human cancer," *Biochimica et Biophysica Acta*, vol. 1473, no. 1, pp. 247–266, 1999.

[6] F. Bizik and I. Tvaroska, "Conformational analysis of disaccharide fragments of blood group determinants in solution by molecular modelling," *Chemical Papers*, vol. 49, no. 4, pp. 202–214, 1995.

[7] A. Imberty, E. Mikros, J. Koca, R. Mollicone, R. Oriol, and S. Perez, "Computer simulation of histo-blood group oligosaccharides: energy maps of all constituting disaccharides and potential energy surfaces of 14 ABH and Lewis carbohydrate antigens," *Glycoconjugate Journal*, vol. 12, no. 3, pp. 331–349, 1995.

[8] T. Kozar and C.-W. von der Lieth, "Efficient modelling protocols for oligosaccharides: from vacuum to solvent," *Glycoconjugate Journal*, vol. 14, no. 8, pp. 925–933, 1997.

[9] D. Case, T. Darden, T. Cheatham et al., "AMBER," University of California, CA, USA, 10th edition, 2006.

[10] R. J. Woods, R. A. Dwek, C. J. Edge, and B. Fraser-Reid, "Glycam biomolecule builder," http://www.glycam.com/CCRC/biombuilder/biomb_index.jsp.

[11] C. Gourmala, Y. Luo, F. Barbault et al., "Elucidation of the LewisX-LewisX carbohydrate interaction with molecular dynamics simulations: a glycosynapse model," *Journal of Molecular Structure*, vol. 821, no. 1–3, pp. 22–29, 2007.

[12] J. P. Ryckaert, G. Ciccotti, and H. J. C. Berendsen, "Numerical integration of the cartesian equations of motion of a system with constraints: molecular dynamics of n-alkanes," *Journal of Computational Physics*, vol. 23, no. 3, pp. 327–341, 1977.

[13] H. J. C. Berendsen, J. P. M. Postma, W. F. van Gunsteren, A. Dinola, and J. R. Haak, "Molecular dynamics with coupling to an external bath," *The Journal of Chemical Physics*, vol. 81, no. 8, article 3684, 7 pages, 1984.

[14] E. Y. Yoon, "A conformational study of linkage positions in oligosaccharides investigated by 2-D NMR spectroscopy and molecular modeling," *Bulletin of the Korean Chemical Society*, vol. 24, no. 3, pp. 339–344, 2003.

[15] J. A. McCammon and S. C. Harvey, *Dynamics of Proteins and Nucleic Acids*, Cambridge University Press, Cambridge, UK, 1987.

[16] E. B. Philip and W. Helge, *Structural Boinformatics*, 1st edition, 2003.

20

Preparation and Characterization of Chitosan/Zinc Oxide Nanoparticles for Imparting Antimicrobial and UV Protection to Cotton Fabric

M. M. AbdElhady

Textile Research Division, National Research Centre, Dokki, Egypt

Correspondence should be addressed to M. M. AbdElhady, marwa_abdalhady@yahoo.com

Academic Editor: R. J. Linhardt

Synthesis of chitosan/ZnO nanoparticles was performed using different concentrations of ZnO at different temperatures. Nanoparticles of ZnO/chitosan were prepared in rod form with average length 60 nm and average width 5–15 nm. Thus, obtained nanoparticles of ZnO/chitosan were characterized using UV spectrophotometer, FTIR, TEM, X-ray, and SEM. Size and shape of chitosan/ZnO nanoparticles relied on conditions of their synthesis. Notably, chitosan/ZnO in rod form with average length of 60 nm and average width 5–15 nm could be achieved. Application of chitosan/ZnO nanoparticles to cotton fabric conferred on the latter antibacterial and UV protection properties. Cotton fabric was characterized using SEM, ultraviolet protection factor (UPF) rating, and antibacterial (gram-positive and gram-negative) characteristics. Finished cotton fabric exhibited good antibacterial properties against gram-positive and gram-negative bacteria. The UV testes indicated a significant improvement in UV protection of finished cotton fabric which is increasing by increasing the concentration of nanoparticles of ZnO/chitosan.

1. Introduction

Chitin, the second most abundant biopolymer, widely distributes in nature as the principal component of exoskeletons of crustaceans and insects as well as of cell walls of some bacteria and fungi. It is a glucose-based unbranched polysaccharide. It differs from cellulose at the C-2 carbon where an acetamido residue locates instead of a hydroxyl group. Chitosan is a partially deacetylated polymer of acetyl glucosamine obtained through alkaline deacetylation of chitin. It is a compound polymer of glucosamine and N-acetyl glucosamine. The term chitosan refers to a group of polymers varying in molecular weight upward to several million Daltons [1]. The structure of chitosan is very similar to that of cellulose; it consists of β(1-4)-linked D-glucosamine residue with the 2-hydroxyl group being substituted by an amino or acetylated amino group. The primary amine groups endow chitosan with many special properties, making it applicable in many areas and readily available for chemical reactions, for example, salt formation with acids. Chitosan is positively charged, making it able to adhere to the negatively charged surface. Chitosan is soluble in diverse acids and able to interact with polyanions to form complexes and gels. It holds antibacterial and antifungal properties. It is safe and nontoxic [2, 3]

With advent of nanotechnology, semiconductor nanoparticles have attracted much attention due to their novel optical, electrical, and mechanical properties. Among various semiconductor nanoparticles, nanosized zinc oxide (ZnO) particles are the most frequently studied because of their interest in fundamental study and also their applied aspects such as in solar energy conversion, varistors, luminescence, photocatalysis, electrostatic dissipative coating, transparent UV protection films, and chemical sensors. Previously, ZnO nanoparticles have been prepared by techniques including the sol-gel method [4, 5], precipitation [6], hydrothermal synthesis [7], and spray pyrolysis [8].

Over recent years, hybrid materials based on chitosan have been developed, including conducting polymers, metal nanoparticles, and oxide agents, due to excellent properties of individual components and outstanding synergistic effects simultaneously [9]. Currently, the research on the combination of chitosan and metal oxide has focused on titanium dioxide, as titanium dioxide has excellent photocatalytic performance and is stable in acidic and alkaline solvents. Compared with titanium dioxide, zinc oxide has similar band-gap and antibacterial activity [9].

Sun protection creams and textiles are common choices to protect against UV radiation although UV radiation also weathers and degrades textiles. So many UV blocking agents are being developed to add or to improve the UV protection function of the textiles [10]. There are both organic and inorganic UV blockers. The organic blockers are also known as UV absorbers as they absorb the UV rays, whereas the inorganic blockers such as TiO_2 and ZnO efficiently scatter both UVA and UVB, the main cause of skin cancer. Compared with organic UV absorbers, inorganic blockers are now preferred due to their properties such as nontoxicity and chemical stability under UV radiation [10–12].

Since chitosan has tremendous ability to form metal complexes with zinc metal [13, 14] because of its amine groups and hydroxyl groups currently, chitosan-ZnO complex attracted great interests for its potential use as UV protector and medicament. This work is focused on (i) preparation and characterization of chitosan/zinc oxide nanorods complex using precipitation method and (ii) application of chitosan/ZnO nanoparticles on cotton fabric to impart antibacterial properties and UV protection.

2. Experimental

2.1. Materials. Mill-bleached 100% cotton fabric was kindly supplied by Misr Company for spinning and weaving Mehalla El Kobra, Egypt. Chitosan is poly [β-2-amino-2-deoxy-(1-4)-D glucopyranose] with a molecular weight of 60, 000 Da, and a deacetylation degree of 85% was supplied by Fluka. Zinc oxide (ZnO), sodium hydroxide (NaOH), and acetic acid were of laboratory grade chemicals.

2.2. Preparation of Chitosan/Zinc Oxide Nanoparticles. The accurate weight of ZnO powder (0.1–1 gm) was dissolved in 100 mL of 1% acetic acid where it changed to zinc cations. To this solution 1 gm of chitosan was added. The mixture was sonicated for 30 min. After magnetic stirring, 1 M NaOH drop by drop was added until the solution attained pH 10. The solution was heated in water bath at (40–80°C) for about 3 hr. It was then filtered and washed with distilled water several times then dried in an oven at 50°C for 1 hr.

2.3. Application Chitosan/Zinc Oxide Nanoparticles to Cotton Fabric. 0.5–4 wt % Chitosan/ZnO nanoparticles powder was suspended in water and was sonicated for 10 min. In these suspensions bleached cotton fabric samples were padded in two dip and nip and then squeezed to a wet pick-up of 100%. The samples were then dried at 100°C for 10 min and cured at 170°C for 5 min. The so treated cotton fabrics were washed with distilled water and finally dried at ambient conditions.

2.4. Characterization

2.4.1. UV-Vis Spectrum. UV-vis spectrum was recorded on Perkin Elmer Lambda 3B UV-vis spectrometer. Ultraviolet protection factor (UPF) was measured using UV Shimadzu 3101 PC spectrophotometer.

2.4.2. FTIR Spectroscopy. FTIR spectroscopy was measured using FT-IR, model: Nexus 670 (Nicollet-Madison-WI-USA). The samples were mixed uniformly with potassium bromide at 1 : 5 (sample : KBr) ratio, respectively. The KBr discs were prepared by compressing the powders (mixture of sample and KBr) at pressure of 5 tons for 5 min in a hydraulic press. The discs were scanned in the range of 400–4000 cm^{-1} to obtain FT-IR spectra.

2.4.3. Scanning Electron Microscope (SEM). Fabric samples under investigation were examined by a JEOL-840X scanning electron microscope, from Japan, magnification range 35–10,000, resolution 200 Å, acceleration voltage 19 kV. All the samples were coated with gold before SEM testing.

2.4.4. Transmission Electron Microscope (TEM). Transmission electron microscope (TEM) observation was performed on a JEOL JEM-1230 electron microscope at accelerating voltage of 120 kV. Specimens for TEM measurements were prepared by depositing a drop of colloid solution on a 400 mesh copper grid coated by an amorphous carbon film and evaporating the solvent in air at room temperature.

2.4.5. X-Ray Diffraction (XRD). XRD patterns recorded on a Philips PW 3050/10 model. The samples were recorded on a Philips X-Pert MMP diffractometer. The diffractometer was controlled and operated by a PC computer with the programs P Rofit and used a MoK (source with wavelength 0.70930 Å, operating with Mo-tube radiation at 50 kV and 40 mA.

2.4.6. Antibacterial Activity. Antibacterial activity towards *Staphylococcus aureus* (*S. aureus*, gram-positive bacteria) and *Escherichia coli* (*E. coli*, gram-negative bacteria) was evaluated using Agar Plate Method (AATCC Test Method 147-1988).

3. Results and Discussion

3.1. Preparation of Chitosan/ZnO Nanoparticles (Effect of Process Parameters). The prepared Chitosan/ZnO nanoparticles at 40–80°C in powder form were suspended in water and sonicated for 10 min then characterized by UV-vis absorbance spectra.

3.1.1. Effect of Concentration of Zinc Oxide. Figure 1 shows UV-vis absorbance spectra of chitosan and prepared chitosan/ZnO nanoparticles at 60°C. Obviously, (Figure 1, curve a), chitosan exhibits characteristic peak at 226 nm. After incorporation of different concentrations of ZnO in chitosan, this peak undergoes a characteristic peak at range 360–348 nm is which observed.

In detail, at a concentration of 0.1% ZnO (Figure 1, curve b), there is an absorption peak observed at 360 nm,

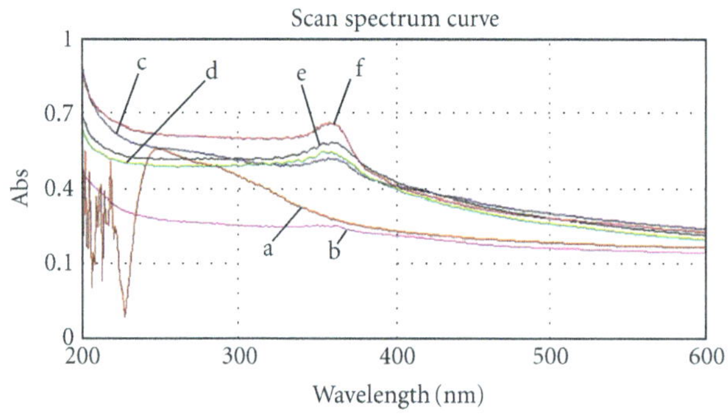

Figure 1: UV-vis absorbance spectra of chitosan and chitosan/ZnO nanoparticles at different concentrations. (a) chitosan, (b) chitosan/ZnO (0.1%), (c) chitosan/ZnO (0.25%), (d) chitosan/ZnO (0.5%), (e) chitosan/ZnO (0.75%) and (f) chitosan/ZnO (1%).

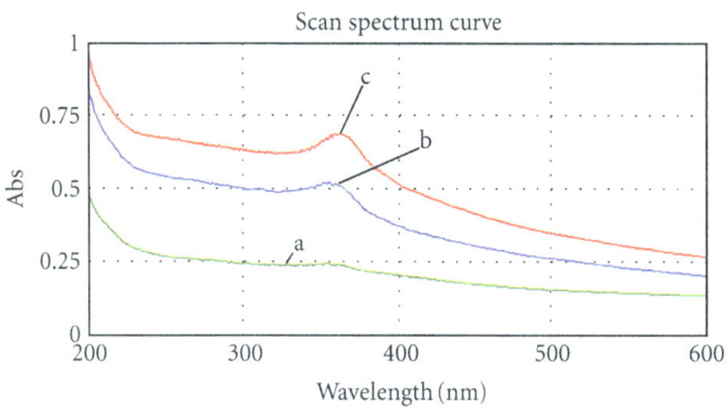

Figure 2: UV-vis absorbance spectra of chitosan and chitosan/ZnO (0.75%) nanoparticles at different temperatures (a) at 40°C, (b) at 60°C, and (c) at 80°C.

indicating the formation of nanoparticles of chitosan/ZnO in larger size. By increasing the concentration of ZnO to 0.25, 0.5, and 0.75% (Figure 1, curves c, d, and e), the absorption peaks are shifted to 356, 350, and 348 nm, respectively. Increasing the concentration of ZnO further to 1% (Figure 1, curve f), the absorption peak changes to about 350 nm. This could be attributed to formation of smaller nanoparticles of chitosan/ZnO complex. So that higher concentrations of ZnO are recommended.

3.1.2. Effect of Temperature. Figure 2 shows UV-vis spectra of different chitosan/ZnO nanoparticles samples prepared at 0.75%wt of ZnO and different temperatures. It is seen that, at temperature 40°C (Figure 2, spectrum a), the absorption peak is observed at 356 nm. Increasing the preparation temperature to 60°C and 80°C results in shifting towards lower absorption bands of 348 and 353 nm (Figure 2, spectra b and c). This could be attributed to complete conversion of chitosan-Zn $(OH)_2$ complexes to ZnO nanoparticles under the influence of higher temperature.

Based on the above results, it is clear that all the concentrations of ZnO used in the synthesis of chitosan/ZnO produce complexes which assume nanocrystalline form, in contrast with the macrocrystalline form with an absorption band at 372 nm when ZnO alone was used [9].

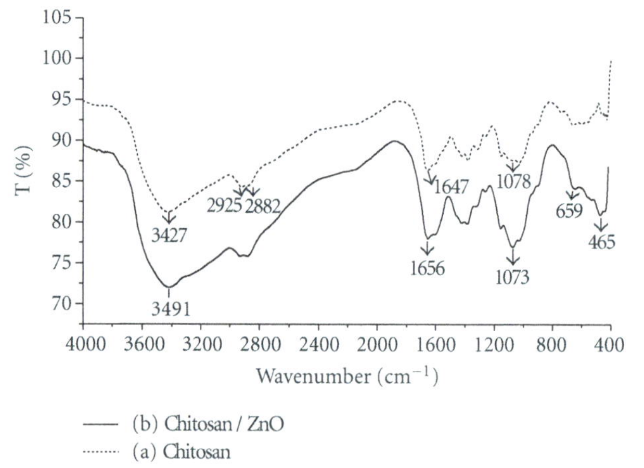

Figure 3: FTIR of chitosan (a) and chitosan/ZnO nanaoparticles (b).

The optimum concentration of ZnO for formation of chitosan/ZnO nanoparticles is 0.75%, and the optimum temperature is 60°C; under this condition an absorption peak at 348 nm is and shifted by 22 nm than macrocrystalline ZnO. This indicates that ZnO within chitosan complex is in the nanoscale and in full agreement with IR and X-ray analysis as will be shown latter.

3.2. Mechanism of Formation of Chitosan/ZnO Nanoparticles. When chitosan and zinc oxide are dissolved in diluted acetic acid; thereby; adjusting the acidity of solution, Zn^{2+} ions were formed. At pH = (6 to 9) the unstable compound $Zn[(OH)]_2$ was formed, according to the following equation:

$$Zn^{2+} + 2OH \rightleftharpoons Zn(OH)_2 \quad (1)$$

Since chitosan NH_2 and OH groups can form co-ordination bond with metal ions [15], by increasing pH of the solution to pH = 10 dropwise addition of NaOH, the stable complex of chitosan/ZnO nanoparticles is formed.

3.3. IR Analysis. (Figure 3, curves a and b) depicts the FTIR of chitosan and chitosan/ZnO nanaoparticles. For chitosan (Figure 3(a)) shows absorption peak at 3427 cm^{-1}. This attributed to the combined peaks of the NH_2 and OH group stretching vibration [16]. Compared with chitosan, (curve 3-b), the broader and stronger peak moved noticeably to lower wave number at 3419 cm^{-1} which indicated the strong interaction between these groups and ZnO [17]. The absorption peaks at 2925, 2882 cm^{-1} are attributed to asymmetric stretching of CH_3 and CH_2 of chitosan polymer [16]. While the absorption peaks at 1647 and 1078 cm^{-1} are ascribed to bending vibration of $-NH_2$ group and C–O stretching group, compared with chitosan, there are new absorption peaks at 659 cm^{-1} and 465 cm^{-1} which are due to the attachment of amide group and stretching mode of ZnO [18]. In addition to these results, the characteristic peaks of (Figure 3(b)) are shifted to lower wavenumber, the wide peak at 3427 cm^{-1}, corresponding to the stretching vibration of hydroxyl, amino and amide groups, moved noticeably

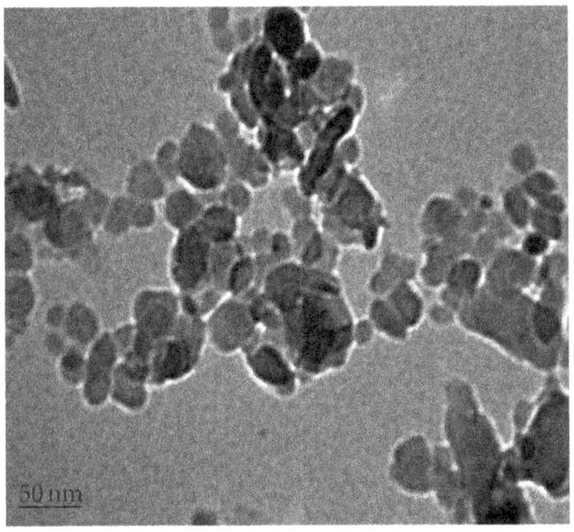

(a) Chitosan/ZnO nanoparticals (0.5%) at 60°C

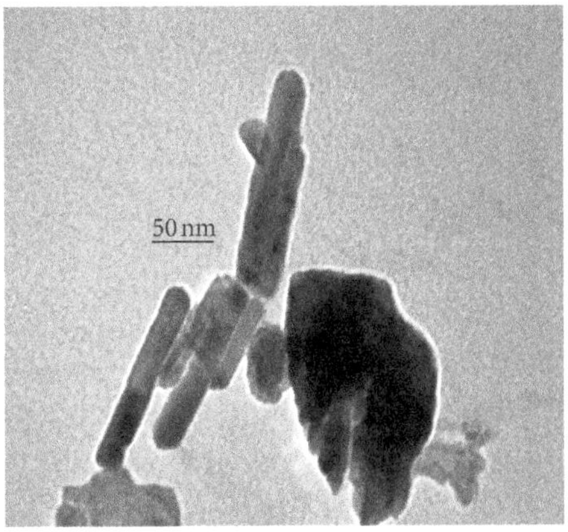

(b) Chitosan/ZnO nanoparticals (0.75%) 60°C

(c) Chitosan/ZnO nanoparticals (0.75%) 80°C

FIGURE 4: TEM photographs of chitosan/ZnO nanoparticles.

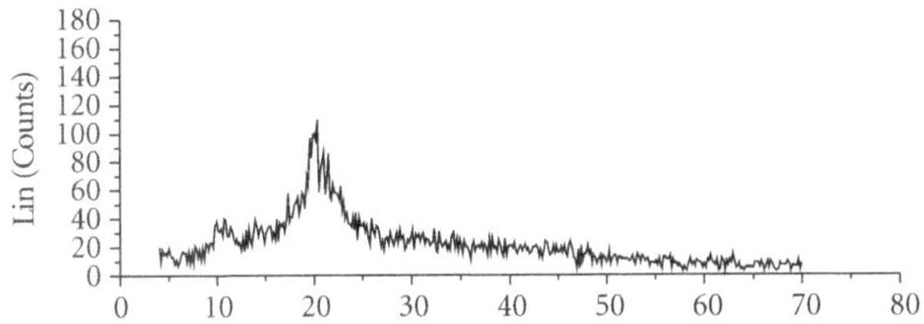

(a)

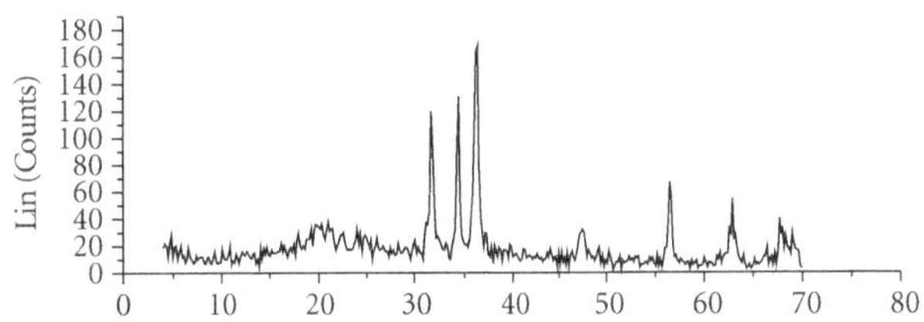

(b)

FIGURE 5: X-ray diffraction patterns of chitosan (a) and chitosan/ZnO nanoparticles (b).

to lower wavenumbers 3419 cm^{-1}, and became broader and stronger, which indicated the strong interaction between these groups and ZnO, compared with Figure 3(a), a point which could be explained in terms of strong attachment of ZnO to the amide groups of chitosan molecules.

3.4. TEM Analysis. Figure 4 represents TEM photograph of chitosan/ZnO nanoparticles. It is evident that at concentration 0.5%, nanospheres ZnO are formed with average sizes 10–20 nm. While increasing the concentration of ZnO up to 0.75%, the average length of ZnO nanorods increases to 60 nm keeping the average width at 5–15 nm. This could be interpreted in terms of lower concentration of ZnO (0.5% at 60°C) which leads to incomplete growth of nanoparticles of chitosan/ZnO complex. On the other hand, ZnO concentrations as high as 0.75% give complete growth of nanorods. It is also evident that increasing the temperature from 60°C to 80°C is accompanied by breakingdown of nanorods to nanospheres (Figure 4(c)) with higher agglomeration of smaller size nanoparticles having 5–10 nm.

3.5. X-Ray Analysis. Figure 5 shows the X-ray diffraction patterns of chitosan and chitosan/ZnO nanoparticles. The typical peaks of chitosan (Figure 5(a)) appeared at 10.67° and 19.99° [9], while these peaks become weak in the XRD pattern of chitosan/ZnO nanoparticals (Figure 5(b)). Other

(a) Untreated cotton fabric

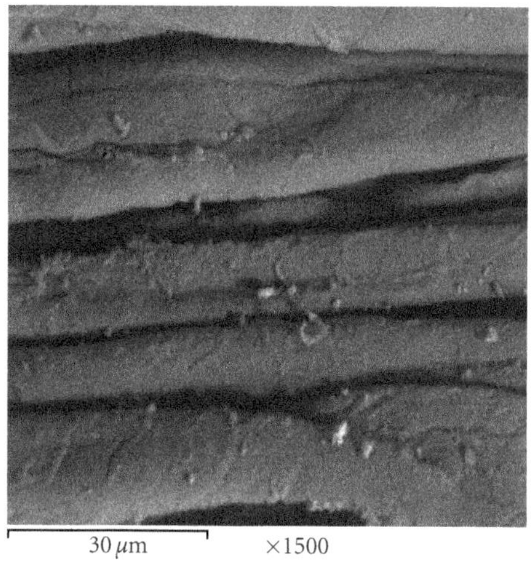

(b) Cotton fabric treated with chitosan/zinc oxide nanoparticles

FIGURE 6: SEM of cotton fabric treated with chitosan/ZnO nanoparticles.

TABLE 1: The antibacterial activity and UPF rating of bleached cotton fabric treated by chitosan/ZnO nanoparticles.

Concentration of chitosan/ZnO nanoparticles (%)	Inhibition zone (mm)		UPF rating
	S. aureus	*E. coli*	
0.0[a]	—	—	1.5
0.5	5	7	3.5
1	9	10	4.8
2	11	12	6.0
4	11	13	8.3

[a] Represents the untreated bleached cotton fabric.
Condition used: cotton fabric treated by chitosan/ZnO nanoparticles (0.75%) at pickup 100%, dried at 100°C for 10 min, and then cured at 170°C for 5 min.

diffraction peaks in Figure 5(b) are sharper and stronger at 31.7°, 34.36°, 36.2°, 56.59°, 62.7°, and 67.90° and were assigned to the (1 0 0), (0 0 2), (1 0 1), (1 1 0), (1 0 3), and (1 1 2) planes of hexagonal zinc oxide can be indexed to the wurtzite ZnO with high crystallinity. All the diffraction peaks are in good agreement with those of hexagonal wurtzite structure of ZnO (JCPDS card 36-1451) [16]. This, indeed, revealed that it is successful formation of nanosized chitosan/ZnO complex [18].

3.6. Scan Electron Microscope. The surface characteristics of untreated and chitosan/ZnO nanoparticals treated cotton fabric were examined by scanning electron microscopy, and the images obtained are shown in Figure 6. In Figure 6(a), the untreated cotton fabric seems to be a smooth fibers. While with cotton fabric treated with chitosan/ZnO nanoparticles Figure 6(b), nanoparticles are dispersed on the surface of the fiber.

3.7. Antibacterial Activity. Different cotton swatches were treated with different concentrations of chitosan/ZnO nanoparticles. These swatches were monitored for antibacterial activities. Results obtained are set out in Table 1. It is seen that the circular inhibition zone increases with increasing the concentrations of chitosan/ZnO nanoparticles in the range studied. It is also observed (Table 1) that the inhibition zone for *E. coli* is greater than the inhibition zone for *S. aureus*. This could be associated with differences in cell wall structure. At any event, however, the antibacterial activity could be attributed to the antibacterial effect of chitosan [19] and photocatalysis of ZnO.

Photocatalysis mechanism was as follows: when ZnO nanoparticles ($Eg = 3.37$ eV) were under light irradiation, the electron transition from the valence band to the conduction band resulted in the electron-hole pair in which electron (e^-) was reductive, and hole (h^+) was oxidative. The hole (h^+) reacted with OH^- on the surface of ZnO nanoparticles, generating hydroxyl radicals (OH^-), superoxide anion (O_2^-), and perhydroxyl radicals (HO_2^-). These highly active free radicals and actual lethal agent hydrogen peroxide (H_2O_2), produced from OH^- and O_2^-, could damage the cells of bacteria, leading to decomposition, complete destruction of its internal structure, eventually, to achieve germicidal and antibacterial effects [20].

3.8. UV Protection. Table 1 shows also UV protection of bleached cotton fabric treated by chitosan/ZnO nanoparticles. It is clearly seen from that ultraviolet protection factor (UPF) of the treated samples is higher than that of untreated cotton fabric. Also UPF rating increases by increasing the concentration of chitosan/ZnO nanoparticles from 0.5% to 4%. This reflects the capacity of UV absorption of ZnO/nanoparticles on the surface of the cotton fabric [21].

4. Conclusion

Synthesis and characterization of chitosan/ZnO nanaoparticles were investigated. Chitosan/ZnO nanoparticles were characterized using UV-vis spectrophotometer at different concentrations of ZnO and different temperature. The optimum concentration of ZnO for formation of nanoparticles is 0.75%, and the optimum temperature is 60°C. The obtained nanoparticles were investigated using FTIR, TEM, and X-ray analysis. Based on TEM results, it is shown nanoparticles in rod form with average length 60 nm and average width 5–15. Coating of cotton fabric by thus obtained chitosan/ZnO nanoparticles enhances the UV absorbing activity of treated fabric. Also the treated fabric has significant improvement at antibacterial properties. From the above results, this innovation is important because it may allow its practical use for industrial applications.

References

[1] S. Sevda and S. J. McClure, "Potential applications of chitosan in veterinary medicine," *Advanced Drug Delivery Reviews*, vol. 56, no. 10, pp. 1467–1480, 2004.

[2] S. A. Agnihotri, N. N. Mallikarjuna, and T. M. Aminabhavi, "Recent advances on chitosan-based micro- and nanoparticles in drug delivery," *Journal of Controlled Release*, vol. 100, no. 1, pp. 5–28, 2004.

[3] S. K. Kim and N. Rajapakse, "Enzymatic production and biological activities of chitosan oligosaccharides (COS): a review," *Carbohydrate Polymers*, vol. 62, no. 4, pp. 357–368, 2005.

[4] N. B. Hubbard, M. L. Culpepper, and L. L. Howell, "Actuators for micropositioners and nanopositioners," *Applied Mechanics Reviews*, vol. 59, no. 1–6, pp. 324–334, 2006.

[5] H. J. Lee, S. Y. Yeo, and S. H. Jeong, "Antibacterial effect of nanosized silver colloidal solution on textile fabrics," *Journal of Materials Science*, vol. 38, no. 10, pp. 2199–2204, 2003.

[6] L. Wang and M. Muhammed, "Synthesis of zinc oxide nanoparticles with controlled morphology," *Journal of Materials Chemistry*, vol. 9, no. 11, pp. 2871–2878, 1999.

[7] H. Y. Xu, H. Wang, Y. C. Zhang et al., "Hydrothermal synthesis of zinc oxide powders with controllable morphology," *Ceramics International*, vol. 30, no. 1, pp. 93–97, 2004.

[8] T. Tani, L. Mdler, and S. E. Pratsinis, "Homogeneous ZnO nanoparticles by flame spray pyrolysis," *Journal of Nanoparticle Research*, vol. 4, no. 4, pp. 337–343, 2002.

[9] L. H. Li, J. C. Deng, H. R. Deng, Z. L. Liu, and L. Xin, "Synthesis and characterization of chitosan/ZnO nanoparticle composite membranes," *Carbohydrate Research*, vol. 345, no. 8, pp. 994–998, 2010.

[10] G. Reinert, F. Fuso, R. Hilfiker, and E. Schmidt, "UV-protecting properties of textile fabrics and their improvement," *Textile Chemist and Colorist*, vol. 29, no. 12, pp. 36–43, 1997.

[11] H. Yang, S. Zhu, and N. Pan, "Studying the mechanisms of titanium dioxide as ultraviolet-blocking additive for films and fabrics by an improved scheme," *Journal of Applied Polymer Science*, vol. 92, no. 5, pp. 3201–3210, 2004.

[12] T. Ohno, K. Sarukawa, K. Tokieda, and M. Matsumura, "Morphology of a TiO_2 photocatalyst (Degussa, P-25) consisting of anatase and rutile crystalline phases," *Journal of Catalysis*, vol. 203, no. 1, pp. 82–86, 2001.

[13] R. A. A. Muzzarelli and L. Sipos, "Chitosan for the collection from seawater of naturally occurring zinc, cadmium, lead and copper," *Talanta*, vol. 18, no. 9, pp. 853–858, 1971.

[14] R. A. A. Muzzarelli and O. Tubertini, "Chitin and chitosan as chromatographic supports and adsorbents for collection of metal ions from organic and aqueous solutions and sea-water," *Talanta*, vol. 16, no. 12, pp. 1571–1577, 1969.

[15] A. Higazy, M. Hashem, A. ElShafei, N. Shaker, and M. M. Abdel Hady, "Development of antimicrobial jute packaging using chitosan and chitosan-metal complex," *Carbohydrate Polymers*, vol. 79, no. 4, pp. 867–874, 2010.

[16] M. Guo, P. Diao, and S. Cai, "Hydrothermal growth of well-aligned ZnO nanorod arrays: dependence of morphology and alignment ordering upon preparing conditions," *Journal of Solid State Chemistry*, vol. 178, no. 6, pp. 1864–1873, 2005.

[17] R. Salehi, M. Arami, N. M. Mahmoodi, H. Bahrami, and S. Khorramfar, "Novel biocompatible composite (Chitosan-zinc oxide nanoparticle): preparation, characterization and dye adsorption properties," *Colloids and Surfaces B*, vol. 80, no. 1, pp. 86–93, 2010.

[18] P. Bhadra, M. K. Mitra, G. C. Das, R. Dey, and S. Mukherjee, "Interaction of chitosan capped ZnO nanorods with *Escherichia coli*," *Materials Science and Engineering C*, vol. 31, no. 5, pp. 929–937, 2011.

[19] I. M. Helander, E.-L. Nurmiaho-Lassila, R. Ahvenainen, J. Rhoades, and S. Roller, "Chitosan disrupts the barrier properties of the outer membrane of Gram-negative bacteria," *International Journal of Food Microbiology*, vol. 71, no. 2-3, pp. 235–244, 2001.

[20] B. Halliwell and J. M. C. Gutteridge, "Oxygen toxicity, oxygen radicals, transition metals and disease," *Biochemical Journal*, vol. 219, no. 1, pp. 1–14, 1984.

[21] S. Kathirvelu, D. Souzaa, and B. Dhurai, "UV protection finishing of textiles using ZnO nanoparticals," *Indian Journal of Fiber and Textile Research*, vol. 34, no. 3, pp. 267–273, 2009.

Effect of Molecular Sizes of Chondroitin Sulfate on Interaction with L-Selectin

Naoko Igarashi,[1] Atsuko Takeguchi,[1] Shinobu Sakai,[2] Hiroshi Akiyama,[2] Kyohei Higashi,[1] and Toshihiko Toida[1]

[1] *Graduate School of Pharmaceutical Sciences, Chiba University, 1-8-1 Inohana, Chuo, Chiba 260-8675, Japan*
[2] *National Institute of Health Sciences, 1-18-1 Kamiyoga, Setagaya, Tokyo 158-8501, Japan*

Correspondence should be addressed to Toshihiko Toida; toida@faculty.chiba-u.jp

Academic Editor: R. J. Linhardt

Chondroitin sulfate (CS) is a glycosaminoglycan (GAG) side chain of proteoglycans (PGs) which are widely distributed in the extracellular matrix and at cell surface. CS shows a highly structural diversity in not only molecular weight (MW) but sulfonation pattern. CS has been reported to exert anti-inflammatory activity by having effects on cytokine production by helper T cells. In this study, we focused on the structures of CS chains, especially MW of CS, and investigated effect of the different MW of CS on binding affinity with L-selectin and cytokine production by murine splenocytes. Firstly, we fractionated CS by employing gel filtration chromatography and obtained several CS fractions with different MW. Then the interaction between fractionated CS and L-selectin was analyzed by surface plasmon resonance (SPR). Finally, the influence of MW of CS on cytokine production by murine splenocytes was investigated *in vitro*. The results showed that interferon-gamma production was significantly increased by mouse splenocytes cocultivated with CS. On the contrary, CS inhibited interleukin 5 production by murine splenocytes depending on MW of the cocultivated CS. These results strongly indicate the existence of the optimal molecular size for an anti-inflammatory effect of CS through cytokine production by murine splenocytes.

1. Introduction

Chondroitin sulfate (CS) belongs to a family of glycosaminoglycans (GAG), which is a linear, sulfonated polysaccharide, such as heparin (HP), heparan sulfate (HS), and dermatan sulfate (DS). CS is composed of a repeating disaccharide unit, [-4) GlcA (β1-3) GalNAc (β1-]$_n$, where GlcA is glucuronate and GalNAc is *N*-acetylgalactosamine [1]. CS chains show structural diversity because sulfonate groups can be bound to variety positions of the disaccharide units, and its molecular weight (MW) is also highly variable due to different numbers of the disaccharide unit [2–4]. On the other hand, it is well known that CS and other GAGs form conjugates with the specific proteins called proteoglycans (PGs) and PGs are ubiquitously distributed in not only extracellular matrix but also on cell surface in animal tissues. PGs are thought to be participating in cell-cell communications such as a cell adhesion and signalling events *via* proteins, especially; HS chains can control a signalling by interacting with cell growth factors and HP has been widely used as an anticoagulant agent [5–7]. Correspondingly, CS has also a similar function and it has been reported that CS contributes to various bioactivities, for example, development of central nervous system, wound healing, and signalling via cell growth factors [8–11]. Furthermore, there have been a large number of reports on anti-inflammatory activity of CS [12–14]. Kawashima et al. revealed that versican, a large CSPG, in the kidney reacts with leukocyte adhesion molecule, L-selectin, by interacting with CS chains, and the binding affinity of CS against L-selectin differs with the sulfonate positions of CS chains [15–18].

Generally, immune response is classified into two types, innate and adaptive immunity systems. Innate immunity system works as the first defense against invading pathogens and then differentiated lymphocytes, for example, killer and helper T cells, and antibody-secreting cells, play a role in antigen-specific immune response. The encounter with antigen can induce the differentiation of helper T cells to at least two functional classes of cell types, Th1 cells and Th2 cells, which secrete specific cytokines such as interferon-gamme (IFN-γ) or interluekin-4 (IL-4), respectively [19, 20]. Once an undifferentiated T cell called naive T cell begins to develop along with one pathway, it tends to become progressively polarized in that direction. This polarization is regulated by cytokines which Th1 cell or Th2 cell secreted. We have already reported that CS upregulates the antigen-specific Th1 immune response on murine splenocytes sensitized with ovalbumin (OVA) *in vitro*, and that CS suppresses the antigen-specific IgE responses [12]. Moreover, we showed that the content, position, and number of *O*-sulfo groups in CS are of importance in immunological activity *in vitro* [13]. Furthermore, Th1-promoted and Th2-inhibitory activity of CS on murine splenocytes could be associated with binding to L-selectin [13]. These findings might have been clarified the contribution of *O*-sulfo groups to immunological activity of CS *via* L-selectin.

In this study, we have further investigated the effect of MW of CS on the immune response, especially cytokine production by murine splenocytes sensitized with OVA, *in vitro*.

2. Materials and Methods

2.1. Chemicals. CS-A, prepared from bovine tracheal cartilage and having an average molecular weight of 15 kDa, was a gift from Shin-Nippon Yakugyo Co. (Tokyo, Japan). CS-C was kindly provided by San-Ei Gen FFI Co. (Osaka, Japan). CS-A from whale cartilage having an average molecule weight of 21 kDa was purchased from Seikagaku Kogyo Co. (Tokyo, Japan). CS-B, CS-D, CS-E, unsaturated disaccharides, and chondroitinase ABC were also purchased from Seikagaku Kogyo Co. (Tokyo, Japan).

2.2. Preparation of Fractionated CS. Solution of CS (10 mg/mL) dissolved in purified water was fractionated by using gel filtration column (100 cm × 4.6 cm i.d.) filled with Sephadex G-100 superfine (GE Healthcare, UK). The mobile phase consisted of 0.2 M NaCl. One hundred drops of eluate solution were collected in each glass vial. Then fractions were detected by UV absorbance at 204 nm. Twenty fractions were collected in one and concentrated into one-twentieth by evaporation followed by desalting on a HiTrap Desalting Column.

The separation of desalted fractionated CS was confirmed by using gradient polyacrylamide gel electrophoresis (PAGE). Ten μg of fractionated CS was dissolved in 50 μL of 50% sucrose solution containing 0.01% bromophenol blue and loaded on a gradient PAGE gel (PAGEL NPG-1020 L, 10–20%, 90 mm × 78 mm × 1 mm gel thickness, ATTO Co. Ltd., Tokyo). The electrophoresis was conducted by using a 0.2 M tris-1.2 M glycine mixture as a buffer under 100 V for 60 min.

The molecular weight of each fractionated CS was determined with a high-performance gel chromatography. The average molecular weight of fractionated CS was calculated from standard curve prepared by using molecular weight known CS sample. The structures of fractionated CS were confirmed by ^{1}H NMR (500 MHz) [21].

2.3. Disaccharide and Composition Analyses of Fractionated CS. The determination of unsaturated disaccharides (ΔDi-0S, ΔDi-4S, ΔDi-6S, ΔDi-diSE, ΔDi-diSB, ΔDi-diSD, and ΔDi-triS), prepared from the fractionated CS, was performed on the lyase-digested samples using high-performance liquid chromatography [22, 23]. Briefly, chondroitinase digestion was performed using tris-acetate buffer (5 μL, 0.2 M, pH 8.0) and 10 μL of an aqueous solution containing chondroitinase ABC (50 mIU) and chondroitinase ACII (50 mIU) were added to a 20 μL portion of the sample solution and incubated at 37°C for 3 h. An 8 μL portion of this mixture was analyzed by high-performance liquid chromatography.

2.4. Estimation of the Molecular Weight of CS. The molecular weights of the fractionated CS were estimated by an HPSEC system that consisted of a Hitachi L-6000 pump (Hitachi Co., Japan), Rheodyne 7725i loop injector (IDEX Health & Science, USA), and a YRD-880 refractive index detector (Shimamura Instruments Co., Japan). The column used was an Asahipak 510 HQ column (7.6 mm, i.d. × 300 mm) (Shodex, Showa Denko Co. Ltd., Tokyo) and eluted with 10 mM NH_4HCO_3 at a flow rate of 0.3 mL/min. The molecular weight standards of CS (15.4, 22.4, 31.0, and 50.0 kDa from shark cartilage) were kindly gifted from Seikagaku Co. Ltd. (Tokyo).

2.5. Kinetic Analysis Using BIAcore between CS and L-Selectin. Recombinant mouse L-selectin was purchased from Sigma-Aldrich Japan (Tokyo, Japan). The interaction between fractionated CS and L-selectin was analyzed by a BIAcore biosensor (GE healthcare, Tokyo). All experiments were performed at 25°C. The running buffer which was provided from the manufacture consisted of 10 mM HEPES pH 7.4, 0.15 M NaCl, 3 mM EDTA, and 0.005% Surfactant P20. First, L-selectin was covalently immobilized on the CM5 sensor chip covered with carboxymethyl dextran via its primary amine groups using an amino coupling kit according to the instructions provided from the manufacturer. The remaining activated groups were blocked with 1 M ethanolamine. One control cell was prepared by blocking with ethanolamine. CS samples were dissolved in 50 mM tris-HCl buffer. The different concentrations of the each fractionated CS were injected at 20 μL/min. The sensor chip surface was regenerated with 1 M NaCl. Affinity kinetic parameters were determined with BIAevaluation 3.0 software using a single-site binding model.

2.6. Animals. Inbred specific pathogen-free BALB/c mice (female, 6 weeks of age) were purchased from Nihon SLC Inc. (Shizuoka, Japan). The mice were maintained in a temperature (25°C), humidity (55%), and light-controlled environment with free access to Rodent diet EQ (Nihon SLC Inc., Shizuoka, Japan) and water. They were acclimatized for at least 1 week before the start of study.

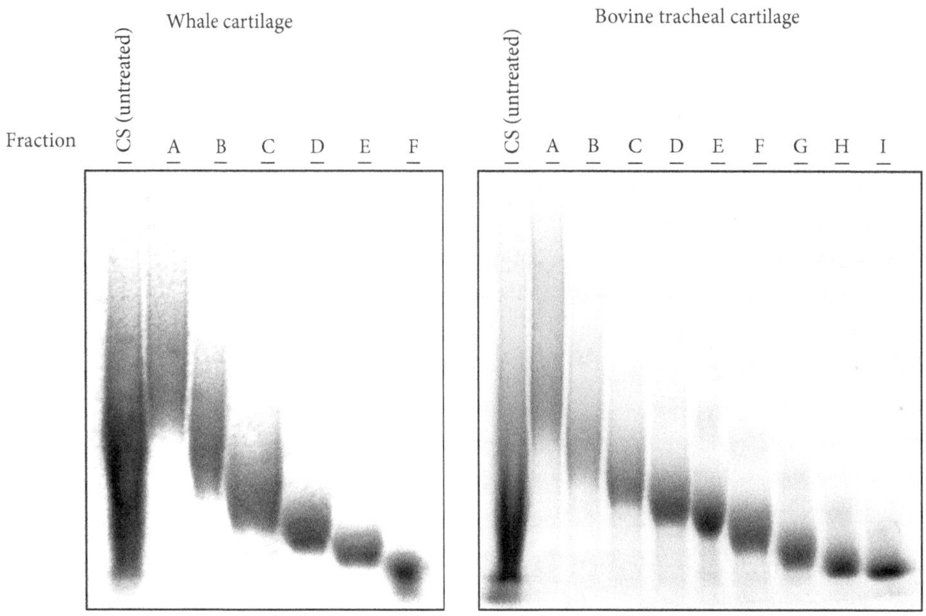

FIGURE 1: Gradient PAGE profiles of original and fractionated chondroitin sulfate.

2.7. Sensitization Protocol. The mice ($n = 5$) were intraperitoneally sensitized with 20 mg of OVA and 2 mg of an aluminium hydroxide hydrate [$Al(OH)_3 \cdot xH_2O$] gel suspension (LSL Co. Ltd., Japan) in a total volume of 400 μL of saline. On 14th day after the first sensitization, the mice were boosted using the same doses of the antigen. On the following day, the mice were sacrificed and their spleens were harvested.

2.8. Preparation and Stimulation of Murine Splenocytes In Vitro. The spleens of mice were removed and cell suspensions were made by passing them through a sterile cell strainer (Becton Dikcinson, CA, USA). The cell suspensions were washed twice in RPMI 1640 medium supplemented with penicillin-streptomycin, 0.1 M sodium pyruvate, 0.2 M L-glutamine, 0.05 mM 2-mercaptoethanol, and FBS, counted by Trypan Blue exclusion staining and adjusted to 5×10^6 cells/mL. The cells were plated at 2 mL/well in a 24-well cell culture cluster (Corning, NY, USA) and challenged with OVA at a final concentration of 100 μg/mL. Furthermore, the samples (each at a final concentration of 0.5 μM) or saline *in vitro* was added to the 24-well culture cluster at 37°C in triplicate in a CO_2 incubator for 72 hrs. The supernatants were then harvested for cytokine production analysis after 72 hrs of incubation.

2.9. Production of Cytokine In Vitro. The cytokines examined in this study were IFN-γ, IL-2, IL-5, and IL-10. The amount of several cytokines in the culture medium after incubation was measured using an ELISA kit (OptEIA TM set, Pharmingen, San Diego, CA, USA). The principle of all ELISA kits was to employ the quantitative sandwich enzyme immunoassay technique. The absorbance was measured at 450 nm with UV-VIS microplate reader (SUNRISE THERMO, TECAN Japan Co. Ltd., Tokyo).

3. Results

3.1. Fractionation of CS. CS has several isomers which have different degrees of sulfonation and binding sites of sulfonate groups [24–26]. Generally, intact CS chains show various MW and a range of their MW is extremely wide from thousands to hundreds of thousands [2–4]. Based on this fact, we firstly fractionated commercially available CS by low-pressure gel chromatography to prepare CS within a strictly limited range of average MW. Two different types of CS samples were applied: one was derived from whale cartilage which is mainly composed of CS-A structure [GlcA (β1-3) GalNAc4S], and the other was from bovine tracheal cartilage which is mainly composed of CS-A structure [GlcA (β1-3) GalNAc4S] [27, 28]. Sephadex G100 Superfine, by which the available fractionation range is from 1 to 100 kDa, was employed for preparation of CS as the average MW of CS from whale cartilage and bovine tracheal cartilage was 21 kDa and 15 kDa, respectively. The structure of each fractionated CS was confirmed by the HPLC for disaccharide analysis and ^{1}H NMR (data not shown). The extent of fractionation of CS was confirmed by applying gradient polyacrylamide gel electrophoresis (PAGE), and we confirmed that we could obtain CS with strictly limited range of MW (Figure 1 and Table 1). Six fractionated CSs from whale cartilage, of which original average MW was 21 kDa, were estimated from 7 to 29 kDa. As for CS from bovine tracheal cartilage, the range of MW was from 6 to 33 kDa.

3.2. Disaccharide Analysis for Fractionated CS. Since it has already been reported that the difference of sulfonation pattern/degree has an impact on a bioactivity of CS, we analyzed the disaccharide composition of the fractionated CS chains. To be concrete, unfractionated and fractionated CSs were degraded by the action of chondroitin lyases and then

TABLE 1: Molecular weight of fractionated chondroitin sulfate.

Fraction	Mean molecular weight (kDa)	
	Whale cartilage	Bovine tracheal cartilage
CS (untreated)	21	15
A	29	33
B	21	26
C	15	21
D	12	17
E	9	15
F	7	13
G	—	9
H	—	7
I	—	6

the compositions of these disaccharides were determined by the reverse-phase ion-pair chromatography with the postcolumn fluorogenic reactions. As a result, it was revealed that the contents of major disaccharide units, which are ΔDi-4S and ΔDi-6S, were quite similar as that of the original CS after fractionation (Table 2). The contents of ΔDi-4S were three and two times higher than those of ΔDi-6S in whale cartilage and bovine tracheal cartilage, respectively. In order to compare the immunoreactivity of both CS from whale and bovine cartilage, 7, 15, and 21 kDa fractionated CS were used for further investigation.

3.3. Binding of CS to L-Selectin. To investigate the binding affinity between CS and L-selectin, and the effect of MW of CS on the binding affinity with L-selectin, we analyzed the binding affinity between CS and L-selectin by surface plasmon resonance (SPR) measurement employed with BIAcore, which is able to detect a biomolecular interaction without labels and analyzes an affinity and kinetics parameters quantitatively in real time. Diluted commercially available recombinant murine L-selectin in 10 mM acetic buffer (pH 4.0) was immobilized to a CM5 sensor chip by the amino-coupling reaction and 11.7 ng/mm^2 of the L-selectin could be immobilized. Subsequently, the interaction between the fractionated CS and L-selectin was investigated. Figure 2 shows both untreated and fractionated CS could interact with L-selectin through dose dependence. As compared with the association and dissociation phases of sensorgrams of the fractionated CS from bovine tracheal cartilage with L-selectin, those of CS from whale cartilage were steeper; that is, the interaction with CS from whale cartilage with L-selectin is tighter. Kinetic parameters such as association constant (K_A), dissociation constant (K_D), association rate constant (k_a), and dissociation rate constant (k_d) were calculated as the Langmuir formula. Evaluation of the kinetic parameters (Table 3) indicated that K_D of the fractionated CS with 21 kDa in both CS was the smallest and there was the tendency of K_D to increase in a molecular weight-dependent manner. Although untreated CS and fractionated CS had the same average MW, K_D of fractionated CS was smaller in comparison to that of untreated CS. It suggests that CS with narrower molecular distribution has a higher affinity with L-selectin compared to CS with a wide range of MW. Furthermore, there was considerable difference for untreated CS from whale cartilage and bovine tracheal cartilage in the affinity with L-selectin, even though the disaccharide compositions of these CS samples showed no significant difference (Tables 2 and 3). These results suggest that not only sulfonation pattern including disaccharide sequence but also MW of CS may affect the affinity with L-selectin.

The many studies on bioactivity of CS have been focused on its sulfonation to date. However, our findings suggest that MW of CS could have effect on such a bioactivity of CS as well as sulfonation pattern of CS.

3.4. Effects of Fractionated CS on Cytokine Production of Mice Splenocytes. To examine the effects of CS with various MWs on cytokine production, splenocytes from OVA-sensitized BALB/c mice were challenged with OVA in the presence and absence of untreated and fractionated CS, and the cytokine levels in the medium of the cultured cells were measured. As shown in Figure 3(a), the levels of IFN-γ, Th1-type cytokine, produced by the splenocytes cultured with fractionated CS from whale cartilage were significantly higher than the levels produced by splenocytes cultured with OVA alone and untreated CS. The production of IFN-γ was the most strongly induced by being cultured with fractionated CS with 21 kDa. As for fractionated CS from bovine tracheal cartilage, it has a tendency to increase the level of IFN-γ in MW dependent. These results coincided with the result of study on the interaction between CS and L-selectin. IL-2 also belongs to Th1-type cytokines but no significant effect on IL-2 production like an IFN-γ production was observed in both CS from whale cartilage and bovine tracheal cartilage (data not shown). In contrast to Th1 cytokines, as shown in Figure 3(b), the levels of Th2 cytokine, IL-5 produced by the splenocytes cultured with fractionated CS from bovine tracheal cartilage were significantly lower than the levels produced by splenocytes cultured with OVA alone. There was a weak tendency to decrease the level of IL-5 production in CS from whale cartilage-treated splenocytes. It seems that fractionated CS has stronger effect on the inhibition of IL-5 production compared to untreated CS. As for the production of IL-10 which is also Th2-type cytokine, decreasing in its production level produced by the splenocytes cultured with fractionated CS was observed (data not shown).

4. Discussion

Selectin is one of cell surface receptors which is engaged in the adhesion between white blood cells and vascular endothelial cells, and it is involved with penetration of white blood cells into extravascular as an adhesion molecule [26]. Presently, three types of selectins, that is, E-, P-, and L-selectin, are reported. E- and P-selectins are expressed in vascular endothelial cell surface and platelets, respectively, and L-selectin is expressed on cell membranes of monocytes, neutrophils, part of natural killer cells, and T and B lymphocytes. Selectins are known as a type I membrane glycoprotein with a C-type lectin domain in its N-terminal domain [29, 30].

TABLE 2: Disaccharide compositions of unfractionated and fractionated chondroitin sulfate samples from whale and bovine cartilage.

Fraction	ΔDi-0S	ΔDi-4S	ΔDi-6S	ΔDi-diS$_E$	ΔDi-diS$_B$	ΔDi-diS$_D$	ΔDi-triS
Whale cartilage							
CS (untreated)	1.2	76.8	20.9	0.3	N.D.	0.7	Trace
A	1.2	75.7	22.8	0.1	N.D.	0.2	Trace
B	1.3	75.4	22.7	0.2	N.D.	0.4	Trace
C	1.5	75.7	22.4	0.1	N.D.	0.3	Trace
D	1.5	76.3	21.9	0.1	N.D.	0.2	Trace
E	1.5	76.7	21.0	0.2	N.D.	0.5	Trace
F	3.4	76.0	20.4	0.1	N.D.	0.1	Trace
Bovine tracheal cartilage							
CS (untreated)	8.1	56.5	33.8	0.6	N.D.	1.0	Trace
A	6.7	61.3	30.9	0.5	N.D.	0.7	Trace
B	7.5	62.3	29.0	0.5	N.D.	0.7	Trace
C	7.3	61.7	29.7	0.5	N.D.	0.8	Trace
D	6.6	59.1	33.1	0.4	N.D.	0.7	Trace
E	6.4	55.7	36.9	0.3	N.D.	0.6	Trace
F	6.7	52.1	40.3	0.3	N.D.	0.6	Trace
G	5.4	50.0	43.4	0.4	N.D.	0.8	Trace
H	8.6	50.1	40.2	0.3	N.D.	0.9	Trace
I	8.2	48.7	41.6	0.4	N.D.	1.0	Trace

Disaccharide compositions (%) of untreated and fractionated CS were determined by HPLC on an amino-bound silica column after chondroitinase ABC digestion as described previously. ΔDi-0S, $\Delta^{4,5}$HexA(αl-3)GalNAc; ΔDi-4S, $\Delta^{4,5}$HexA(αl-3)GalNAc(4-O-sulfate); ΔDi-6S, $\Delta^{4,5}$HexA(αl-3)GalNAc(6-O-sulfate); ΔDi-diSE, $\Delta^{4,5}$HexA(αl-3)GalNAc(4,6-O-disulfate); ΔDi-diSB, $\Delta^{4,5}$HexA(2-O-sulfate) (αl-3)GalNAc(4-O-sulfate); ΔDi-diSD, $\Delta^{4,5}$HexA(2-O-sulfate)(αl-3)GalNAc(4-O-sulfate); ΔDi-tirS, $\Delta^{4,5}$HexA(2-O-sulfate)(αl-3)GalNAc(4,6-O-disulfate); N.D.: not determined.

TABLE 3: Kinetic parameters for the interaction of fractionated chondroitin sulfate with immobilized L-selectin.

	Fraction	k_a (M^{-1} s^{-1})	k_d (s^{-1})	K_A (M^{-1})	K_D (μM)
Whale cartilage	CS (untreated)	6.16×10^2	7.52×10^{-3}	8.18×10^4	12.2
	B (21 kDa)	7.58×10^3	2.27×10^{-2}	3.34×10^5	3.0
	C (15 kDa)	2.31×10^3	2.05×10^{-2}	1.13×10^5	8.8
	E (9 kDa)	6.05×10^2	3.42×10^{-2}	1.77×10^4	56.4
Bovine tracheal cartilage	CS (untreated)	3.32×10^2	1.30×10^{-2}	2.56×10^4	39.1
	C (21 kDa)	7.97×10^2	3.48×10^{-3}	2.23×10^5	4.5
	E (15 kDa)	1.22×10^2	2.30×10^{-3}	5.29×10^4	18.9
	G (9 kDa)	4.52×10^2	2.37×10^{-2}	2.00×10^4	49.9

The association rate constant (k_a), dissociation (k_d) rate constant, equilibrium association constant (K_A) and equilibrium dissociation constant (K_D) for the interaction of CS with immobilized L-selectin were determined by BIAcore. Standard errors of kinetic parameters obtained were within 15% by tripricate experiments.

The C-type lectin domain of L-selectin is known to interact with several glycan ligands directly. MAdCAM-1 [31], CD 34 [32], podocalyxin [33], and GlyCAM-1 [34] have been identified as L-selectin ligands which are expressed on cell surface of endothelial venule. Recently, it has been reported that penetration of white blood cell into an inflammatory tissue was inhibited by administration of anti-L-selectin antibody, and that penetration of white blood cell was suppressed significantly in L-selectin knock-out mouse [35, 36]. Consequently, it is assumed that L-selectin and the ligand play a role in penetration of white blood cell to nonlymphoid tissue, but the ligand is still unknown. Kawashima et al. revealed that L-selectin binding substance except for Sialyl-Lewis X exists in white matter and distal tubule which are nonlymphoid tissue and versican, a large chondroitin/dermatan sulfate (CS/DS) proteoglycan, interacts through its CS/DS chains with L- and P-selectin and CD 44 [15, 16, 18]. And we have reported that CS impacts intracellular transcription system through contact with L-selectin in OVA-sensitized mouse [12].

It has been shown that CSPG and CS are involved in anti-inflammatory reaction [37–39]. We also reported for the immunochemical reactivities of CS on murine splenocytes sensitized with OVA in order to demonstrate that CS could be associated with immune system. And we revealed that CS affects the pattern of cytokine production, differentiation of lymphocyte subsets, and level of IFN-γ mRNA in

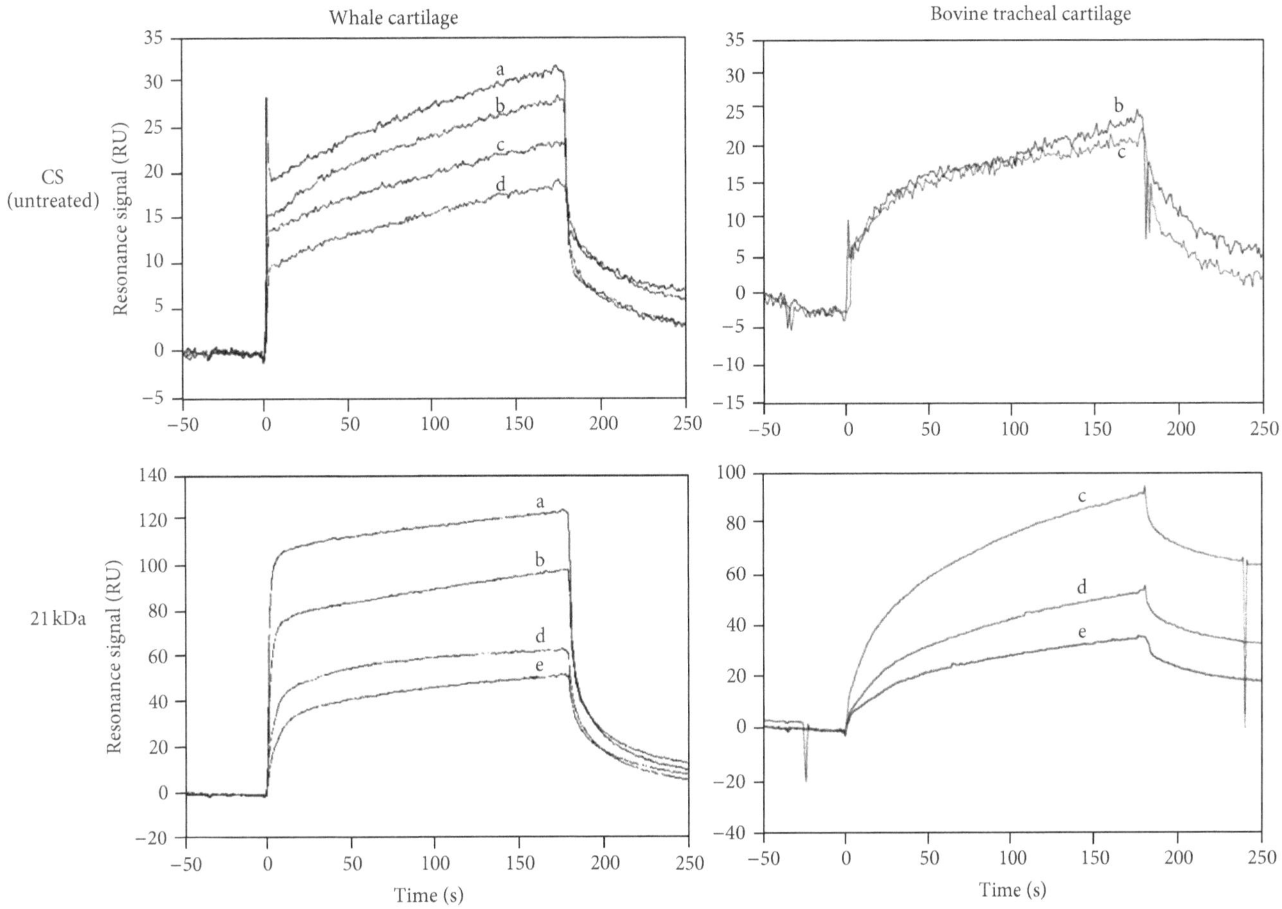

FIGURE 2: Sensorgrams of recording the interaction of fractionated chondroitin sulfate with immobilized L-selectin in the BIAcore. Various concentrations of original CS and fractionated CS (a, 50 μg/mL; b, 100 μg/mL; c, 250 μg/mL; d, 500 μg/mL; e, 1000 μg/mL) were continuously injected over sensor chip surfaces coupled to L-selectin at a 20 μL/min; after this, running buffer was injected and the response in resonance units was recorded as a function time.

the splenocytes of the mice immunized with OVA. Furthermore, it was shown that CS induced these reactions by interacting with L-selectin in sulfonation degree-dependent manner, and these findings suggest that CS could impact transcription system [12, 13]. Additionally, Zhou and his group have reported on the immune modulation by the animal model of experimental autoimmune encephalomyelitis (EAE) using CS and its disaccharide. It was shown that CS-A and disaccharide degradation of CSPG had an opposite effect on EAE [40]. Thus, it is suggested that CS has different immunochemical reactivity due to its structural variation. In this study, we focused on a variability of MW of CS in order to investigate the immunochemical reactivities of CS in detail and investigated the effect of CS with different MW on systemic immune systems.

5. Conclusion

In the past, many studies on the bioactivities of CS have been focused on the degree of its sulfonation patterns and they revealed that effects of CS on bioactivities differ with the degree and positions of sulfonation CS chains. However, we clarified here that the immunoreactivity of CS is affected by the MW, length of leaner polysaccharide chains, of CS. Furthermore, even though the MW of CS was similar, the effect on the production of IFN-γ and IL-5 by mouse splenocytes was more augmented by CS with a narrow range of MW compared to CS with a broad range of MW. Namely, it is suggested that the length of polysaccharide chains of CS has also the impact on the immunoreactivity.

Conflict of Interests

There is no ethical problem or conflict of interests with regard to this paper.

Acknowledgments

This work was supported in part by both Special Funds for Education and Research (Development of SPECT Probes for Pharmaceutical Innovation) from the Ministry of Education, Culture, Sports, Science, and Technology, Japan, and by

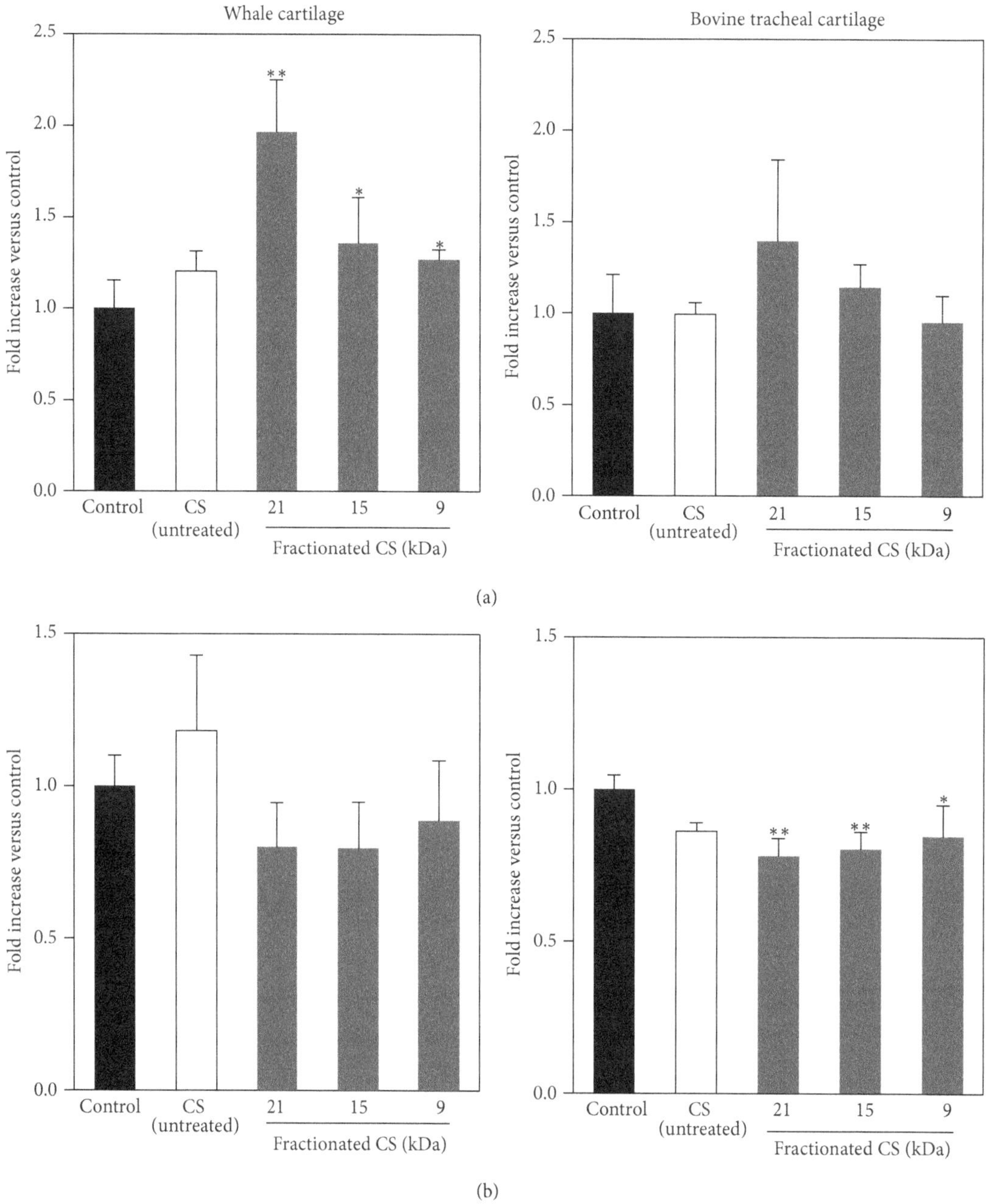

FIGURE 3: Effect of fractionated chondroitin sulfate on the cytokine productions of murine splenocytes *in vitro*. (a) IFN-γ production. (b) IL-5 production. BALB/c mice ($n = 5$) were intraperitoneally injected on day 0 and day 14 with 20 μg of OVA and 2 mg of Al $(OH)_3$ in a total volume of 400 μL. Splenocytes (5.0×10^6 cells/mL) were collected on day 15 and were cocultured with OVA (final concentration 100 μg/mL). The samples (each at a final concentration of 0.5 μM) or saline *in vitro* was added to the 24-well culture cluster in triplicate at 37°C in a CO_2 incubator for 48 hrs. The amounts of cytokines in the supernatant were measured by ELISA. $^*P < 0.05$, $^{**}P < 0.01$ (significantly different from control value). Error bars represent means ± S.D. for 6 wells.

Grants from the Ministry of Health, Labour, and Welfare of Japan and from the Japanese Health Sciences Foundation (KHB1212, TT).

References

[1] L. Roden, "Structure and metabolism of connective tissue proteoglycans," in *The Biochemistry of Glycoproteins and Proteoglycans*, pp. 267–371, Plenum Press, New York, NY, USA, 1980.

[2] K. Kimata, M. Okayama, A. Oohira, and S. Suzuki, "Cytodifferentiation and proteoglycan biosynthesis," *Molecular and Cellular Biochemistry*, vol. 1, no. 2, pp. 211–228, 1973.

[3] T. Toida, H. Toyoda, and T. Imanari, "High-resolution proton nuclear magnetic resonance studies on chondroitin sulfates," *Analytical Sciences*, vol. 9, pp. 53–58, 1993.

[4] J. E. Scott, "Structure and function in extracellular matrices depend on interactions between anionic glycosaminoglycans," *Pathologie Biologie*, vol. 49, no. 4, pp. 284–289, 2001.

[5] K. Sugahara, T. Mikami, T. Uyama, S. Mizuguchi, K. Nomura, and H. Kitagawa, "Recent advances in the structural biology of chondroitin sulfate and dermatan sulfate," *Current Opinion in Structural Biology*, vol. 13, no. 5, pp. 612–620, 2003.

[6] A. Salustri, A. Camaioni, M. di Giacomo, C. Fulop, and V. C. Hascall, "Hyaluronan and proteoglycans in ovarian follicles," *Human Reproduction Update*, vol. 5, no. 4, pp. 293–301, 1999.

[7] M. Yoneda, S. Shimizu, Y. Nishi, M. Yamagata, S. Suzuki, and K. Kimata, "Hyaluronic acid-dependent change in the extracellular matrix of mouse dermal fibroblasts that is conducive to cell proliferation," *Journal of Cell Science*, vol. 90, no. 2, pp. 275–286, 1988.

[8] G. S. Kelly, "The role of glucosamine sulfate and chondroitin sulfates in the treatment of degenerative joint disease," *Alternative Medicine Review*, vol. 3, no. 1, pp. 27–39, 1998.

[9] T. Omata, Y. Segawa, Y. Itokazu, N. Inoue, and Y. Tanaka, "Effects of chondroitin sulfate-C on bradykinin-induced proteoglycan depletion in rats," *Arzneimittel-Forschung/Drug Research*, vol. 49, no. 7, pp. 577–581, 1999.

[10] J. Beren, S. L. Hill, M. Diener-West, and N. R. Rose, "Effect of pre-loading oral glucosamine HCl/chondroitin sulfate/manganese ascorbate combination on expierimental arthritis in rats," *Proceedings of the Society for Experimental Biology and Medicine*, vol. 226, no. 2, pp. 144–151, 2001.

[11] G. Rovetta, "Galactosaminoglycuronoglycan sulfate (Matrix) in therapy of tibiofibular osteoarthritis of the knee," *Drugs under Experimental and Clinical Research*, vol. 17, no. 1, pp. 53–57, 1991.

[12] S. Sakai, H. Akiyama, N. Harikai et al., "Effect of chondroitin sulfate on murine splenocytes sensitized with ovalbumin," *Immunology Letters*, vol. 84, no. 3, pp. 211–216, 2002.

[13] H. Akiyama, S. Sakai, R. J. Linhardt, Y. Goda, T. Toida, and T. Maitani, "Chondroitin sulphate structure affects its immunological activities on murine splenocytes sensitized with ovalbumin," *Biochemical Journal*, vol. 382, no. 1, pp. 269–278, 2004.

[14] S. Haylock-Jacobs, M.B. Keough, L. Lau, and V. W. Yon, "Chondroitin sulphate proteoglycans: extracellular matrix proteins that regulate immunity of the central nervous system," *Autoimmunity Reviews*, vol. 10, no. 12, pp. 766–772, 2011.

[15] H. Kawashima, Y. F. Li, N. Watanabe, J. Hirose, M. Hirose, and M. Miyasaka, "Identification and characterization of ligands for L-selectin in the kidney. I. Versican, a large chondroitin sulfate proteoglycan, is a ligand for L-selectin," *International Immunology*, vol. 11, no. 3, pp. 393–405, 1999.

[16] H. Kawashima, M. Hirose, J. Hirose, D. Nagakubo, A. H. K. Plaas, and M. Miyasaka, "Binding of a large chondroitin sulfate/dermatan sulfate proteoglycan, versican, to L-selectin, P-selectin, and CD44," *Journal of Biological Chemistry*, vol. 275, no. 45, pp. 35448–35456, 2000.

[17] H. Kawashima, K. Atarashi, M. Hirose et al., "Oversulfated chondroitin/dermatan sulfates containing GlcAβ1/IdoAα1-3GalNAc(4,6-O-disulfate) interact with L- and P-selectin and chemokines," *Journal of Biological Chemistry*, vol. 277, no. 15, pp. 12921–12930, 2002.

[18] J. Hirose, H. Kawashima, O. Yoshie, K. Tashiro, and M. Miyasaka, "Versican interacts with chemokines and modulates cellular responses," *Journal of Biological Chemistry*, vol. 276, no. 7, pp. 5228–5234, 2001.

[19] A. K. Abbas, K. M. Murphy, and A. Sher, "Functional diversity of helper T lymphocytes," *Nature*, vol. 383, no. 6603, pp. 787–793, 1996.

[20] K. M. Murphy and S. L. Reiner, "The lineage decisions of helper T cells," *Nature Reviews Immunology*, vol. 2, no. 12, pp. 933–944, 2002.

[21] T. Toida, K. Sato, N. Sakamoto, S. Sakai, S. Hosoyama, and R. J. Linhardt, "Solvolytic depolymerization of chondroitin and dermatan sulfates," *Carbohydrate Research*, vol. 344, no. 7, pp. 888–893, 2009.

[22] H. Toyoda, K. Shinomiya, S. Yamanashi, I. Koshiishi, and T. Imanari, "Microdetermination of unsaturated disaccharide's produced from chondroitinsulfates in rabbit plasma by high-performance liquid chromatography with fluorometric detection," *Analytical Sciences*, vol. 4, pp. 381–384, 1988.

[23] H. Toyoda, K. Motoki, M. Tanikawa, K. Shinomiya, H. Akiyama, and T. Imanari, "Determination of human urinary hyaluronic acid, chondroitin sulphate and dermatan sulphate as their unsaturated disaccharides by high-performance liquid chromatography," *Journal of Chromatography*, vol. 565, no. 1-2, pp. 141–148, 1991.

[24] J. A. Santos, B. Mulloy, and P. A. S. Mourao, "Structural diversity among sulfated α-L-galactans from ascidians (tunicates)—studies on the species Ciona intestinalis and Herdmania monus," *The European Journal of Biochemistry*, vol. 204, no. 2, pp. 669–677, 1992.

[25] A. P. Alves, B. Mulloy, J. A. Diniz, and P. A. S. Mourão, "Sulfated polysaccharides from the egg jelly layer are species-specific inducers of acrosomal reaction in sperms of sea urchins," *Journal of Biological Chemistry*, vol. 272, no. 11, pp. 6965–6971, 1997.

[26] W. R. L. Farias, A. P. Valente, M. S. Pereira, and P. A. S. Mourão, "Structure and anticoagulant activity of sulfated galactans. Isolation of a unique sulfated galactan from the red algae Botryocladia occidentalis and comparison of its anticoagulant action with that of sulfated galactans from invertebrates," *Journal of Biological Chemistry*, vol. 275, no. 38, pp. 29299–29307, 2000.

[27] Y. M. Michelacci and C. P. Dietrich, "Structure of chondroitin sulphate from whale cartilage: distribution of 6- and 4-sulphated oligosaccharides in the polymer chains," *International Journal of Biological Macromolecules*, vol. 8, no. 2, pp. 108–113, 1986.

[28] D. Vestweber and J. E. Blanks, "Mechanisms that regulate the function of the selectins and their ligands," *Physiological Reviews*, vol. 79, no. 1, pp. 181–213, 1999.

[29] M. P. Bevilacqua and R. M. Nelson, "Selectins," *Journal of Clinical Investigation*, vol. 91, no. 2, pp. 379–387, 1993.

[30] S. D. Rosen and C. R. Bertozzi, "The selectins and their ligands," *Current Opinion in Cell Biology*, vol. 6, no. 5, pp. 663–673, 1994.

[31] E. L. Berg, L. M. McEvoy, C. Berlin, R. F. Bargatze, and E. C. Butcher, "L-selectin-mediated lymphocyte rolling on MAdCAM-1," *Nature*, vol. 366, no. 6456, pp. 695–698, 1993.

[32] S. Baumhueter, M. S. Singer, W. Henzel et al., "Binding of L-selectin to the vascular sialomucin CD34," *Science*, vol. 262, no. 5132, pp. 436–438, 1993.

[33] C. Sassetti, K. Tangemann, M. S. Singer, D. B. Kershaw, and S. D. Rosen, "Identification of Podocalyxin-like protein as a high endothelial venule ligand for L-selectin: parallels to CD34," *Journal of Experimental Medicine*, vol. 187, no. 12, pp. 1965–1975, 1998.

[34] Y. Imai, L. A. Lasky, and S. D. Rosen, "Sulphation requirement for GlyCAM-1, an endothelial ligand for L-selectin," *Nature*, vol. 361, no. 6412, pp. 555–557, 1993.

[35] M. L. Arbonés, D. C. Ord, K. Ley et al., "Lymphocyte homing and leukocyte rolling and migration are impaired in L-selectin-deficient mice," *Immunity*, vol. 1, no. 4, pp. 247–260, 1994.

[36] T. R. Mosmann, H. Cherwinski, M. W. Bond, M. A. Giedlin, and R. L. Coffman, "Two types of murine helper T cell clone. I. Definition according to profiles of lymphokine activities and secreted proteins," *Journal of Immunology*, vol. 136, no. 7, pp. 2348–2357, 1986.

[37] F. Petersen, L. Bock, H. D. Flad, and E. Brandt, "A chondroitin sulfate proteoglycan on human neutrophils specifically binds platelet factor 4 and is involved in cell activation," *Journal of Immunology*, vol. 161, no. 8, pp. 4347–4355, 1998.

[38] P. Morreale, R. Manopulo, M. Galati, L. Boccanera, G. Saponati, and L. Bocchi, "Comparison of the antiinflammatory efficacy of chondroitin sulfate and diclofenac sodium in patients with knee osteoarthritis," *Journal of Rheumatology*, vol. 23, no. 8, pp. 1385–1391, 1996.

[39] F. Ronca, L. Palmieri, P. Panicucci, and G. Ronca, "Anti-inflammatory activity of chondroitin sulfate," *Osteoarthritis and Cartilage*, vol. 6, pp. 14–21, 1998.

[40] J. Zhou, P. Nagarkatti, Y. Zhong, and M. Nagarkatti, "Immune modulation by chondroitin sulfate and its degraded disaccharide product in the development of an experimental model of multiple sclerosis," *Journal of Neuroimmunology*, vol. 223, no. 1-2, pp. 55–64, 2010.

Synthesis and Characterization of Acrylic Acid-g-(κ-Carrageenan) Copolymer and Study of Its Application

Arti Srivastava[1] and Rajesh Kumar[2]

[1] *Department of Chemistry, School of Physical Sciences, Guru Ghasidas Vishwavidyalaya, Bilaspur, Chhattisgarh 495009, India*
[2] *Organic Polymer Laboratory, Department of Chemistry, Centre of Advanced Studies in Chemistry, Banaras Hindu University, Varanasi, Uttar Pradesh 221 005, India*

Correspondence should be addressed to Rajesh Kumar; rkr_bhu@yahoo.com

Academic Editor: R. J. Linhardt

The effects of concentrations of peroxymonosulphate, mercaptosuccinic acid, hydrogen ion, acrylic acid, and carrageenan on grafting parameters were studied to find out the maximum grafting ratio. The corresponding values were found to be 14.0×10^{-3}, 32×10^{-3}, 15.0×10^{-3}, 18.0×10^{-2} mol dm^{-3}, and 1.0×10^{-2} g dm^{-3}, respectively. The optimal reaction time duration and reaction temperature were found to be 120 min and 40°C, respectively. The effect of hydrogen ion variation from 5.0 up to 25.0×10^{-4} mol dm^{-3} shows prompt changes on grafting parameter. The grafted copolymer was found more thermally stable than the ungrafted substrate. Also the grafted copolymer absorbed more water, namely hazardous metal cations. Hg^{2+}, Pb^{2+}, and Cd^{2+}, showed better flocculation behavior than carrageenan.

1. Introduction

Currently the commercial application of natural polymers and their derivatives is most captivating and immeasurably investigated. For this purpose, biodegradable natural polymers, such as polysaccharides and protein, have been widely used [1, 2]. Amongst the natural polysaccharides, carrageenan is widely applied due to its anionic nature. It is extracted from red seaweed and obtained from the cell wall of *Rhodophyceae* (red algae) belonging mainly to the genera *Chondrus, Gigartina, Kappaphycus, and Eucheuma* [3, 4]. Carrageenan is a linear polymer of about 25,000 galactose derivatives with regular but imprecise structures, depending on the source and extraction conditions. Carrageenans are usually classified according to their sulphate content. From an industrial application point of view, the most important configurations are the κ (kappa), ι (iota), and λ (lambda) forms; their structures differ mostly in the number of sulphated groups per disaccharide: one, two, or three for κ, ι, and λ, respectively. They are composed of alternation-(1 → 3)-β-D-galactopyranose-4-sulfate-(1 → 4)-3,6-anhydro-α-D-galactopyranose-(1 → 3) (Figure 1) which are well documented [5–9]. The presence of 3,6-anhydro-D-galactose and conformation of the pyranoside ring are responsible for their interesting rheological properties. κ-carrageenan is widely used as a thickening and gelling agent in food industry [10–12]. κ-carrageenan undergoes a coil (disordered state) to helix (ordered) transition, depending on temperature and ionic environment. Gelation is achieved through junction zones formed by the helices, leading to a three-dimensional network [5].

Gelation of ι-carrageenan is enhanced mainly by calcium, and κ-carrageenan is enhanced by potassium. It is reported [13] that cations $Ca^{2+} > K^{+} > Na^{+}$ bind to the carrageenan, blocking the sulphate groups and thereby preventing interaction with thaumatin (a basic protein). In recent years, Yuan et al. [14]. and Zhang et al. [15, 16] have demonstrated their significant role in antioxidant activities and explored them as effective excipients in controlled drug release delivery systems [17, 18].

Despite their numerous applications, κ-carrageenan suffers from drawbacks like easier susceptibility of microbial attack, which is destroying the gelation. Free radical initiated

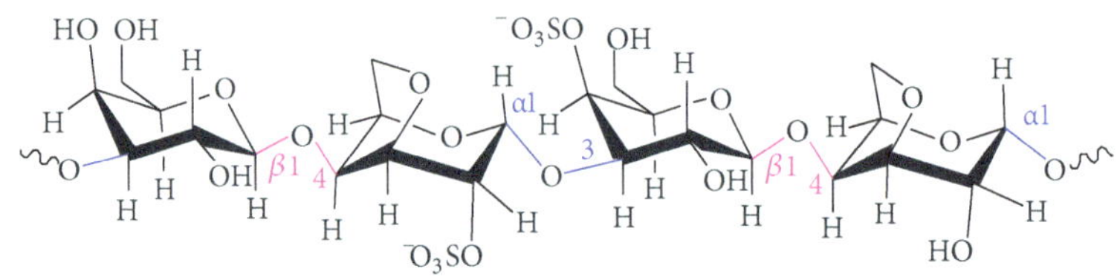

FIGURE 1: Structure of κ-carrageenan[-(1,3)-β-D-galactopyranose-4-sulphate-(1,4)-3,6-anhydro-α-D-galactopyranasoe(1,3)-].

graft copolymerization provides green and efficient approach for removing the drawback of easier susceptibility of microbial attack due to use of water, which is an easily decomposable oxidant and reductant and needs fairly lower temperature than any other radical initiated polymerization process like ATRP, RAFT, NMP, and so forth, which are generally carried out in nonaqueous media. Up to date many investigations were carried out on graft copolymerization reactions in view of preparing biopolymers-based materials [19–23]. A Few recent reports highlighted the grafting of κ-carrageenan with vinyl monomers [24–30], but no report showed the effect of grafting on gelation of carrageenan. In view of gelation properties of κ-carrageenan, this work has been carried for the first time out with an aim to tailor κ-carrageenan based hybrid materials by grafting of acrylic acid using peroxymonosulphate/mercaptosuccinic acid redox pair. The optimal grafting conditions were obtained for gelation, and studies were carried out on swelling capacity, metal ion sorption, and flocculating ability.

2. Experimental

2.1. Materials. Acrylic acid (Sigma-Aldrich, USA) was distilled under reduced pressure in inert atmosphere at 14°C. Peroxymonosulphate was purchased from Sigma-Aldrich, USA, and mercaptosuccinic acid, sulphuric acid, and methanol were purchased from Merck, India and used as received. κ-carrageenan was purchased from Sigma-Aldrich, USA.

2.2. Synthesis of Acrylic Acid-g-(κ-Carrageenan) Copolymer. The grafted copolymer of carrageenan and acrylic acid was synthesized by using peroxymonosulphate/mercaptosuccinic acid redox initiator in inert atmosphere. In each case carrageenan solution was prepared by adding a precalculated amount of carrageenan to 100 mL triple distilled water in a three-neck reactor to achieve the desired concentration of 1.0 g dm^{-3}. To determine the optimal conditions for the grafting of acrylic acid onto carrageenan, using PMS/mercaptosuccinic acid as redox initiator, graft copolymerization was carried out under various reaction times, temperatures, and concentrations of PMS, mercaptosuccinic acid, carrageenan, hydrogen ion, and acrylic acid. A definite concentration of sulphuric acid (15.0×10^{-4} mol dm^{-3}) was found essential to start the reaction. A definite amount of mercaptosuccinic acid and acrylic acid solutions, that is, 32.0×10^{-4} mol dm^{-3} and 18.0×10^{-2} mol dm^{-3}, respectively, was added to the reaction mixture. The stream of nitrogen gas was passed into the solution and in peroxymonosulphate solution separately. After the desired time interval (30 min), the reaction was initiated by addition of 14.0×10^{-3} mol dm^{-3} peroxymonosulphate solution. The reaction was allowed to continue for 120 minute, after which the reaction mixture was poured into water-methanol mixture (58 : 42 v/v). The grafted copolymer precipitates out, whereas polyacrylic acid remained in the solution. The graft copolymer was separated by filtration and washed with water-methanol mixture (two times) so that any homopolymer stuck to the grafted copolymer sample pass into the filtrate. The grafted copolymer thus obtained was dried in vacuum oven at 40°C and weighed.

2.3. Separation of Homopolymer. For the separation of homopolymer from the filtrate, a pinch of hydroquinone was added to the filtrate, and the homopolymer solution was concentrated by distillation under reduced pressure at room temperature. Thereafter, polyacrylic acid was precipitated by acidifying the reduced amount as given by [31], filtered, dried, and weighed.

2.4. Characterization

2.4.1. Determination of Grafting Parameters. The grafting parameters were calculated according to Fanta's (1973) definition [32] in terms of grafting ratio (%G), grafting efficiency (%E), add on (%A), conversion (%C), and homopolymer (%H):

$$\%G = \frac{\text{Weight of grafted copolymer}}{\text{Weight of } \kappa\text{-carrageenan}} \times 100,$$
$$\%A = \frac{\text{Weight of synthetic polymer}}{\text{Weight of grafted copolymer}} \times 100, \quad (1)$$

where synthetic polymer/polymer formed is a polymer as homopolymer and incorporated into the κ-carrageenan chain:

$$\%C = \frac{\text{Weight of polymer formed}}{\text{Weight of monomer charged}} \times 100,$$
$$\%E = \frac{\text{Weight of grafted copolymer}}{\text{Weight of polymer formed}} \times 100, \quad (2)$$
$$\%H = 100 - \%E.$$

TABLE 1: Grafting ratio and swelling behavior of acrylic acid-g-carrageenan as a function of the concentration of acrylic acid. [PMS] = 14.0×10^{-3} mol dm^{-3}; [MSA] = 32.0×10^{-4} mol dm^{-3}; [H^+] = 15.0×10^{-4} mol dm^{-3}; [CGN] = 1.0 g dm^{-3}; Time = 120 min.; Temp = 40°C.

Sample Code	[Acrylic acid] × 10^2 mol dm^{-3}	%G	P_S	S_R
C_{AA1}	8.0	330.0	310	3.1
C_{AA2}	12.0	336.0	328	3.3
C_{AA3}	16.0	380.0	342	3.4
C_{AA4}	18.0	510.0	390	3.9
C_{AA5}	20.0	535.0	396	4.0
C_{AA6}	22.0	494.0	382	3.8
C_{AA7}	24.0	450.0	360	3.6

[PMS] = concentration of potassium peroxymonosulphate, [MSA] = concentration of mercaptosuccininc acid, [CGN] = concentration of carrageenan.

Besides the above parameters, the rate of grafting (Rg) was calculated according to the following formula [33]:

$$\text{Rg} = \frac{\text{Weight of grafted polymer } (W)}{\text{Vol. } (V) \times \text{time } (T) \times \text{m.wt. of AA } (M)} \times 1000 \left(\text{mol L}^{-1}\,\text{s}^{-1}\right), \tag{3}$$

where W = Weight of grafted gum − weight of ungrafted gum.

To determine the optimal condition for the grafting of acrylic acid onto carrageenan, using PMS/mercaptosuccinic acid as redox initiator, graft copolymerization was carried out under various reaction times, temperatures, and concentrations of PMS, mercaptosuccinic acid, carrageenan, hydrogen ion, and acrylic acid.

2.4.2. Techniques. FT-IR spectra on KBr disc were recorded on a Varian Excalibur 3000 (Palo Alto, CA) spectrophotometer in the region 4000–400 cm^{-1}. The electronic absorption spectra was recorded on Perkin Elmer-Lambda 35 UV-VIS spectrometer connected with PTP-1 Peltier system at 25°C. The optical path length of measurement cell was 10 mm. The thermal behavior of pure carrageenan and acrylic acid-g-carrageenan was recorded on a Perkin Elmer and Diamond TG/DTA thermal analyzer within a temperature range 25–1000°C at a heating rate of 15°C/min in N_2.

2.5. Studies on Properties

2.5.1. Swelling Test. The swelling behavior of polyacrylic acid was carried out in water for one hour, and the sample was prepared by taking the different grafting ratios (Table 1) of acrylic acid. These studies were carried out by adding 0.100 g of poly(acrylic acid) into 20 mL triple distilled water and keeping undisturbed for 120 minutes at constant temperature to get the swelling equilibrium. The surface water on the swollen polyacrylic acid was removed by safely pressing between the folds of the moisten filter paper, and weights were taken. Percent swelling (P_S) and swelling ratio (S_R) have been calculated by using the following expressions [34]:

$$P_S = \frac{\text{Weight of swollen polymer} - \text{Weight of the dry polymer}}{\text{Weight of the dry polymer}} \times 100,$$

$$S_R = \frac{P_S}{100}. \tag{4}$$

2.5.2. Viscosity Measurement. Viscosity measurements of the aqueous solutions of carrageenan and the grafted copolymers (acrylic acid-g-carrageenan) were carried out with an Ubbelohde viscometer at 30 ± 0.1°C.

2.5.3. Absorption of Metal Ions: Removal of Hazardous Metal Ions. The metal-ion sorption studies were carried out on the different grafted copolymers obtained by varying the concentration of acrylic acid (Table 2). For this, 0.05 g acrylic acid-g-carrageenan with different molecular weight were placed in a 5 mL solution of a metal ion with a known concentration and kept for 24 h. After 24 h absorption, the concentration of the metal solution was determined by a standard method [35]. For the metal-ion sorption studies, five metal ions were selected, namely, Cu^{2+}, Ni^{2+}, Zn^{2+}, Pb^{2+}, and Hg^{2+}. The sorption behavior of the polymer for those five metal ions was investigated in terms of the following parameters [36]:

$$\text{Percent ion uptake } (P_u) = \frac{\text{Amount of metal ion in the polymer}}{\text{Amount of metal ion in the feed}} \times 100,$$

$$\text{Partition coefficient } (K_d) = \frac{\text{Amount of metal ion in the polymer}}{\text{Amount of the metal ion left in the solution}} \times \frac{\text{Volume of the solution (mL)}}{\text{Weight of the dry polymer (mg)}}, \tag{5}$$

$$\text{Retention Capacity } (Q_r) = \frac{\text{Amount of metal ion in the polymer}}{\text{Weight of the dry polymer}}.$$

2.5.4. Flocculation Test. Into a 1 L beaker, 200 mL of 1 wt% coal suspension was added. The beaker was placed on flocculator (Sonar Jar Set Digital Apparatus), dipping the stirrer blade in the suspension. At a 100 rpm stirring condition, a required quantity of polymer solution was added to the beaker to make a predetermined dose with respect to suspension volume. After the addition of the polymer solution, the suspension was stirred at a constant rpm (600) for 15 min. The flocks were allowed to settle down for one hour. Clean supernatant liquid was drawn from a depth of 1.0 cm, and its turbidity was measured using a digital

Table 2: Metal ion sorption.

Sample Code	%G	Percent ion uptake (P_u)					Partition coefficient (K_d)					Retention capacity (Q_r)				
		Cu^{2+}	Ni^{2+}	Zn^{2+}	Pb^{2+}	Hg^{2+}	Cu^{2+}	Ni^{2+}	Zn^{2+}	Pb^{2+}	Hg^{2+}	Cu^{2+}	Ni^{2+}	Zn^{2+}	Pb^{2+}	Hg^{2+}
C	—	3.1	2.1	1.8	4.1	1.0	16.1	5.4	4.8	10.8	5.2	1.5	0.50	0.5	1.0	0.5
C_{AA1}	330	5.18	4.73	3.80	4.16	3.27	27.35	24.83	19.76	21.73	16.91	2.5	2.3	2.0	2.1	1.6
C_{AA2}	336	7.88	6.79	5.32	6.15	4.49	42.79	36.42	28.11	32.76	23.55	3.8	3.3	2.8	3.1	2.2
C_{AA3}	380	9.54	8.43	7.60	8.13	6.33	52.75	46.06	41.15	44.27	33.84	4.6	4.1	4.0	4.1	3.1
C_{AA4}	510	8.09	7.40	5.13	6.94	4.90	44.01	40.00	27.05	37.31	25.80	3.9	3.6	2.7	3.5	2.4
C_{AA5}	535	4.97	4.11	2.66	3.57	2.45	26.20	21.46	13.67	18.51	12.57	2.4	2.0	1.4	1.8	1.2

Nephelometer (Eutech Instruments Digital Turbidity Meter TN-100 supplied by Merck, India). The turbidity is expressed in nephelometric unit (NTU).

3. Results and Discussion

3.1. FTIR Analysis. The FTIR spectrum of acrylic acid-g-carrageenan showed the peaks of poly(acrylic acid) as well as carrageenan (Figure 2). The following additional peaks/bands in the spectrum of acrylic acids-g-carrageenan were observed. The broad band of the hydrogen bonding with carboxylic group -OH stretching disappeared from the region 3000–3500 cm^{-1} in graft copolymer spectrum, while the band at 2730–3050 cm^{-1} is attributed to stretching vibration of nonhydrogen bonded OH stretching. This suggests that the polymeric network is arranged in such a way that the possibility of hydrogen bonding is diminished. The peak at 1740 cm^{-1} is attributed to stretching vibration of carbonyl group of acrylic acid. The broad and intense bands centered at 1240 cm^{-1} and 1450 cm^{-1} are attributed to C–O stretching and C–O–H bending vibrations. The presence of these bands/peaks confirms the grafting of poly(acrylic acid) onto carrageenan backbone. Thus, the presence of these bands and disappearance of O–H bending vibration in the spectrum of carrageenan prove that acrylic acid was grafted onto O–H site of the carrageenan macromolecule.

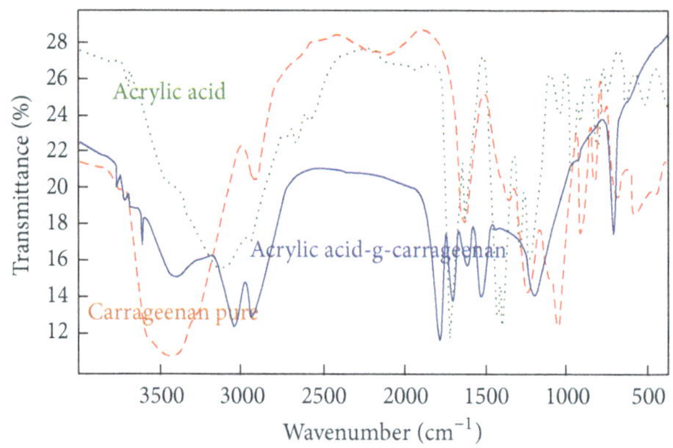

Figure 2: FTIR spectra of acrylic acid, carrageenan, and acrylic acid-g-carrageenan.

3.2. UV Analysis. Figure 3 shows UV visible spectra of acrylic acid and carrageenan-g-acrylic acid in the region 200 to 400 nm. UV-visible spectrum of carrageenan shows the sharp peak at 223 nm and a broad shoulder between 300 and 340 nm, while no absorption has been seen in the spectrum of acrylic acid. The spectrum of carrageenan-g-acrylic acid clearly indicates the incorporation of the above peak in the graft copolymer with major shift of the peak appeared at 223 nm to 280 nm and shoulder between 340 and 403 nm. These peaks and shoulders are due to carbonyl chromophore of acrylic acid, which confirms the grafting of acrylic acid onto carrageenan.

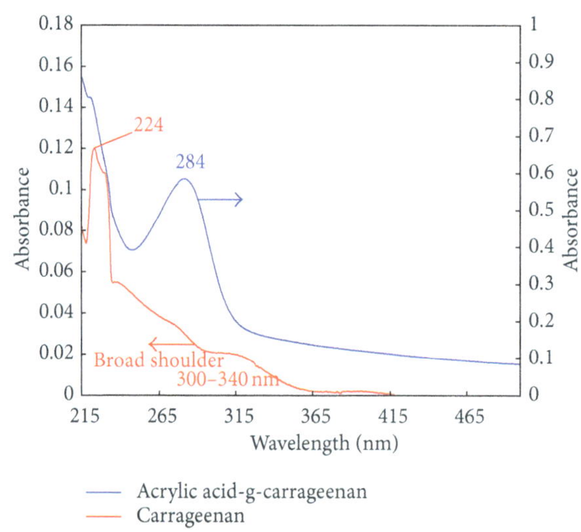

Figure 3: UV-visible spectra of acrylic acid-g-carrageenan and carrageenan.

3.3. Studies on Properties

3.3.1. Swelling Studies. The effect of swelling behavior on grafting ratio was studied by varying the concentration of acrylic acid from 8.0 to 24.0 × 10^{-2} mol dm^{-3} (Table 1). It was observed that percent swelling and swelling ratio of the grafted copolymer increase with increasing the grafting ratio. This is due to the hydrophilic nature of acrylic acid. The increase in percentage of poly (acrylic acid) chain on to carrageenan provides more hydrophilicity to graft copolymer thereby increases in percent swelling and swelling ratio were observed.

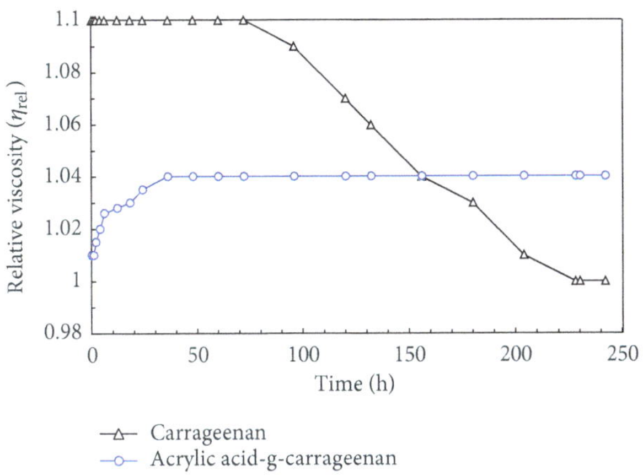

FIGURE 4: Relative viscosity of pure carrageenan and acrylic acid-g-carrageenan as a function of time.

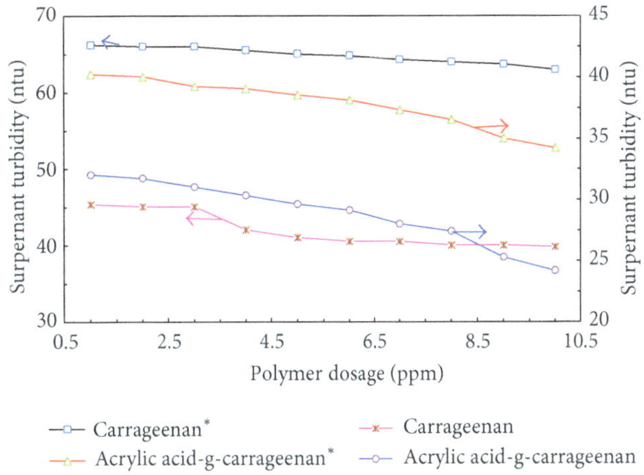

FIGURE 5: Effect of polymer dosage on turbidity for coking coal and non-coking coal (shown with *).

3.3.2. Resistance to Biodegradation. If two polymers have a similar molecular weight, a branched polymer's hydrodynamic volume will be lower, and hence it will have a lower viscosity than the linear polymer. It was observed that relative viscosity of acrylic acid-g-carrageenan is lower than the carrageenan (Figure 4). This might be due to the presence of grafted chains, which make the molecule more flexible and reduce the viscosity drastically [37]. Initially up to 40 hours, viscosity of the grafted copolymer solution was increased that was due the swelling nature of polyacrylic acid chains, which absorbs water resulting in swelling in grafted copolymer. Carrageenan, like other polysaccharide solutions, is highly prone to biodegradation, and it was found that its solution after 70 h of preparation started to degrade, and after 10 days, the solution showed considerable loss of viscosity (Figure 4). The graft copolymer solution was subjected to the same type of study for biodegradation, and it was observed that graft copolymer solution showed no loss of viscosity up to 10 days (Figure 4). These results show that the grafted copolymer is less susceptible to biodegradation, and similar results were reported by others [37] also. This is in agreement with the fact that by incorporating relatively poly(acrylic acid) chains onto carrageenan, the carrageenan can be made less susceptible to bacterial attack [38]. Thus, it can be concluded that, by incorporation of poly(acrylic acid) onto carrageenan through graft copolymerization, the resistance to biodegradation of carrageenan can be improved.

3.3.3. Metal Ion Sorption: Removal of Hazardous Metal Ions. Hydrophilic-hydrophobic balance and nature of monomer are major factors that affect the metal ion uptake [39]. These functionalities can be tailored and incorporated by grafting to retain metal ions by polymer analogous reaction or by simple chelation absorption due to the opening of polymer matrix or pores created. Out of these possibilities, all may contribute in the metal ion sorption, although the extent of contribution of possibilities may vary [40]. However, functional group incorporated with the help of grafting by its ability to interact with metal ions may play more important role in determination of selectivity and quantum of metal ion absorption [41]. Five metal ions were chosen (Cu^{2+}, Ni^{2+}, Zn^{2+}, Pb^{2+}, and Hg^{2+}) for the study of metal ion sorption. The amount of metal ion absorption was calculated by titrating the unabsorbed metal ions. The increases in the values of P_u, K_d, and Q_r up to 380%G (Table 2) might have been caused by the increase in poly(acrylic acid) chains, which accounts for the higher absorption of these divalent cations through a chelating reaction [42]. The highest percent ion uptake was found for Copper ions, while the lowest percent ion uptake was for mercury ions. The decreases in the values of P_u, K_d, and Q_r when the grafting ratio was higher than 380 are due to the increase in viscosity of the medium.

3.3.4. Flocculation Behaviour. Flocculation efficiencies of carrageenan backbone and its poly(acrylic acid) grafted copolymer were measured by the turbidity value of the supernatant liquid. Plots of polymer dosage versus supernatant turbidity for coking and noncoking coals are presented in Figure 5. It was found that grafted copolymer showed better performance with low turbidity than the pure substrate. In the grafted copolymer, the hanging of poly(acrylic acid) chains led to better accessibility to the contaminant coal particles and hence resulted in higher flocculation potential. Here, the bridging mechanism operates which involves binding or bridging individual particles. Poly(acrylic acid) has stronger bridging mechanism than carrageenan. By forming good flocks, the flocculation potential of carrageenan was improved. By grafting of poly(acrylic acid) onto carrageenan, efficient flocculants were obtained, and they might be used for the treatment of coal waste water.

3.4. Mechanism. A mechanism has been proposed for the reaction containing carrageenan, acrylic acid, hydrogen ion, mercaptosuccinic acid, and peroxymonosulphate. It was observed that potassium peroxymonosulphate is not able to produce grafted copolymer alone. Hence, $R^{\bullet}$ free radicals are generated by the interaction of peroxymonosulphate and

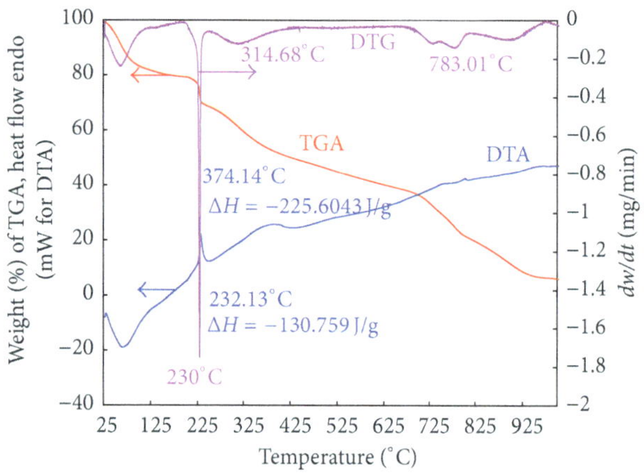

Figure 6: Thermal curves of TGA, DTG, and DTA of pure carrageenan.

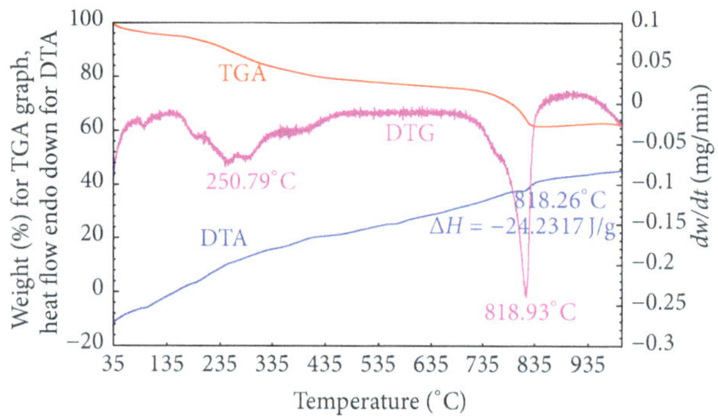

Figure 7: Thermal curves of TGA, DTG, and DTA of acrylic acid-g-carrageenan.

mercaptosuccinic acid. These free radicals abstract hydrogen atoms from the carrageenan molecule producing carrageenan macroradicals. The monomer radicals which are at the close proximity of the reaction sites become acceptors of the carrageenan radicals, resulting in chain initiation, and thereafter, themselves become free radical donors to neighbouring molecules to give grafted chain. These grafted chains are terminated by coupling to give grafted copolymer.

3.5. Thermal Analysis

3.5.1. Carrageenan. Figure 6 presents the thermograms of carrageenan, which revealed that decomposition of carrageenan had a two-step degradation process. The 18% weight loss at 100°C might be due to the loss of absorbed water. It started to degrade at 200–240°C. The polymer decomposition temperature (PDT) was found to be 125°C. The rate of weight loss increased with the increase in temperature from 125°C to 240°C and thereafter decreased and attained a maximum value at about 230°C. T_{max}, temperature at which maximum degradation occurred, was 230°C. The final decomposition temperature (FDT) and integral procedural decomposition temperature (IPDT) were found to be 800°C and 305°C, respectively.

3.6. Acrylic Acid-g-Carrageenan. Thermogram of acrylic acid-g-carrageenan (Figure 7) showed a two-step degradation process. Acrylic acid-g-carrageenan started to degrade at about 150°C; however, 4% weight loss was observed around 100°C, which was due to the absorbed water. The rate of weight loss was very slow up to 730°C. Only 25% weight loss was observed up to this temperature, while pure substrate degraded by almost 80% at 700°C, but after 730°C, acrylic acid-g-carrageenan degraded rapidly (Figure 7). The maximum weight loss (T_{max}) appeared at two temperatures, that is, 251°C and 819°C. And PDT, FDT, and IPDT of acrylic acid-g-carrageenan were found to be 180°C, 825°C, and 376.93°C, respectively. The PDT and FDT indicated that grafting of acrylic acid increased the initial decomposition temperature by 55°C while increasing the final decomposition temperature by 20°C than that of carrageenan. The higher values of FDT, IPDT, and char yield of grafted carrageenan compared to those of carrageenan indicated an overall improvement in thermal stability of the grafted copolymer.

4. Conclusion

The successful synthesis of high grafting ratio for the grafted copolymer was achieved. Grafted copolymer showed very good water swelling ability and resistance to biodegradation. Grafting is further supported by enhanced properties like metal ion uptake and flocculation efficiency. The thermal analysis data showed that the grafted copolymer was more thermally stable than the ungrafted polymer because of the higher final decomposition temperature and integral procedural decomposition temperature for the grafted copolymer as compared to substrate.

Acknowledgments

The authors thanks the Department of Chemistry, BHU, Varanasi (India), for providing spectral and analytical facilities and DST, New Delhi, for financial support in the form of Fast Track Project (SR/FTP/CS-107/2005).

References

[1] B. Jeong, Y. H. Bae, D. S. Lee, and S. W. Kim, "Biodegradable block copolymers as injectable drug-delivery systems," *Nature*, vol. 388, no. 6645, pp. 860–862, 1997.

[2] H. Younes and D. Cohn, "Morphological study of biodegradable PEO/PLA block copolymers," *Journal of Biomedical Materials Research*, vol. 21, no. 11, pp. 1301–1316, 1987.

[3] D. A. Rees, E. R. Morris, D. Thorn, and J. K. Madden, *The Polysaccharides*, vol. 1 of *Edited by G. O. Aspinall*, Academic Press, New York, NY, USA, 1982.

[4] O. Smidsrød and H. Grasdalen, "Some physical properties of carrageenan in solution and gel state," *Carbohydrate Polymers*, vol. 2, no. 4, pp. 270–272, 1982.

[5] G. A. De Ruiter and B. Rudolph, "Carrageenan biotechnology," *Trends in Food Science and Technology*, vol. 8, no. 12, pp. 389–395, 1997.

[6] L. Hilliou, F. D. S. Larotonda, A. M. Sereno, and M. P. Gonçalves, "Thermal and viscoelastic properties of κ/ι-hybrid carrageenan

gels obtained from the Portuguese seaweed Mastocarpus stellatus," *Journal of Agricultural and Food Chemistry*, vol. 54, no. 20, pp. 7870–7878, 2006.

[7] S. E. Harding, K. Day, R. Dhami, and P. M. Lowe, "Further observations on the size, shape and hydration of kappa-carrageenan in dilute solution," *Carbohydrate Polymers*, vol. 32, no. 2, pp. 81–87, 1997.

[8] M. Layahe, "Chemistry and physico-chemistry of phycocolloids," *Chhiers de Biologie Marine*, vol. 42, pp. 137–157, 2001.

[9] T. T. T. Thanh, Y. Yuguchi, M. Mimura et al., "Molecular characteristics and gelling properties of carrageenan family, preparation of novel carrageenan and their dilute solution properties," *Macromolecular Chemical Physics*, vol. 203, pp. 15–23, 2002.

[10] A. D. Anderson, C. R. Daubert, and B. E. Farkas, "Rheological characterization of skim milk stabilized with carrageenan at high temperatures," *Journal of Food Science*, vol. 67, no. 2, pp. 649–652, 2002.

[11] L. González-Tomás and E. Costell, "Relation between consumers' perceptions of color and texture of dairy desserts and instrumental measurements using a generalized Procrustes analysis," *Journal of Dairy Science*, vol. 89, no. 12, pp. 4511–4519, 2006.

[12] S. Y. Hsu and H. Y. Chung, "Comparisons of 13 edible gum-hydrate fat substitutes for low fat Kung-wan (an emulsified meatball)," *Journal of Food Engineering*, vol. 40, no. 4, pp. 279–285, 1999.

[13] S. Ohashi, F. Ura, T. Ochi, H. Lida, and S. Ukai, "Interaction of thaumatin with carrageenans. I. effects of pH, temperature and competing cations," *Food Hydrocolloids*, vol. 4, no. 2, pp. 105–119, 1990.

[14] H. Yuan, J. Song, W. Zhang, X. Li, N. Li, and X. Gao, "Antioxidant activity and cytoprotective effect of κ-carrageenan oligosaccharides and their different derivatives," *Bioorganic and Medicinal Chemistry Letters*, vol. 16, no. 5, pp. 1329–1334, 2006.

[15] Q. Zhang, N. Li, X. Liu, Z. Zhao, Z. Li, and Z. Xu, "The structure of a sulfated galactan from Porphyra haitanensis and its in vivo antioxidant activity," *Carbohydrate Research*, vol. 339, no. 1, pp. 105–111, 2004.

[16] Q. Zhang, P. Yu, Z. Li, H. Zhang, Z. Xu, and P. Li, "Antioxidant activities of sulfated polysaccharide fractions from Porphyra haitanesis," *Journal of Applied Phycology*, vol. 15, no. 4, pp. 305–310, 2003.

[17] C. Michona, C. Chapuisa, V. Langendorffb, P. Boulenguerb, and G. Cuveliera, "Structure evolution of carrageenan/milk gels: effect of shearing, carrageenan concentration and nu fraction on rheological behavior," *Food Hydrocolloids*, vol. 19, pp. 541–547, 2005.

[18] R. Lazzarini, P. C. Maiorka, J. Liu, V. Papadopoulos, and J. Palermo-Neto, "Diazepam effects on carrageenan-induced inflammatory paw edema in rats: role of nitric oxide," *Life Sciences*, vol. 78, no. 26, pp. 3027–3034, 2006.

[19] R. Kumar, A. Srivastava, and K. Behari, "Graft copolymerization of methacrylic acid onto xanthan gum by Fe^{2+}/H_2O_2 redox initiator," *Journal of Applied Polymer Science*, vol. 105, no. 4, pp. 1922–1929, 2007.

[20] R. Kumar, A. Srivastava, and K. Behari, "One-pot synthesis of a polysaccharide-based graft copolymer with an efficient redox pair (Fe^{2+}/BrO_3^-)," *Journal of Applied Polymer Science*, vol. 107, no. 5, pp. 2883–2891, 2008.

[21] R. Kumar, A. Srivastava, and K. Behari, "Synthesis and characterization of polysaccharide based graft copolymer by using potassium peroxymonosulphate/ascorbic acid as an efficient redox initiator in inert atmosphere," *Journal of Applied Polymer Science*, vol. 112, no. 3, pp. 1407–1415, 2009.

[22] A. Srivastava, V. Mishra, S. K. Singh, and R. Kumar, "One pot synthesis and characterization of industrially important graft copolymer (GOH-g-ACM) by using peroxymonosulphate/mercaptosuccinic acid redox pair," *E-Polymers*, pp. 1–14, 2009.

[23] A. Srivastava, V. Mishra, S. K. Singh, and R. Kumar, "Vanadium(V)/Mandelic acid initiated graft copolymerization of acrylamide onto Guar Gum in an aqueous medium," *Journal of Applied Polymer Science*, vol. 115, no. 4, pp. 2375–2385, 2010.

[24] A. Pourjavadi, A. M. Harzandi, and H. Hosseinzadeh, "Modified carrageenan 3. Synthesis of a novel polysaccharide-based superabsorbent hydrogel via graft copolymerization of acrylic acid onto kappa-carrageenan in air," *European Polymer Journal*, vol. 40, no. 7, pp. 1363–1370, 2004.

[25] S. Francis, M. Kumar, and L. Varshney, "Radiation synthesis of superabsorbent poly(acrylic acid)-carrageenan hydrogels," *Radiation Physics and Chemistry*, vol. 69, no. 6, pp. 481–486, 2004.

[26] A. Pourjavadi, H. Hosseinzadeh, and R. Mazidi, "Modified carrageenan. 4. Synthesis and swelling behavior of crosslinked KC-g-AMPS superabsorbent hydrogel with antisalt and pH-responsiveness properties," *Journal of Applied Polymer Science*, vol. 98, no. 1, pp. 255–263, 2005.

[27] A. Pourjavadi, S. Barzegar, and F. Zeidabadi, "Synthesis and properties of biodegradable hydrogels of κ-carrageenan grafted acrylic acid-co-2-acrylamido-2-methylpropanesulfonic acid as candidates for drug delivery systems," *Reactive and Functional Polymers*, vol. 67, no. 7, pp. 644–654, 2007.

[28] D. K. Mishra, J. Tripathy, and K. Behari, "Synthesis of graft copolymer (k-carrageenan-g-N,N-dimethylacrylamide) and studies of metal ion uptake, swelling capacity and flocculation properties," *Carbohydrate Polymers*, vol. 71, no. 4, pp. 524–534, 2008.

[29] M. M. Mishra, A. Sand, D. K. Mishra, M. Yadav, and K. Behari, "Free radical graft copolymerization of N-vinyl-2-pyrrolidone onto k-carrageenan in aqueous media and applications," *Carbohydrate Polymers*, vol. 82, no. 2, pp. 424–431, 2010.

[30] H. L. A. El-Mohdy and H. A. A. El-Rehim, "Radiation-induced kappa carrageenan/acrylic acid graft-copolymers and their application as catalytic reagent for sucrose hydrolysis," *Chemical Engineering Journal*, vol. 145, no. 1, pp. 154–159, 2008.

[31] P. Elayaperumal, T. Balakrishnan, M. Santappa, and R. W. Lenz, "A comparison of the absolute reactivity of vinyl monomers. I. Kinetic studies on radical polymerization of vinyl monomers initiated by a Mn^{3+}–diglycolic acid redox system," *Journal of Polymer Science A*, vol. 20, no. 12, pp. 3325–3336, 1982.

[32] G. F. Fanta, "Block and graft copolymerization," in *Synthesis of Graft and Block Copolymers of Starch*, R. J. Ceresa, Ed., pp. 1–27, Wiley-Interscience, New York, NY, USA, 1973.

[33] A. Srivastava and K. Behari, "Synthesis and characterization of graft copolymer (guar gum-g-N-vinyl-2-pyrrolidone) and investigation of metal ion sorption and swelling behavior," *Journal of Applied Polymer Science*, vol. 100, no. 3, pp. 2480–2489, 2006.

[34] H. A. Abd EL Rehim, A. Hegazy EL-Sayed, and A. M. Ali, "Selective separation of some heavy metals by poly(vinyl-alcohol)-grafted membranes," in *Journal of Applied Polymer Science*, vol. 76, pp. 125–132, 2000.

[35] J. Basset, R. C. Denny, G. H. Jeffery, and J. Mendham, *Vogel's Text Book of Quantitative Analysis*, William Clowes and Sons, London, UK, 4th edition, 1978.

[36] B. L. Rivas, H. A. Maturana, M. Jesús Molina, M. R. Gómez-Antón, and I. F. Piérola, "Metal ion binding properties of poly(N-vinylimidazole) hydrogels," *Journal of Applied Polymer Science*, vol. 67, no. 6, pp. 1109–1118, 1998.

[37] S. Ungeheuer, H. W. Bewersdorff, and R. P. Singh, "Turbulent drag effectiveness and shear stability of xanthan-gum-based graft copolymers," *Journal of Applied Polymer Science*, vol. 37, no. 10, pp. 2933–2948, 1989.

[38] J. K. Seaman, "Locust bean gum," in *Handbook of Water Soluble Gums and Resins*, R. L. Davidson, Ed., pp. 6–9, McGraw-Hill, New York, NY, USA, 1980.

[39] J. Lehto, K. Vaaramaa, E. Vesterinen, and H. Tenhu, "Uptake of zinc, nickel, and chromium by N-isopropyl acrylamide polymer gels," *Journal of Applied Polymer Science*, vol. 68, no. 3, pp. 355–362, 1998.

[40] K. Inou, Y. Baba, and K. Yoshizuka, "Adsorption of metal ions on chitosan and crosslinked copper(II)-complex chitosan," *Bulletin of the Chemical Society of Japan*, vol. 66, pp. 2915–2921, 1993.

[41] P. X. Sheng, Y. P. Ting, J. P. Chen, and L. Hong, "Sorption of lead, copper, cadmium, zinc, and nickel by marine algal biomass: characterization of biosorptive capacity and investigation of mechanisms," *Journal of Colloid and Interface Science*, vol. 275, no. 1, pp. 131–141, 2004.

[42] E. Onsoyen and O. Skaugrud, "Metal recovery using chitosan," *Journal of Chemical Technology and Biotechnology*, vol. 49, no. 4, pp. 395–404, 1990.

23

Miscibility Studies of Hydroxypropyl Cellulose/Poly(Ethylene Glycol) in Dilute Solutions and Solid State

K. Sudharsan Reddy,[1] M. N. Prabhakar,[1] P. Kumara Babu,[1] G. Venkatesulu,[1] U. Sajan Kumarji Rao,[2] K. Chowdoji Rao,[1] and M. C. S. Subha[2]

[1] *Department of Polymer Science and Technology, Sri Krishnadevaraya University, Andhra Pradesh Anantapur 515003, India*
[2] *Department of Chemistry, Sri Krishnadevaraya University, Andhra Pradesh Anantapur 515003, India*

Correspondence should be addressed to K. Chowdoji Rao, chowdojirao@gmail.com

Academic Editor: Yuriy A. Knirel

The miscibility of Hydroxypropyl cellulose (HPC)/poly(ethylene glycol) (PEG) blends over an extended range of concentrations in water. The viscosity, ultrasonic velocity, and refractive index of the above blend solutions have been measured at 30°C. The interaction parameters such as $\Delta\beta$ and μ proposed by Chee and α proposed by Sun have been obtained using the viscosity data to probe the miscibility of the polymer blends. The values indicated that the blends were miscible when HPC content is more than 40% in the blend. The obtained results have been confirmed by the ultrasonic velocity and refractive index studies. The films of the blends were prepared by solution casting method using water as a solvent. The prepared films have been characterized by analytical techniques such as FTIR, DSC, X-RD, and SEM to probe the miscibility of HPC/PEG blends. The compatibility in the above compositions may be due to the formation of H-bonding between hydroxyl groups of HPC and etheric oxygen atom of PEG molecules.

1. Introduction

Usage of blended polymers is the most effective way to produce new multipurpose materials. During the past few years, researchers have paid considerable attention to the study of polymer blending [1–3].

A combination of synthetic and natural polymers results in new materials, which have specific properties of synthetic components (good mechanical properties, easy possibilities, low production, and transformation costs) and biocompatibility nature of biopolymers [4, 5]. These blends have already been utilized as biodegradable biomaterials [6], drug delivery systems [7], membranes [8], materials for agricultural applications [9], and so forth.

Blending of polymers is one of the simplest methods to obtain a variety of physical and chemical properties from the constituent polymers at a molecular level. The gain in newer properties depends on the degree of compatibility or miscibility of the polymers. There have been number of techniques to study the miscibility of polymer blends such as FTIR, DSC, X-RD, and SEM. Most of these techniques may be complicated, costly, and time consuming. Chee [10] and Sun et al. [11] suggested viscometry for the study of the polymer-polymer miscibility. Palladhi and Singh [12, 13] showed that the variation of ultrasonic velocity and viscosity with blend composition is linear for miscible and nonlinear for immiscible blends.

Hydroxypropyl cellulose (HPC) is a cellulose derivative employed as coatings, excipients, encapsulations, binding materials, foaming agents, protection colloids, flocculants, and so forth, for a wide variety of applications in food, drugs, paper, ceramics, plastics, and so forth [14]. Hydroxypropyl cellulose films are used as a food additive, HPC is used as a thickener, a low level binder and as an emulsion stabiliser with E number E463. In pharmaceuticals it is used as a disintegrant, and a binder, in tablets [15]. They act as acceptable barriers to moisture and oxygen [16, 17]. Blends based on HPC with natural or synthetic polymers were proposed to design new materials with enhanced properties and with wide range of applications.

Poly(ethylene glycol) is used in all applications of aqueous phase partitioning. PEG is a biocompatible and biodegradable polymer and has adhesive and flocculent properties [18]. For biochemical separations on the laboratory scale, the most commonly used aqueous phase system is composed of Dextron and PEG [19].

In the present study HPC and PEG polymer blend solutions were prepared in water and their miscibility was studied by using viscosity, ultrasonic velocity, and refractive index measurements. By solution casting method these polymer blend films were also prepared and their miscibility studied using FTIR, DSC, X-RD, and SEM techniques. The compatibility of these polymer blends is discussed and it is concluded that the presence of hydrogen bonding groups [20, 21] in these two polymers is responsible for the polymer-polymer interactions [22]. These results are presented here.

2. Materials and Experimental Techniques

2.1. Materials. Hydroxypropyl cellulose (M_W = 1,40,000) was purchased from Sigma Aldrich, USA, poly(ethylene glycol) (M_W = 6,000) was purchased from Qualigens, Mumbai, (India) and both were used without further purification. Double distilled water having zero conductivity (distilled-water usually runs between 2 and 10 S/cm) was used as a solvent.

2.2. Preparation of Blend Solutions. Two kinds of polymer powders (HPC and PEG) were dried in a vacuum oven for 24 hrs at 50°C to remove the free water from the samples before use. HPC and PEG were separately dissolved in water to form 1% (w/v) solution. It is worth mentioning that in the preparation of HPC solution, the HPC powder must be slowly added to the solvent with gently stirring for about one week at room temperature to obtain a homogenous solution; otherwise HPC tends to form sticky clumps when large amounts of samples were added to the solvent. The blends of stock solutions (20/80, 40/60, 50/50, 60/40, and 80/20) of HPC/PEG were prepared by stirring the mixtures at room temperature for about 45 minutes. Using the above pure and blended stock solutions different blend solutions (0.1, 0.3, 0.5, 0.7, and 0.9 (w/v) concentrations) were prepared.

2.3. Preparation of Blend Films. Thin films of the polymers and their blends were prepared by solution casting method. Separate aqueous solutions of HPC and PEG were prepared. A solution of HPC was added to that of PEG with constant stirring. The mixtures were stirred for 45 min at room temperature to ensure complete mixing. The total polymer concentration was kept at 1% (w/v). Stock solutions of HPC and PEG and their different blend compositions were then casted onto a Teflon-coated clean glass plate and dried using IR lamp in a dust-free atmosphere. The dried thin films were peeled off from the glass plate and were found to be transparent. The prepared thin films of pure polymers and blends were characterized using FTIR, DSC, X-RD, and SEM methods.

2.4. Techniques. Viscosity and density measurements were made at 30°C using Ubbelohde suspended level viscometer (with the flow time of 95 sec for distilled water) and specific gravity bottle, respectively. The required temperature (30°C) was maintained within ±0.05°C. The ultrasonic velocities of the blend solutions with different compositions, namely, 0/100, 20/80, 40/60, 50/50, 60/40, 80/20, and 100/0 by weight, were measured at 30°C using ultrasonic interferometer. The constant temperature was maintained by circulating water from a thermostat with a thermal stability of ±0.05°C through the double-walled jacket of ultrasonic experimental cell. The experimental frequency was 2 MHz, and the velocity measurements were accurate to better than ±0.5%. The refractive indices of blend solutions with different compositions were measured directly with an Abbe refractometer (digital) with thermostated water circulation system at 30°C. The accuracy of the refractive index measurement is ±0.02%.

2.5. Fourier Transform Infrared Spectroscopy. Fourier transform infrared (FTIR) spectra of HPC, PVP, and their blend films were taken using Bomen MB-3000 FTIR spectrometer. Blend films were characterized at room temperature from 4000 to 400 cm^{-1} under an N_2 atmosphere at a scan rate of 21 cm^{-1}.

2.6. Differential Scanning Calorimetry. DSC curves of HPC, PVP, and their blend films of different compositions were recorded using TA instruments Differential Scanning Calorimeter (Model: SDT Q600, USA). The analysis of samples was performed at heating rate of 20°C/min under N_2 atmosphere at a purge speed of 100 mL/min.

2.7. X-Ray Diffraction. The X-ray Diffraction (XRD) patterns of the blend samples were obtained with an Intel diffractometer (Paris, France) with monochromatized Cu Kα radiation (scan speed of 1°/min in a 2θ range of 5°–40°) at room temperature.

3. Results and Discussion

3.1. Viscosity Studies. The absolute viscosity versus concentration curves for the blends of HPC and PEG of different compositions at 30°C in water is shown in Figure 1. It was well established earlier by many workers [23, 24] that the variations of viscosity versus concentration of blend composition plots are linear for compatible blends and nonlinear for incompatible blends. On this basis in the present study it is noticed that absolute viscosity varied nonlinearly with the concentration of blend composition up to 40/60 HPC/PEG and linearly beyond this composition. This indicates that the polymer blends of HPC/PEG are found to be immiscible when HPC content is up to 40% and then miscible beyond this composition.

From Figure 1, it is also observed that as the concentration of HPC increases in the blend composition, the absolute viscosity also increases. With the increasing percentage of HPC in the HPC/PEG blend, PEG can interact with several chains of HPC and can form highly cross-linked structure,

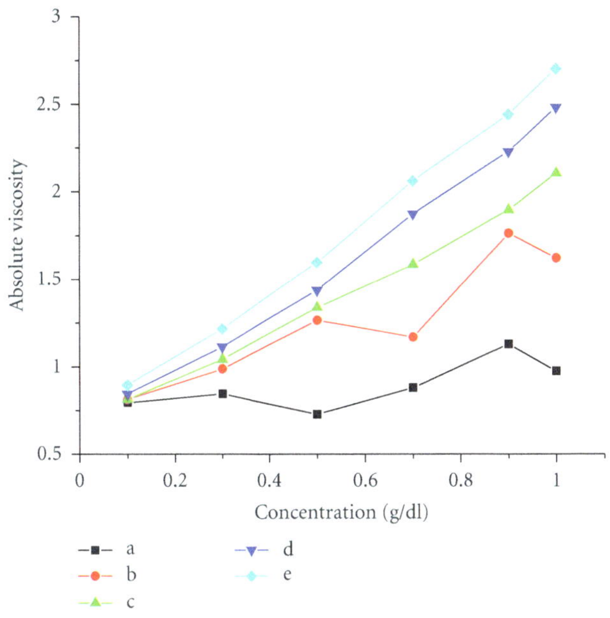

FIGURE 1: Plots of absolute viscosity versus concentration for 1% w/v HPC-PEG blends of 20/80 (a), 40/60 (b), 50/50 (c), 60/40 (d), 80/20 (e).

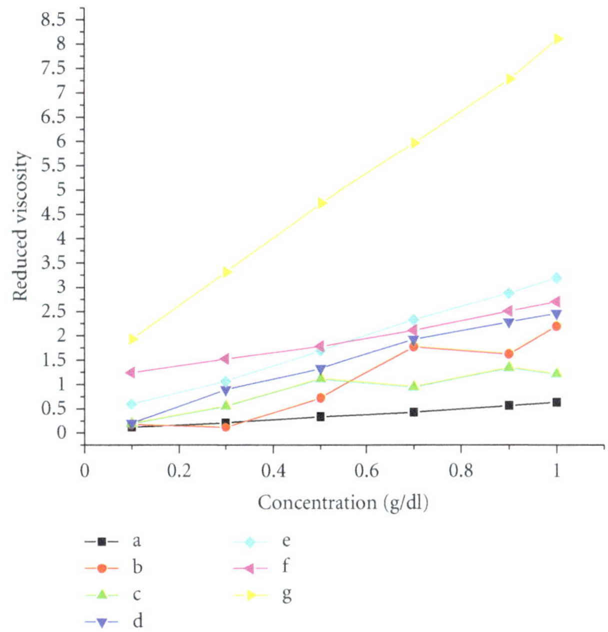

FIGURE 2: Reduced viscosity versus concentration curves for pure PEG (a), HPC/PEG blends 20/80 (b), 40/60 (c), 50/50 (d), 60/40 (e) 80/20 (f) and pure HPC (g).

which causes high solution viscosity. Further, with the increase in proportion of HPC in the blend along with cross-linking effect, the effect of hydrogen bonding between the OH groups of HPC and etheric oxygen atom of PEG also increases. These effects at higher contents of HPC of these blends lead to the high viscosity of the blends as observed in Figure 1.

The Huggins plots of reduced viscosity against concentration of different compositions of 1% (w/v) HPC/PEG blends pure HPC and pure PEG in water at 30°C are shown in Figures 2(a)–2(g). From these graphs on extrapolating to zero concentration, the intrinsic viscosities are determined, and the values of HPC, PEG, and HPC/PEG blends of different compositions (20/80, 40/60, 50/50, 60/40, 80/20) are 1.3258, 0.1373, 0.0426, 0.1956, 0.555, 0.5184, and 0.7529 dL/g, respectively. The intrinsic viscosity of HPC/PEG blends for different compositions is in between the values of pure polymers. The reasonable interpretation may be that the attractive interactions are existing between HPC and PEG in water. These attractive interactions may lead to decrease the intermolecular excluded volume effect and decrease the hydrodynamic volume. As a consequence, the HPC/PEG blend coils expand, which causes an increase of the intrinsic viscosity of HPC/PEG blends in water.

From these graphs (Figures 2(a)–2(e)), it is also clearly evident that the Huggins curves composed of two regions with varying slopes for the polymer blends up to 40/60 HPC/PEG blend composition. The change in the slope of the curve may be attributed to the mutual repulsion of macromolecules in solution which favors the polymer immiscibility up to 40% of HPC content in the blend. But later, that is, above 40% HPC content in the blend composition, the reduced viscosity versus concentration of

TABLE 1: Chee and Sun's interaction parameters for different composition of 1% (w/v) HPC/PEG blends in water at 30°C.

HPC/PEG blend composition	Chee's differential interaction parameters ΔB	Chee's differential interaction parameters μ	Sun's miscibility parameter $\alpha \times 10^2$
20/80	−0.0821	−1.1578	−2.4609
40/60	−0.1609	−0.3186	−0.3344
50/50	0.1238	8.1672	7.7292
60/40	0.0504	0.5192	0.054
80/20	0.0184	0.9732	0.0252

blends shows linear variation which is an indication of mutual attraction of molecules in solution which favors the polymer miscibility. A similar observation was made by Vasanthakumari et al. [25] from their viscosity studies on miscibility of poly(2-hydroxyethylmethocrylate) and hydroxyl propyl methyl cellulose blends.

3.2. Chee and Sun Interaction Parameters. In order to quantify the miscibility or immiscibility nature of the present polymer blend studies, the interaction parameters ΔB and μ suggested by Chee and Sun's interaction parameter α are calculated with the help of (1), (4), and (5) for the HPC/PEG blend systems in water at 30°C and are presented in the Table 1.

Chee suggested that the general expression for interaction parameter when polymers are mixed in weight fractions w_1 and w_2 is as follows:

$$\Delta B = \frac{b - \bar{b}}{2w_1 w_2}, \qquad (1)$$

where $\overline{b} = w_1 b_{11} + w_2 b_{22}$ in which b_{11} and b_{12} are the slopes of the viscosity curves for the pure components. The coefficient b is related to the Huggins coefficient K_H as

$$b = K_H [\eta]^2. \tag{2}$$

For ternary systems, the coefficient b is also given by

$$b = w_1^2 b_{11} + w_2^2 b_{22} + 2 w_1 w_2 b_{12}, \tag{3}$$

where b_{12} is the slope for the blend solution. Using these values, Chee defined a more effective parameter as follows:

$$\mu = \frac{\Delta B}{\left\{[\eta]_2 - [\eta]_1\right\}^2}, \tag{4}$$

where η_1 and η_2 are the intrinsic viscosities for the pure component solutions. Recently, Sun et al. has suggested a new formula for the determination of polymer miscibility as follows:

$$\alpha = K_m - \frac{K_1 [\eta]_1^2 w_1^2 + K_2 [\eta]_2^2 w_2^2 + 2\sqrt{K_1 K_2} [\eta_1][\eta_2] w_1 w_2}{\left\{[h]_1 w_1 + [h]_2 w_2\right\}^2}, \tag{5}$$

where K_1, K_2 and K_m are the Huggins constants for individual components 1, 2 and the blend, respectively.

It is observed from Table 1 that the values of ΔB, μ, and α are negative up to 40/60 HPC/PEG blend composition and positive beyond this composition.

In general [23, 24] if ΔB, μ, and α are positive for any polyblend system, these are considered to be as miscible one whereas if these values are negative, the polyblends are considered as immiscible one. Based on this it is concluded from the present study, as per the values given in Table 1, that HPC/PEG blend shows immiscibility when HPC content is up to 40% and then miscible beyond this composition. The probable reason for this may be that up to 40/60 HPC/PEG blends, that is, in the PEG rich content blend composition the etheric "O" atoms are responsible for H-bonding with –OH groups of HPC (Scheme 1).

H-bonding between HPC and PEG can be explained as follows. The available OH-groups of HPC are less in number up to 40/60 HPC/PEG blend; hence the hydrogen bonding possibility is less, and this may be the reason for immiscibility of these blend compositions, that is, up to 40% HPC content. Later as the HPC content increases beyond this composition more no. of HPC molecules are available and are responsible for having more no. of –OH groups which in turn gives a scope to form more H-bonding with etheric oxygen of PEG molecules. This leads to miscibility of this blend beyond 40% HPC + 60% PEG. A similar observation was reported by Ravi Prakash and Rai [26] from their work on the miscibility studies of Sodium Alginate/PEG blends in water by viscosity, ultrasonic velocity, and refractive index methods.

3.3. Ultrasonic Velocity Studies. The variation of ultrasonic velocity (v) with the concentration of different compositions of 1% (w/v) of HPC/PEG blends in water is plotted in Figures 3(a)–3(e).

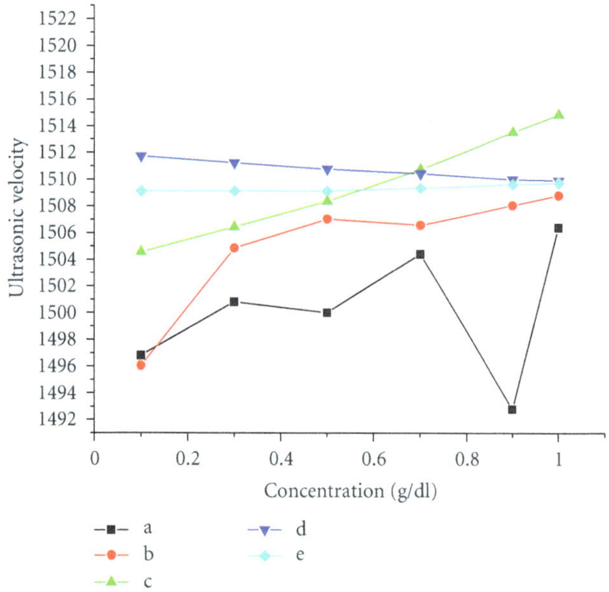

FIGURE 3: Ultrasonic velocity versus concentration curves for HPC/PEG blends of 20/80 (a), 40/60 (b), 50/50 (c), 60/40 (d), 80/20 (e).

It has been observed from Figures 3(a) to 3(e) that ultrasonic velocity varies nonlinearly up to 40/60 (HPC/PEG) blend composition and linearly with polymer blend concentration beyond this blend composition. It was well established [27–29] that for a miscible blend, the ultrasonic velocity varies linearly with polymer blend concentration, and for an immiscible blend it varies nonlinearly with polymer blend concentration. This indicates HPC and PEG form immiscible blend in water up to 40/60 (HPC/PEG) blend composition and miscible beyond this composition.

From these results, we can conclude that the variation of the curves relating the different acoustical parameters indicates immiscibility nature of (HPC/PEG) blend up to 40/60 HPC/PEG blend composition, and beyond this composition these blends behave as miscible. A similar observation was reported by Illiger et al. [24] from their ultrasonic investigation on compatibility of polymer blends.

3.4. Refractive Index Studies. It is evident from Figures 4(a)–4(e) that the refractive index varies non-linearly with the concentration up to 40/60 HPC/PEG blend composition, whereas it varies linearly beyond this composition. This behavior indicates that HPC/PEG blends are immiscible up to 40/60 blend composition, and beyond this composition these blends are miscible. A similar observation was reported by Basavaraju et al. in [30] case of Xanthan gum/PEO blends.

These conclusions are in good agreement with the results obtained for this blend system from viscometric and ultrasonic studies.

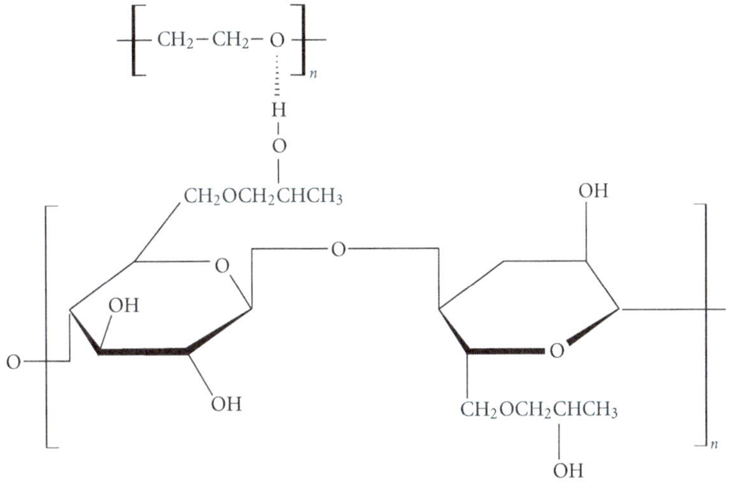

SCHEME 1: Hydrogen bonding between the Hydroxyl groups of PEG and the hydroxyl groups of HPC.

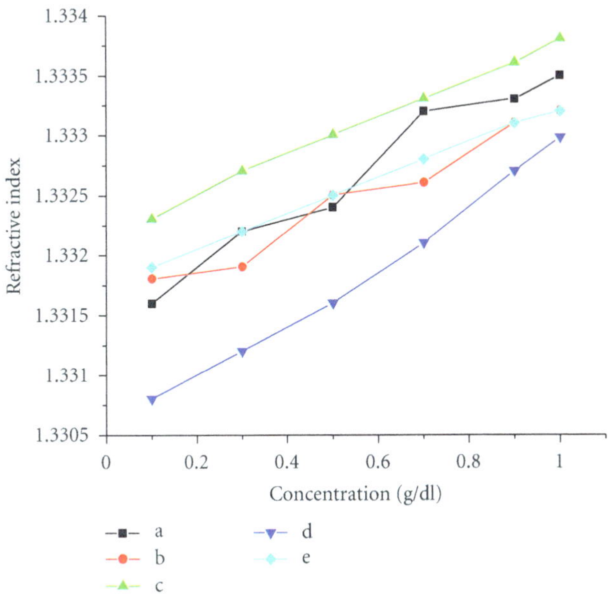

FIGURE 4: Refractive Index versus concentration curves for HPC/PEG blends of 20/80 (a), 40/60 (b), 50/50 (c), 60/40 (d), 80/20 (e).

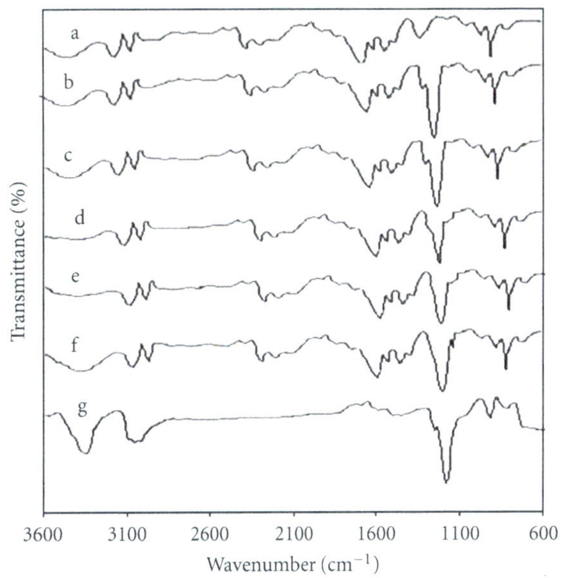

FIGURE 5: FTIR spectra of PEG (a), 20/80 HPC/PEG (b), 40/60 HPC/PEG (c), 50/50 HPC/PEG (d), 60/40 HPC/PEG (e), 80/20 HPC/PEG (f), HPC (g).

4. Miscibility Studies of Blend Films by Characteristic Techniques

4.1. Fourier Transform Infrared Spectroscopy Studies. FTIR spectroscopy has been widely used by many researchers to study the formation of blends [31–33]. FTIR spectrum provides information regarding intermolecular interactions corresponding to stretching or blending vibrations of particular bands, and the positions at which these peaks appear depend directly on the force constant or band strength. Hydrogen bonding or other secondary interactions between chemical groups on the dissimilar polymers should theoretically cause a shift in peak position of the participating groups. This kind of behavior is exhibited by miscible blends that show extensive phase mixing. Hydrogen bonding interactions usually move the stretching frequencies of participating groups, for example, O–H towards lower numbers usually with increased intensity and peak broadening. The shift in peak position will depend on the strength of the interaction.

The FTIR spectra of HPC, PEG, and the blends of HPC/PEG (20/80, 40/60, 50/50, 60/40 and 80/20) are shown in Figure 5. The formation of strong hydrogen bonds between HPC and PEG was demonstrated by FTIR spectroscopy from the shifts of absorption bands showing hydroxyl stretching vibrations, which were sensitive to the hydrogen bonds formed during blending. The broad transmission bands at 3600–3100 cm^{-1} produced by stretching of the hydroxyl groups in the spectrum of HPC can be

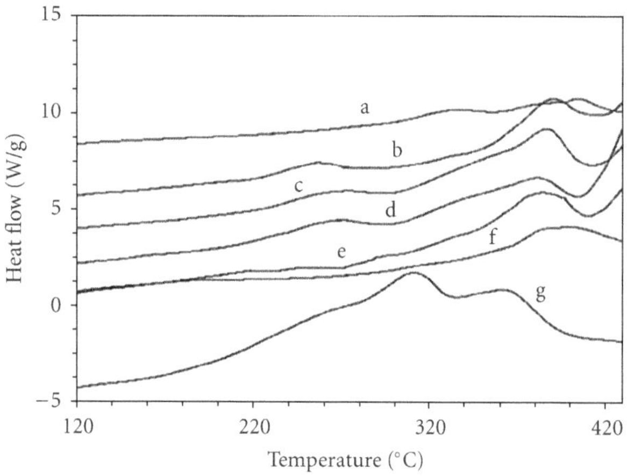

FIGURE 6: DSC thermograms of HPC (a), HPC/PEG 80/20(b), HPC/PEG 60/40 (c), HPC/PEG 50/50 (d), HPC/PEG 40/60 (e), HPC/PEG 20/80 (f), PEG (g).

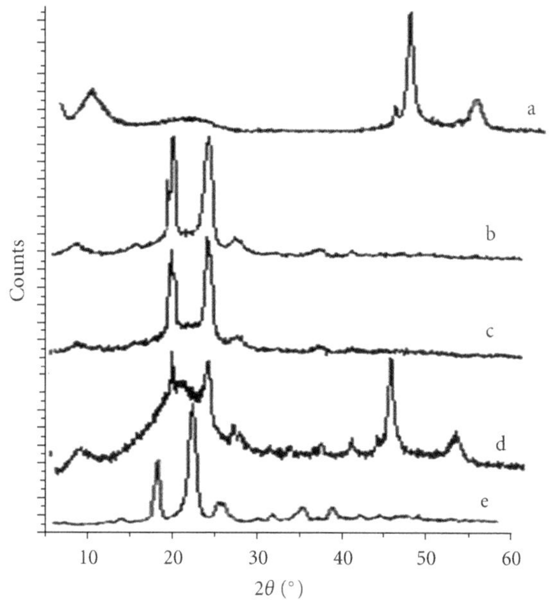

FIGURE 7: X-ray diffraction patterns of HPC (a), HPC/PEG 80/20 (b), HPC/PEG 50/50 (c), HPC/PEG 20/80 (d), PEG (e).

remarkably distinguished. It can be seen from the spectra that the peak intensity and the peak shape were clearly different, and these differences were induced by the different blend ratios. The presence of hydrogen-bond structures in some blends could be inferred from the peak shape and peak intensity of the absorption band of the hydroxyl starching vibrations in the FTIR spectra [34]. The broadband in the pure HPC spectrum at 3600–3100 cm^{-1}, with a maximum at 3460 cm^{-1}, was assigned to stretching vibrations of the –OH groups. The difference among the curves in Figure 5, a little broadening or shifting, or a peak at 3600–3100 cm^{-1} was observed in the transmission band of the HPC/PEG blends when they were compared with that of pure HPC, which suggested that a relative low amount of interaction was presented between the polymers.

From Figures 5(b) and 5(c) it is clear that 20/80, 40/60 HPC/PEG blends, there is no change (or) shift in –OH stretching vibrations. In these blends as the HPC content is lower (40%) and no. of OH groups available also less to form the hydrogen bonds with etheric oxygen of PEG molecules. Hence these blends show immiscible nature. By observing (d, e, f, and g) spectrum it is clear that the hydroxyl stretching vibrations shifted to a lower wave number with increasing amounts of HPC which are responsible for formation of more hydrogen bonding interactions between the OH groups belonging to HPC and etheric oxygen belonging to PEG molecules. This leads to miscibility nature of HPC/PEG blends, where the HPC content is more than 40% in the blend compositions.

4.2. Differential Scanning Calorimetry Studies. Figure 6 shows DSC thermogram of HPC, PEG, and their different blend compositions (20/80, 40/60, 50/50, 60/40, and 80/20). DSC thermograms were carried out to determine the compatibility of the polymer blend. T_m values for pure HPC and Pure PEG are about 353°C and 333°C, respectively. As shown in Figure 6 broad endothermic peaks, observed in case of 50/50, 60/40, and 80/20 HPC/PEG blend films, which indicates hydrogen bonding interactions between HPC and PEG. By careful observation in blends where HPC content is less than 40%, that is, in case of DSC thermograms of 20/80 and 40/60 HPC/PEG blends the narrow peaks are noticed which shows the non-H-bonding nature between HPC and PEG. From this we can conclude that for blend where the content of HPC is more than 40% they are miscible whereas the blends were the content of HPC is less than 40% they are immiscible.

4.3. X-Ray Diffraction Studies. The measurement of the X-RD patters of polymer blend is also used as a criterion to determine its miscibility. If the two components have low compatibility, then each polymer would have its own crystal region in the blend films, and X-ray patterns of the sample would express as simply mixed patterns with the same ratio as those for blending. The typical X-RD patterns of PEG, HPC, and their blend compositions are shown in Figure 7. For the pure HPC (a), there were two peaks around $2\theta = 8.5°$ and 44.6°. The diffraction model of PEG (b) showed at $2\theta = 19.7$ and 23.9. Peak of 50/50 (c), HPC/PEG blend film is $2\theta = 19.9°$ and 23.9°; 80/20 HPC/PEG (d) blend film is $2\theta = 19.5°$ and 23.7°. These peaks are similar to PEG. These evidences further concluded that strong interaction occurred between HPC/PEG molecules in the blends of 50/50 HPC/PEG and 80/20 HPC/PEG which is an indication of miscibility nature. But in case of 20/80 blend film showed $2\theta = 19.4$ and 44.6 which indicates that there is no interaction between HPC and PEG molecules in the 20/80 blend film; each component has shown its own crystal region in the blend film, and X-RD patterns have expressed simple mixed patterns. These evidences further concluded

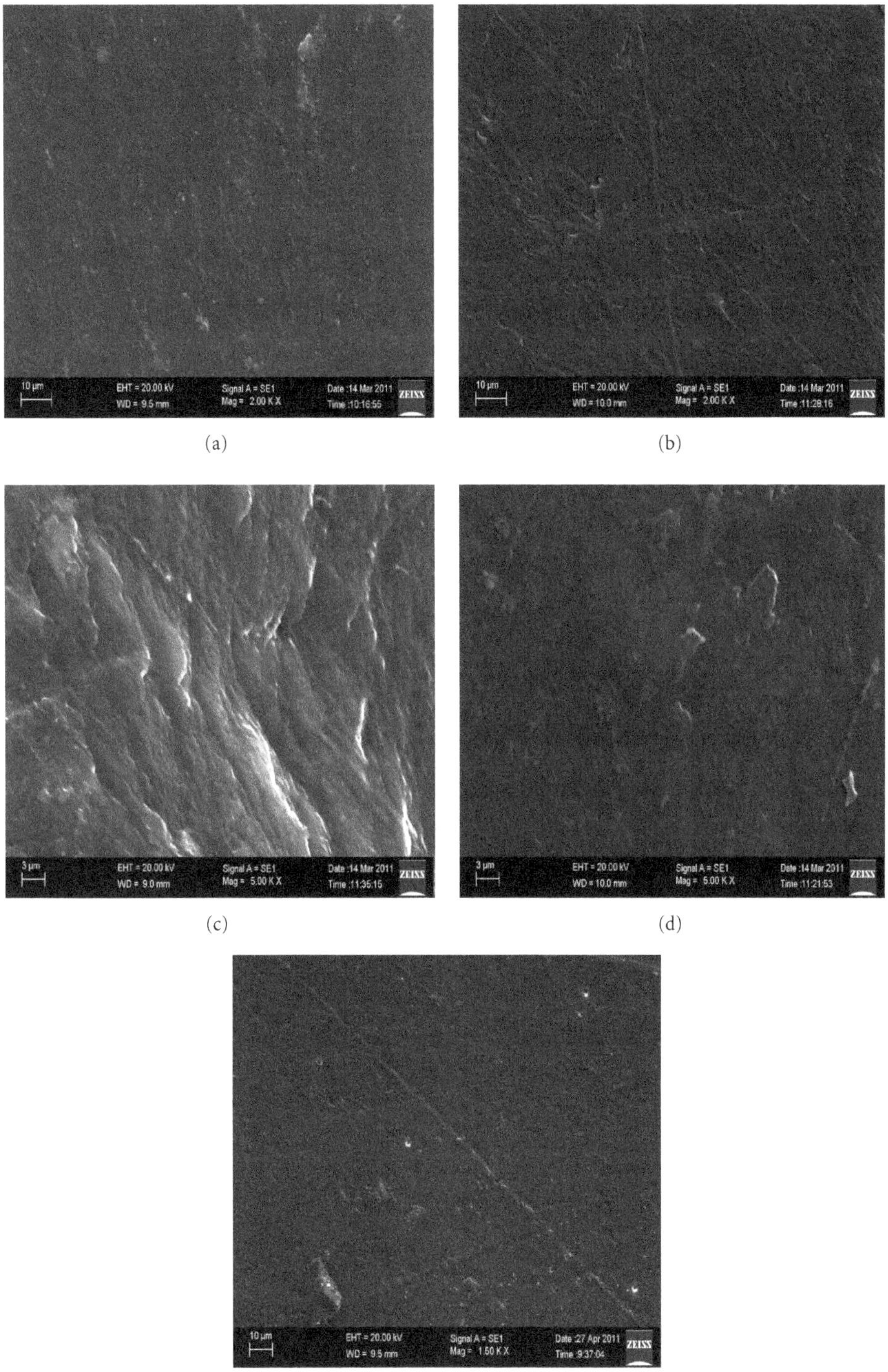

(a) (b) (c) (d) (e)

Figure 8: SEM images of HPC (a), PEG (b), 20/80 HPC/PEG (c), 50/50 HPC/PEG (d), 80/20 HPC/PEG (e).

that 20/80 HPC/PEG is immiscible, and 50.50, 80/20 blend compositions are miscible.

4.4. Scanning Electron Microscopy. Figure 8 shows the SEM images of HPC (a), PEG (b), and their blends of different compositions 20/80 (c), 50/50 (d), and 80/20 (e). The surface morphology of HPC (a) and PEG (b) films was homogeneous. The bright strips presented in the image of pure PEG. With the addition of PEG to HPC, the morphologies of the blended films changed dramatically. There is no obvious phase separation observed in the 50/50 and 80/20 HPC/PEG blend films, but phase separation is visible in the 20/80 HPC/PEG blend. Hence, we conclude that the blends of 50/50, 80/20 are miscible, and 20/80 of

HPC/PEG blend is immiscible in nature. This confirms the HPC/PEG blends where the HPC content is more 40% are miscible and the blends where HPC content is less 40% are immiscible. This further supports the conclusions drawn from other analytical techniques in case of the miscibility nature of this blend.

5. Conclusions

The miscibility of HPC/PEG blends in water has been studied by solution methods such as viscosity, ultrasonic velocity, and refractive index techniques at 30°C. Using the viscosity data, interaction parameters ΔB, μ, and α are were computed. These values revolved that the blend is miscible when HPC content is more than 40% in the blend. This was further confirmed by ultrasonic velocity and refractive index measurements. Below this critical HPC concentration, the blends were found to be immiscible.

The prepared films of the HPC/PEG blends are also characterized for FTIR, DSC, X-RD, and SEM to support the data obtained from simple methods such refractive index, ultrasonic velocity, and viscosity.

References

[1] M. Barsbay and A. Güner, "Miscibility of dextran and poly(ethylene glycol) in solid state: effect of the solvent choice," *Carbohydrate Polymers*, vol. 69, no. 2, pp. 214–223, 2007.

[2] G. S. Guru, P. Prasad, H. R. Shivakumar, and S. K. Rai, "Miscibility studies of polysaccharide xanthan Gum/PVP blend," *Journal of Polymers and the Environment*, vol. 18, no. 2, pp. 135–140, 2010.

[3] T. M. M. Swamy, B. Ramaraj, and Siddaramaiah, "Thermal and morphological properties of SA/HPMC blends," *Journal of Applied Polymer Science*, vol. 112, no. 4, pp. 2235–2240, 2009.

[4] Y. Nishio, N. Hirose, and T. Takahashi, "Miscibility studies of hydroxy propyl cellulose/ poly (ethylene glycol) in dilute solutions and solid state," *Sen' Gakkashi*, vol. 46, p. 441, 1990.

[5] Y. Nishio and R. S. J. Manley, "Blends of cellulose with nylon 6 and poly(ε-caprolactone) prepared by a solution-coagulation method," *Polymer Engineering & Science*, vol. 30, no. 2, pp. 71–82, 1990.

[6] Y. Kikuchi, N. Kubota, H. Fukuma, and H. Mitsuishi, "Preparation and properties of macromolecular complexes consisting of chitosan derivatives, iron(III) hydroxide sulfate, and poly(potassium vinyl sulfate)," *Macromolecular Rapid Communications*, vol. 17, no. 2, pp. 131–136, 1996.

[7] N. Kubota and Y. Kikuchi, "Preparation and properties of macromolecular complexes consisting of chitosan derivatives and potassium metaphosphate," *Molecular Chemistry and Physics*, vol. 193, no. 2, pp. 559–566, 1992.

[8] N. Kubota and Y. Eguchi, "Facile preparation of water-soluble N-acetylated chitosan and molecular weight dependence of its water-solubility," *Polymer Journal*, vol. 29, no. 2, pp. 123–127, 1997.

[9] N. Kubota, G. Konaka, and Y. Eguchi, "Characterization of blend films of chitin regenerated from porous chitosan with poly (vinyl alcohol)," *Sen'i Gakkaishi*, vol. 54, no. 4, pp. 212–218, 1998.

[10] K. K. Chee, "Determination of polymer-polymer miscibility by viscometry," *European Polymer Journal*, vol. 26, no. 4, pp. 423–426, 1990.

[11] Z. Sun, W. Wang, and Z. Feng, "Criterion of polymer-polymer miscibility determined by viscometry," *European Polymer Journal*, vol. 28, no. 10, pp. 1259–1261, 1992.

[12] R. Paladhi and R. P. Singh, "Miscibility and interaction studies on some aqueous polymer blend solutions by ultrasonic and rheological techniques," *Journal of Applied Polymer Science*, vol. 51, no. 9, pp. 1559–1565, 1994.

[13] R. Paladhi and R. P. Singh, "Ultrasonic and rheological investigations on interacting blend solutions of poly(acrylic acid) with poly(vinyl pyrrolidone) or poly(vinyl alcohol)," *European Polymer Journal*, vol. 30, no. 2, pp. 251–257, 1994.

[14] M. G. Cascone, L. Di Silvio, B. Sim, and S. Downes, "Collagen and hyaluronic acid based polymeric blends as drug delivery systems for the release of physiological concentrations of growth hormone," *Journal of Materials Science*, vol. 5, no. 9-10, pp. 770–774, 1994.

[15] Myra L. Weiner and Lois A. Kotkoskie, *Excipient Toxicity and Safety*, 1999.

[16] R. D. Hagenmaier and P. E. Shaw, "Moisture permeability of edible films made with fatty acid and (hydroxypropyl)methylcellulose," *Journal of Agricultural and Food Chemistry*, vol. 38, no. 9, pp. 1799–1803, 1990.

[17] J. F. Hanlon, *Handbook of Package Engineering*, Technomic, Lancaster, UK, 1992.

[18] E. Yilmaz, N. Erdenizci, and O. Yilmaz, "Miscibility of chitosan and poly(ethylene oxide) in dilute Solution," *International Journal of Polymer Analysis and Characterization*, vol. 8, no. 5, pp. 327–338, 2003.

[19] P. A. Albertson, *Partition of Cell Particles and Macromolecules*, Wiley, New York, NY, USA, 3rd edition, 1986.

[20] P. A. A. P. Mamza and F. M. Folaranmi, "Compatibility studies on solution of polystyrene and polyvinyl acetate blend by density and viscometric methods," *European Polymer Journal*, vol. 32, no. 7, pp. 909–912, 1996.

[21] M. Mucha and A. Pawlak, "Thermal analysis of chitosan and its blends," *Thermochimica Acta*, vol. 427, no. 1-2, pp. 69–76, 2005.

[22] C. Bonnet-Gonnet, S. Leikin, S. Chi, D. C. Rau, and V. Adrian Parsegian, "Measurement of forces between hydroxypropylcellulose polymers: temperature favored assembly and salt exclusion," *Journal of Physical Chemistry B*, vol. 105, no. 9, pp. 1877–1886, 2001.

[23] H. M. P. Naveen Kumar, M. N. Prabhakar, C. Venkata Prasad et al., "Compatibility studies of chitosan/PVA blend in 2% aqueous acetic acid solution at 30°C," *Carbohydrate Polymers*, vol. 82, no. 2, pp. 251–255, 2010.

[24] S. R. Illiger, C. Fadnis, T. Demappa, J. Jayaraju, and J. Keshavayya, "Miscibility studies of HPMC/PEG blends in water by viscosity, density, refractive index and ultrasonic velocity method," *Carbohydrate Polymers*, vol. 75, no. 3, pp. 484–488, 2009.

[25] K. Prasnt. Vasanthakumari, S. K. Rai, and H. Shiringara, "Studies on the mechanical properties of unsaturated polyester toughened epoxy/silk composites," *International Journal of Polymer Materials*, vol. 25, no. 1, pp. 33–41, 2006.

[26] S. D. Ravi Prakash and S. K. Rai, "Miscibility studies on sodium alginate/polyethylene glycol blend in solution by refractive index, ultrasonic velocity, density and viscometric

methods," *International Journal of Plastics Technology*, vol. 8, p. 334, 2004.

[27] W. Schenk, D. Reichert, and H. Schneider, "Compatibility in polymer blends of poly(vinyl acetate) and poly(methyl methacrylate) studied by nuclear magnetic relaxation," *Polymer*, vol. 31, no. 2, pp. 329–335, 1990.

[28] R. Mikkonen and A. Savolainen, "Blends of low density polyethylene/plasticized poly(vinyl chloride). The rheological, morphological, and mechanical characterization," *Journal of Applied Polymer Science*, vol. 39, no. 8, pp. 1709–1725, 1990.

[29] M. A. Sidkey, A. M. A. El Fattah, A. A. Yehia, and N. S. A. El All, "Ultrasonic investigation of some rubber blends," *Journal of Applied Polymer Science*, vol. 43, no. 8, pp. 1441–1449, 1991.

[30] K. C. Basavaraju, T. Demappa, and S. K. Rai, "Miscibility studies of polysaccharide Xanthan gum and PEO (polyethylene oxide) in dilute solution," *Carbohydrate Polymers*, vol. 69, no. 3, pp. 462–466, 2007.

[31] G. W. Adams and J. M. G. Cowie, "Blends of rigid and flexible macromolecules: poly(benzimidazole) mixed with poly(vinyl acetate-star-vinyl alcohol) copolymers," *Polymer*, vol. 40, no. 8, pp. 1993–2001, 1999.

[32] M. Mathew, K. N. Ninan, and S. Thomas, "Compatibility studies of polymer-polymer systems by viscometric techniques: nitrile-rubber-based polymer blends," *Polymer*, vol. 39, no. 25, pp. 6235–6241, 1998.

[33] R. K. Wanchoo and P. K. Sharma, "Viscometric study on the compatibility of some water-soluble polymer-polymer mixtures," *European Polymer Journal*, vol. 39, no. 7, pp. 1481–1490, 2003.

[34] T. Cai, Z. B. Hu, B. Ponder, J. St. John, and D. Moro, "Synthesis and study of and controlled release from nanoparticles and their networks based on functionalized hydroxypropylcellulose," *Macromolecules*, vol. 36, no. 17, pp. 6559–6564, 2003.

Brönsted Acidic Ionic Liquid 1-(1-Propylsulfonic)-3-methylimidazolium-Chloride Catalyzed Hydrolysis of D-Cellobiose in Aqueous Medium

Ananda S. Amarasekara and Bernard Wiredu

Department of Chemistry, Prairie View A&M University, Prairie View, TX 77446, USA

Correspondence should be addressed to Ananda S. Amarasekara, asamarasekara@pvamu.edu

Academic Editor: Yuriy A. Knirel

Brönsted acidic ionic liquid 1-(1-propylsulfonic)-3-methylimidazolium chloride (PSMIMCl) shows a higher catalytic activity than sulfuric acid in the hydrolysis of D-cellobiose to D-glucose in water at 90–120°C. This catalytic activity enhancement is more significant at higher temperatures, and at 120°C, PSMIMCl produced 64.5% glucose yield, whereas H_2SO_4 produced only 42.2% after 40 min. reaction, and this is a 52.8% enhancement of catalytic activity due to the alkylimidazolium group attached to the sulfonic acid group. ^{1}H NMR monitoring of the D-cellobiose hydrolysis in PSMIMCl and sulfuric acid mediums failed to reveal intermediates in the hydrolysis reaction, and this is probably due to rapid conversion of the intermediate(s) to a mixture of D-glucose anomers with $\alpha : \beta \approx 1 : 1.6$.

1. Introduction

Hydrolysis of lignocellulosic biomass to fermentable sugars is an essential step in the production of cellulosic-ethanol and is the most challenging step in the whole process [1–5]. Enzymatic hydrolysis is the widely used technology in current pilot plants, but this method requires a drastic pretreatment at high temperature and pressure to disrupt the strong hydrogen bonding network in cellulose, before exposure to the enzyme. Furthermore, commercialization of the enzymatic process is hindered by the prohibitive cost of the currently available enzyme preparations as well [6]. As an alternative to the enzymatic methods, dilute aqueous solution of sulfuric acid can also be used as a catalyst for the direct hydrolysis of cellulose at high temperature and pressure. Even though, the direct dilute acid saccharification gives lower sugar yields compared to enzymatic saccharification, a number of research groups have taken an interest in this old process [7–9] taking a second look at this technology due to its simplicity, and lower cost when compared to enzymatic saccharification, which anyhow requires an energy consuming pretreatment.

Ionic liquids are well known for their unique combination of attractive properties and as green solvents for numerous applications, but currently there is a marked interest in using ionic liquids in a number of other functions such as electrolytes [10], polymeric materials [11, 12], and catalysts [13]. In addition to this, 2002 discovery [14] of the use of ionic liquids as a cellulose dissolving solvent has sparked a new field of research on the use of ionic-liquid-based systems for the depolymerization of cellulose. The initial efforts in this direction were reported by Li et al. in 2007, where they published the use of catalytic amount of H_2SO_4 and controlled small amount of water for the hydrolysis of cellulose dissolved in butylmethylimidazolium chloride under mild conditions [15, 16]. Later, the use of solid acid catalysts [17–19], as well as an immobilized acidic ionic liquid catalyst [20] was reported as well for the hydrolysis of cellulose dissolved in imidazolium ionic liquids. In 2009, our research group developed the use of

Brönsted acidic ionic liquids with a built-in-SO_3H group for the dual role of solvent and acid catalyst [21]. Where we reported that cellulose dissolved in Brönsted acidic ionic liquids 1-(alkyl sulfonic)-3-methylimidazolium chloride can be hydrolyzed by the addition of 2.0 equivalents of water per glucose unit of cellulose and heating the solution at 70°C, and at atmospheric pressure to give glucose along with other reducing sugars in good yields [21].

As these ionic-liquid-based cellulose depolymerization methods require large volumes of ionic liquids as solvents, a complete recovery for reuse of the ionic liquid is essential in any large scale industrial process, and this is quite challenging since both the resulting sugars and these ionic liquids are highly soluble in water. Therefore, the use of an ionic liquid as a glycosidic bond hydrolysis catalyst in aqueous medium is a very attractive proposition [22]. However, the ionic liquid-cellulose system is typically studied under anhydrous conditions, and the use of ionic liquids in water as a catalyst is a relatively unexplored area. In a recent account of the use of ionic liquids in water, Dwiatmoko et al. reported [23] the enhancement of the catalytic activity of Nafion NR50 in the hydrolysis of cellulose model compound D-cellobiose to glucose, with the addition of halogenated ionic liquid 1-butyl-3-methylimidazolium chloride, to the aqueous medium. Our interests [20–22] in the development of an efficient chemical cellulose depolymerization catalyst have led us to study the SO_3H group substituted Brönsted acidic ionic liquid 1-(1-propylsulfonic)-3-methylimidazolium chloride (PSMIMCl) (Figure 1) as a catalyst in aqueous medium for the hydrolysis of cellulose model compound D-cellobiose.

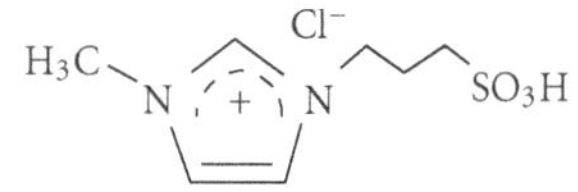

FIGURE 1: Brönsted acidic ionic liquid 1-(1-propylsulfonic)-3-methylimidazolium chloride (PSMIMCl).

2. Experimental

2.1. Materials and Instrumentation. D-cellobiose (99.9%), D_2O (99.9% atom D), sulfuric acid-d_2 (99.9% atom D), 1-methylimidazole, and 1,3-propanesultone were purchased from Aldrich Chemical Co. Brönsted acidic ionic liquid 1 was prepared by condensation of 1-methylimidazole with 1,3-propanesultone and acidification of the resulting salt with conc. HCl according to the literature procedure [24, 25]. D-Cellobiose hydrolysis experiments were carried out in 25 mL glass vials with polycarbonate screw caps. These vials were heated in a preheated mineral oil bath placed on a VWR Scientific 1500 W ceramic hotplate with VT-5 temperature controller of temperature accuracy ±0.1°C. Glucose concentrations in aqueous solutions were determined using a Carey 50 UV-Vis spectrophotometer and 1 cm quartz cells. ^{1}H NMR Spectra in D_2O were recorded on a Varian Mercury plus spectrometer operating at 400 MHz, rd = 1s, spectral width of 6398 Hz, and typically 16 scans were collected for spectra. All NMR spectra were collected at room temperature, 23°C.

2.2. General Experimental Procedures for the Hydrolysis of D-Cellobiose Using Aqueous 1-(1-Propylsulfonic)-3-Methylimidazolium Chloride, and Sulfuric Acid Solutions at 90, 105, and 120°C. Stock solutions of the acids, 1-(1-propylsulfonic)-3-methylimidazolium chloride and sulfuric acid were prepared by dissolving appropriate amounts of these acids in deionized water to give acid concentration of 0.0321 mol H^+/L in each solution. The accuracy of the concentration was checked by titration with standardized aq. NaOH solution using phenolphthalein as the indicator.

D-Cellobiose (0.0300 g, 0.0877 mmol) was dissolved in 10.00 mL of aqueous acid solution in a 25 mL glass vial with a skrew cap. The vial was firmly closed and heated in a preheated mineral oil bath maintained (±0.1°C) at the desired temperature for a specified length of time. Then, 0.30 mL Aliquots was withdrawn from the reaction solution at specific time intervals for glucose analysis, and an aliquot taken at room temperature before heating begins is referred to as zero time.

2.3. Glucose Assay. 0.30 mL of Aliquots taken from the reaction vial at specific time intervals was transferred to glass vials, neutralized by drop wise addition of 0.5 M aq. NaOH, and then diluted to 2.00 mL. At zero time, reaction was started by adding 2.00 mL of glucose oxidase-peroxidase assay reagent [26, 27] to the vial and mixing thoroughly, and the vial was incubated in a water bath at 37°C for 30 min. Then reaction was quenched by adding 2.00 mL of 6 M HCl to give a pink solution. The reagent blank was prepared by mixing 2.00 mL of deionized water and 2.00 mL of assay reagent and was treated similarly. Then the absorbance was immediately measured at 540 nm against the reagent blank, and glucose concentration in the solution was calculated by employing a standard curve prepared using glucose. The changes in percent yields of glucose produced during the hydrolysis of D-cellobiose in aqueous 1-(1-propylsulfonic)-3-methylimidazolium chloride (PSMIMCl), and H_2SO_4 mediums at 90, 105, and 120°C are shown in Figure 2.

2.4. ^{1}H NMR Studies of D-Cellobiose Hydrolysis in D_2O Using1-(1-Propylsulfonic)-3-Methylimidazolium Chloride and Sulfuric Acid-d_2 as Catalysts. Two NMR samples were prepared in 5 mm NMR tubes by dissolving 10.0 mg (0.029 mmol) of D-cellobiose in each tube and using 0.6 mL of D_2O solutions of 1-(1-propylsulfonic)-3-methylimidazolium chloride and sulfuric acid-d_2 of acid strength 0.0321 mol H^+/L. These samples were allowed to stabilize at room temperature for 24 hr, and then ^{1}H NMR spectra were recorded as the baseline data. Then the tubes were heated in a preheated mineral oil bath maintained at 90.0 ± 0.1°C, tubes were taken out of oil bath after specified time intervals, reaction was quenched by immersing the NMR tubes in cold water, and ^{1}H NMR spectra were recorded immediately. Then the tubes were returned to the thermostated

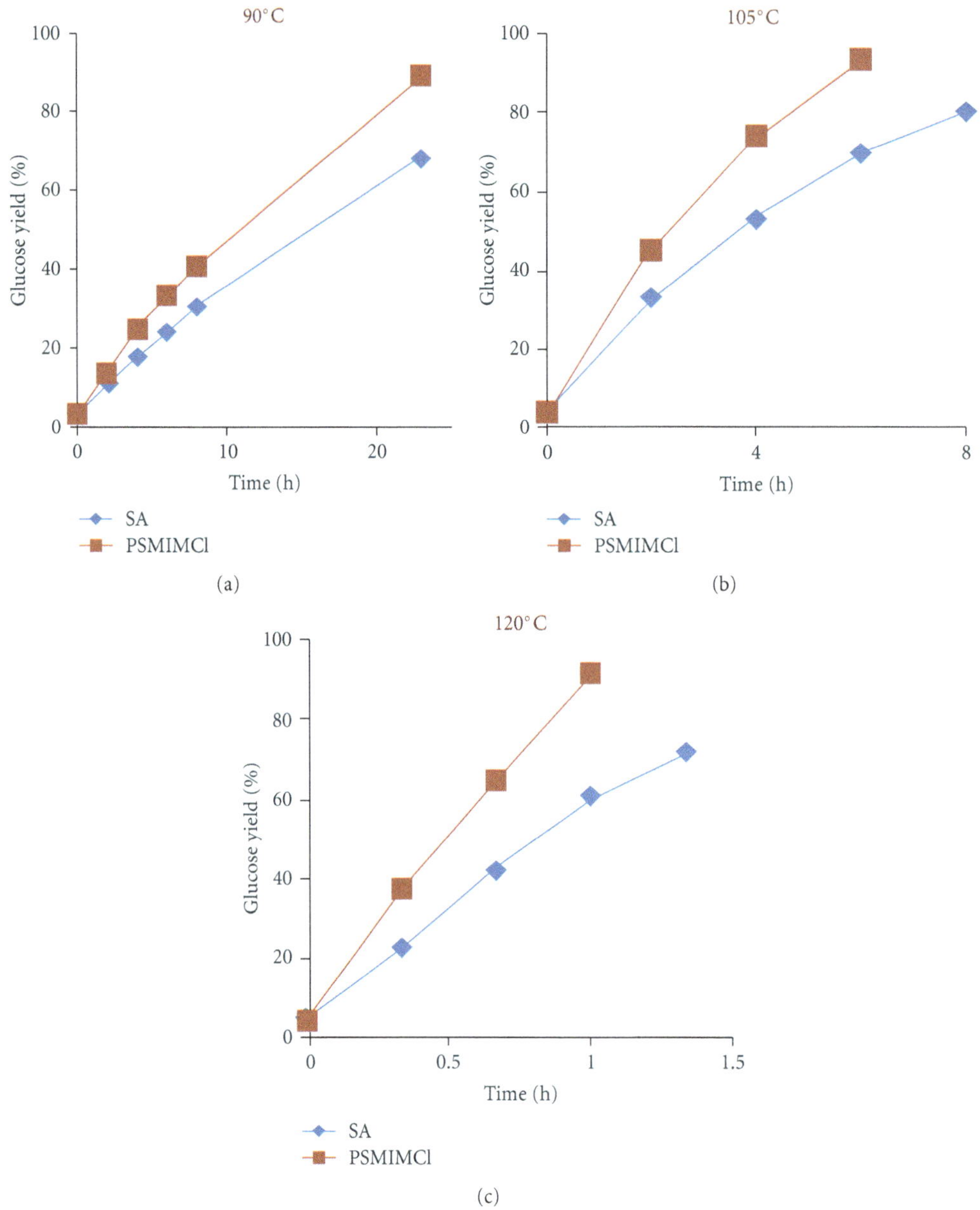

FIGURE 2: Change in percent yields of glucose produced during the hydrolysis of D-cellobiose in aqueous 1-(1-propylsulfonic)-3-methylimidazolium chloride (PSMIMCl) and H_2SO_4(SA) mediums at 90, 105, and 120°C. 30.0 mg of D-cellobiose in 10.0 mL of 0.0321 mol H^+/L acid mediums were used in all experiments.

oil bath for further heating, and NMR spectrums were recorded at regular time intervals to monitor the progress of the reaction. A typical series of spectra recorded for the sample using 0.0321 mol H^+/L of 1-(1-propylsulfonic)-3-methylimidazolium chloride in D_2O medium is shown in Figure 3. D-Cellobiose sample heated in sulfuric acid-d_2 medium also gave a similar series of spectra.

3. Results and Discussion

3.1. Hydrolysis of D-Cellobiose Using Aqueous 1-(1-Propylsulfonic)-3-Methylimidazolium Chloride and Sulfuric Acid Solutions at 90, 105, and 120°C. Catalytic activities of 1-(1-propylsulfonic)-3-methylimidazolium chloride were compared with the activities of sulfuric acid of the same molar H^+ ion concentration, and according to Oscarson and Izatt's expression on temperature dependence of the first and second dissociation constants of sulfuric acid in water, it is assumed that H_2SO_4 completely dissociates to give two H^+ ions in the temperature range of the study [28, 29]. The changes in percent yields of glucose produced during the hydrolysis of D-cellobiose in aqueous PSMIMCl, and H_2SO_4 mediums at 90, 105 and 120°C are shown in Figure 2. These experiments clearly show that Brönsted acidic ionic liquid PSMIMCl has a higher catalytic activity than sulfuric acid

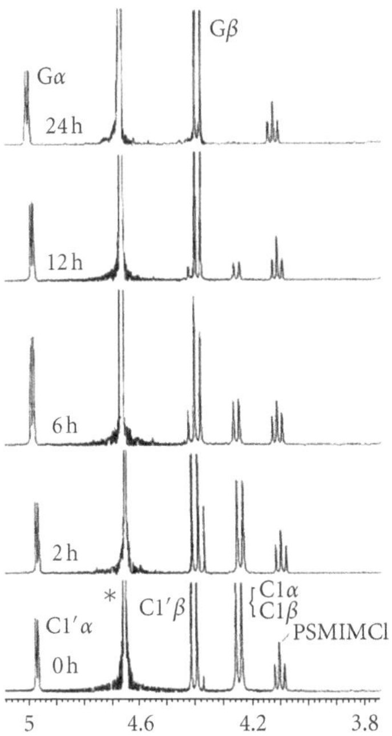

FIGURE 3: A series of ^{1}H NMR spectra recorded during the hydrolysis of D-cellobiose to D-glucose using 0.0321 mol H^+/L 1-(1-propylsulfonic)-3-methylimidazolium chloride (PSMIMCl) in D_2O at 90°C. *= HDO.

of the same acid strength. Furthermore, catalytic activity enhancement is more significant at higher temperatures, for instance at 120°C PSMIMCl produced 64.5% glucose yield after 40 min. reaction, whereas H_2SO_4 produced only 42.2% during the same period, and this is a 52.8% enhancement of catalytic activity due to the alkylimidazolium group attached to the sulfonic acid group. This catalytic activity enhancement may be due to ion-dipole-type interactions of alkyl imidazolium groups and chloride ions with hydroxyl groups of D-cellobiose. In fact, Xiang has observed [30] similar interactions of 1-ethyl-3-methylimidazolium acetate with D-cellobiose in DMSO using ^{13}C NMR spectroscopy. Additionally, the glucose yield shows slight decrease with longer reaction times, especially at higher temperatures, and this could be due to well-known decomposition of glucose to various products like 5-hydroxymethylfurfural, 1,6-anhydroglucose, levulinic acid, and formic acid in the acid medium [31, 32].

3.2. ^{1}H NMR Studies of D-Cellobiose Hydrolysis in D_2O Using1-(1-Propylsulfonic)-3-Methylimidazolium Chloride and Sulfuric Acid-d_2 as Catalysts. A typical series of ^{1}H NMR spectra recorded during the hydrolysis of D-cello-biose in 0.0321 mol H^+/L 1-(1-propylsulfonic)-3-methylimidazolium chloride in D_2O medium is shown in Figure 3. D-Cellobiose sample heated in sulfuric acid-d_2 medium also gave a similar series of spectra. The spectrum recorded at $t = 0$ h. shows a mixture of α and β anomers of D-cellobiose and only a trace amount of D-glucose. The anomeric composition of D-cellobiose in this mixture was calculated as $\alpha := \beta$ 1.00 : 1.60 using the peak area ratio of two doublets at 4.97 ppm ($J = 3.6$ Hz) and 4.40 ppm ($J = 8.4$ Hz, in Figure 3, $t = 0$ h), and this value is compatible with the reported anomeric ratio [33] of D-cellobiose in water. A gradual increase in D-glucose and the disappearance of D-cellobiose is seen in the series of spectra recorded during the course of the reaction. Only the anomeric hydrogens of D-cellobiose and D-glucose could be assigned due to the complexity of the spectra shown in Figure 3. The anomeric ratios of D-glucose formed during the PSMIMCl and sulfuric acid catalyzed hydrolysis were calculated by using the following equations, assuming that anomeric ratio of D-cellobiose remains constant throughout the reaction.

TABLE 1: $\alpha : \beta$ Anomeric ratios of D-glucose produced during the 1-(1-propylsulfonic)-3-methylimidazolium chloride (PSMIMCl) and sulfuric acid catalyzed hydrolysis of D-cellobiose to D-glucose. 10.0 mg of D-cellobiose in 0.60 mL acid solutions (0.0321 mol H^+/L) were used.

Time h.	Catalyst	
	PSMIMCl $\alpha : \beta$	H_2SO_4 $\alpha : \beta$
2	1 : 1.56	1 : 1.62
6	1 : 1.65	1 : 1.58
12	1 : 1.60	1 : 1.59
24	1 : 1.63	1 : 1.65

$$C_\alpha + 2G_\alpha = \text{Peak area at 4.97 ppm}, \quad (1)$$

$$C_\beta + 2G_\beta = \text{Peak area at 4.37–4.42 ppm}, \quad (2)$$

$$C_\alpha + C_\beta = \text{Peak area at 4.25 ppm}, \quad (3)$$

where, C_α = peak area of D-cellobiose 1α or $1'\alpha$ hydrogen. C_β = peak area of D-cellobiose 1β or $1'\beta$ hydrogen. G_α = peak area of D-glucose 1α hydrogen. G_β = peak area of D-glucose 1β hydrogen.

Calculated anomeric ratios of D-glucose produced remains approximately constant during the course of the reactions as seen in Table 1, and this may be due to a rapid equilibration of D-glucose formed. Furthermore, as seen in the array of NMR spectra of PSMIMCl-catalyzed hydrolysis of D-cellobiose in Figure 3, ^{1}H NMR study failed to reveal any reaction intermediates, and similarly NMR spectra of H_2SO_4-catalyzed reaction also showed no reaction intermediates. This can be explained by a mechanism involving a slow protonation of glycosidic oxygen in D-cellobiose and fast attack of water on the anomeric carbon, resulting fast hydrolysis D-cellobiose to D-glucose, as shown in the case of PSMIMCl catalyzed reaction in Figure 4.

4. Conclusion

We have shown that Brönsted acidic ionic liquid1-(1-propylsulfonic)-3-methylimidazolium chloride (PSMIMCl) has

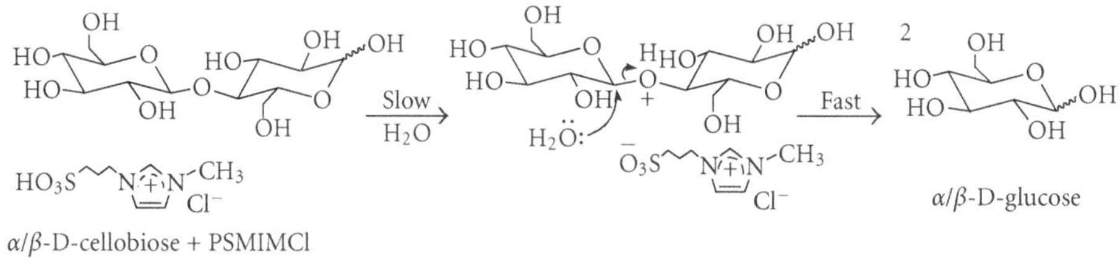

FIGURE 4: Proposed mechanism for Brönsted acidic ionic liquid 1-(1-propylsulfonic)-3-methylimidazolium chloride (PSMIMCl) catalyzed hydrolysis of D-cellobiose to D-glucose in water.

a higher catalytic activity than sulfuric acid in the hydrolysis of D-cellobiose to D-glucose in water at 90–120°C. This catalytic activity enhancement is more significant at higher temperatures and may be due to an interaction of alkylimidazolium group with D-cellobiose. ^{1}H NMR monitoring of the PSMIMCl as well as sulfuric acid catalyzed D-cellobiose hydrolysis reactions at 90°C failed to show any intermediates in the reaction, and this is possibly due to fast conversion of any intermediate(s) to products.

Acknowledgments

Authors would like to thank Center for Environmentally Beneficial Catalysis (CEBC), University of Kansas, American Chemical Society: PRF Grant UR1-49436, NSF Grant CBET-0929970, and USDA Grant CBG-2010-38821-21569 for financial support.

References

[1] J. Goettemoeller and A. Goettemoeller, *Sustainable Ethanol: Biofuels, Biorefineries, Cellulosic Biomass, Flex-fuel Vehicles, and Sustainable Farming for Energy Independence*, Prairie Oak, Maryville, Mo, USA, 2007.

[2] R. Katzen and D. J. Schell, *Biorefineries—Industrial Processes and Products*, vol. 1, Wiley-VCH, Weinheim, Germany, 2006, edited by B. Kamm, P.R. Gruber, M. Kamm.

[3] C. Wyman, *Handbook on Bioethanol: Production and Utilization*, Taylor & Francis, Washington, DC, USA, 1996.

[4] G. W. Huber, S. Iborra, and A. Corma, "Synthesis of transportation fuels from biomass: chemistry, catalysts, and engineering," *Chemical Reviews*, vol. 106, no. 9, pp. 4044–4098, 2006.

[5] M. Balat and H. Balat, "Recent trends in global production and utilization of bio-ethanol fuel," *Applied Energy*, vol. 86, no. 11, pp. 2273–2282, 2009.

[6] R. K. Sukumaran, R. R. Singhania, G. M. Mathew, and A. Pandey, "Cellulase production using biomass feed stock and its application in lignocellulose saccharification for bio-ethanol production," *Renewable Energy*, vol. 34, no. 2, pp. 421–424, 2009.

[7] P. Lenihan, A. Orozco, E. O'Neill, M. N. M. Ahmad, D. W. Rooney, and G. M. Walker, "Dilute acid hydrolysis of lignocellulosic biomass," *Chemical Engineering Journal*, vol. 156, no. 2, pp. 395–403, 2010.

[8] G. Sanchez, L. Pilcher, C. Roslander, T. Modig, M. Galbe, and G. Liden, "Dilute-acid hydrolysis for fermentation of the Bolivian straw material Paja Brava," *Bioresource Technology*, vol. 93, no. 3, pp. 249–256, 2004.

[9] Q. Xiang, Y. Y. Lee, and R. W. Torget, "Kinetics of glucose decomposition during dilute-acid hydrolysis of lignocellulosic biomass," *Applied Biochemistry and Biotechnology*, vol. 113, pp. 1127–1138, 2004.

[10] T. Sato, S. Marukane, T. Narutomi, and T. Akao, "High rate performance of a lithium polymer battery using a novel ionic liquid polymer composite," *Journal of Power Sources*, vol. 164, no. 1, pp. 390–396, 2007.

[11] A. S. Amarasekara and P. Shanbhag, "Synthesis and characterization of polymeric ionic liquid poly(imidazolium chloride-1,3-diylbutane-1,4-diyl)," *Polymer Bulletin*, vol. 67, no. 4, pp. 623–629, 2011.

[12] A. S. Amarasekara, B. Callis, and B. Wiredu, "Synthesis and characterization of branched polymeric ionic liquids with imidazolium chloride segments," *Polymer Bulletin*, vol. 68, no. 4, pp. 901–908, 2011.

[13] H. Olivier-Bourbigou, L. Magna, and D. Morvan, "Ionic liquids and catalysis: recent progress from knowledge to applications," *Applied Catalysis A*, vol. 373, no. 1-2, pp. 1–56, 2010.

[14] R. P. Swatloski, S. K. Spear, J. D. Holbrey, and R. D. Rogers, "Dissolution of cellose with ionic liquids," *Journal of the American Chemical Society*, vol. 124, no. 18, pp. 4974–4975, 2002.

[15] C. Li and Z. K. Zhao, "Efficient acid-catalyzed hydrolysis of cellulose in ionic liquid," *Advanced Synthesis and Catalysis*, vol. 349, no. 11-12, pp. 1847–1850, 2007.

[16] C. Li, Q. Wang, and Z. K. Zhao, "Acid in ionic liquid: an efficient system for hydrolysis of lignocellulose," *Green Chemistry*, vol. 10, no. 2, pp. 177–182, 2008.

[17] R. Rinaldi, R. Palkovits, and F. Schüth, "Depolymerization of cellulose using solid catalysts in ionic liquids," *Angewandte Chemie—International Edition*, vol. 47, no. 42, pp. 8047–8050, 2008.

[18] Z. Zhang and Z. K. Zhao, "Solid acid and microwave-assisted hydrolysis of cellulose in ionic liquid," *Carbohydrate Research*, vol. 344, no. 15, pp. 2069–2072, 2009.

[19] S. J. Kim, A. A. Dwiatmoko, J. W. Choi, Y. W. Suh, D. J. Suh, and M. Oh, "Cellulose pretreatment with 1-*n*-butyl-3-methylimidazolium chloride for solid acid-catalyzed hydrolysis," *Bioresource Technology*, vol. 101, no. 21, pp. 8273–8279, 2010.

[20] A. S. Amarasekara and O. S. Owereh, "Synthesis of a sulfonic acid functionalized acidic ionic liquid modified silica catalyst and applications in the hydrolysis of cellulose," *Catalysis Communications*, vol. 11, no. 13, pp. 1072–1075, 2010.

[21] A. S. Amarasekara and O. S. Owereh, "Hydrolysis and decomposition of cellulose in bronösted acidic ionic liquids under mild conditions," *Industrial and Engineering Chemistry Research*, vol. 48, no. 22, pp. 10152–10155, 2009.

[22] A. S. Amarasekara and B. Wiredu, "Degradation of cellulose in dilute aqueous solutions of acidic ionic liquid

1-(1-propylsulfonic)-3-methylimidazolium chloride, and *p*-toluenesulfonic acid at moderate temperatures and pressures," *Industrial and Engineering Chemistry Research*, vol. 50, no. 21, pp. 12276–12280, 2011.

[23] A. A. Dwiatmoko, J. W. Choi, D. J. Suh, Y.-W. Suh, and H. H. Kung, "Understanding the role of halogen-containing ionic liquids in the hydrolysis of cellobiose catalyzed by acid resins," *Applied Catalysis A*, vol. 387, no. 1-2, pp. 209–214, 2010.

[24] Q. Yang, Z. Wei, H. Xing, and Q. Ren, "Brönsted acidic ionic liquids as novel catalysts for the hydrolyzation of soybean isoflavone glycosides," *Catalysis Communications*, vol. 9, no. 6, pp. 1307–1311, 2008.

[25] G. Y. Zhu, R. Wang, G. H. Liu, L. Q. Xu, B. Zhang, and X. Q. Wu, "Synthesis of multi-hydroxyl and sulfonyl dual-functionalized room temperature ionic liquids," *Chinese Chemical Letters*, vol. 18, no. 6, pp. 633–635, 2007.

[26] H. U. Bergmeyer and E. Bernt, "Glucose determination with glucose oxidase and peroxidase," in *Methods of Enzymatic Analysis*, H. U. Bergmeyer, Ed., pp. 1205–1212, Academic Press, New York, NY, USA, 1974.

[27] D. A. T. Southgate, *Determination of Food Carbohydrates*, Applied Science Publishers, London, UK, 1961.

[28] L. Kupiainen, J. Ahola, and J. Tanskanen, "Comparison of formic and sulfuric acids as a glucose decomposition catalyst," *Industrial and Engineering Chemistry Research*, vol. 49, no. 18, pp. 8444–8449, 2010.

[29] J. L. Oscarson, R. M. Izatt, P. R. Brown, Z. Pawlak, S. E. Gillespie, and J. J. Christensen, "Thermodynamic quantities for the interaction of SO_4^{2-} with H^+ and Na^+ in aqueous solution from 150 to 320°C," *Journal of Solution Chemistry*, vol. 17, no. 9, pp. 841–863, 1988.

[30] J. Zhang, H. Zhang, J. Wu, J. Zhang, J. He, and J. Xiang, "NMR spectroscopic studies of cellobiose solvation in EmimAc aimed to understand the dissolution mechanism of cellulose in ionic liquids," *Physical Chemistry Chemical Physics*, vol. 12, no. 8, pp. 1941–1947, 2010.

[31] Q. Xiang, Y. Y. Lee, and R. W. Torget, "Kinetics of glucose decomposition during dilute-acid hydrolysis of lignocellulosic biomass," *Applied Biochemistry and Biotechnology A*, vol. 115, no. 1–3, pp. 1127–1138, 2004.

[32] X. Huang, H. Duan, and S. A. Barringer, "Effects of buffer and temperature on formation of furan, acetic acid and formic acid from carbohydrate model systems," *LWT—Food Science and Technology*, vol. 44, no. 8, pp. 1761–1765, 2011.

[33] M. U. Roslund, P. Tähtinen, M. Niemitz, and R. Sjöholm, "Complete assignments of the ^{1}H and ^{13}C chemical shifts and $J_{H,H}$ coupling constants in NMR spectra of D-glucopyranose and all D-glucopyranosyl-D-glucopyranosides," *Carbohydrate Research*, vol. 343, no. 1, pp. 101–112, 2008.

Permissions

The contributors of this book come from diverse backgrounds, making this book a truly international effort. This book will bring forth new frontiers with its revolutionizing research information and detailed analysis of the nascent developments around the world.

We would like to thank all the contributing authors for lending their expertise to make the book truly unique. They have played a crucial role in the development of this book. Without their invaluable contributions this book wouldn't have been possible. They have made vital efforts to compile up to date information on the varied aspects of this subject to make this book a valuable addition to the collection of many professionals and students.

This book was conceptualized with the vision of imparting up-to-date information and advanced data in this field. To ensure the same, a matchless editorial board was set up. Every individual on the board went through rigorous rounds of assessment to prove their worth. After which they invested a large part of their time researching and compiling the most relevant data for our readers. Conferences and sessions were held from time to time between the editorial board and the contributing authors to present the data in the most comprehensible form. The editorial team has worked tirelessly to provide valuable and valid information to help people across the globe.

Every chapter published in this book has been scrutinized by our experts. Their significance has been extensively debated. The topics covered herein carry significant findings which will fuel the growth of the discipline. They may even be implemented as practical applications or may be referred to as a beginning point for another development. Chapters in this book were first published by Hindawi Publishing Corporation; hereby published with permission under the Creative Commons Attribution License or equivalent.

The editorial board has been involved in producing this book since its inception. They have spent rigorous hours researching and exploring the diverse topics which have resulted in the successful publishing of this book. They have passed on their knowledge of decades through this book. To expedite this challenging task, the publisher supported the team at every step. A small team of assistant editors was also appointed to further simplify the editing procedure and attain best results for the readers.

Our editorial team has been hand-picked from every corner of the world. Their multi-ethnicity adds dynamic inputs to the discussions which result in innovative outcomes. These outcomes are then further discussed with the researchers and contributors who give their valuable feedback and opinion regarding the same. The feedback is then collaborated with the researches and they are edited in a comprehensive manner to aid the understanding of the subject.

Apart from the editorial board, the designing team has also invested a significant amount of their time in understanding the subject and creating the most relevant covers. They scrutinized every image to scout for the most suitable representation of the subject and create an appropriate cover for the book.

The publishing team has been involved in this book since its early stages. They were actively engaged in every process, be it collecting the data, connecting with the contributors or procuring relevant information. The team has been an ardent support to the editorial, designing and production team. Their endless efforts to recruit the best for this project, has resulted in the accomplishment of this book. They are a veteran in the field of academics and their pool of knowledge is as vast as their experience in printing. Their expertise and guidance has proved useful at every step. Their uncompromising quality standards have made this book an exceptional effort. Their encouragement from time to time has been an inspiration for everyone.

The publisher and the editorial board hope that this book will prove to be a valuable piece of knowledge for researchers, students, practitioners and scholars across the globe.

List of Contributors

Defang Zeng, Xinrong Luo and Renjie Tu
School of Resource & Environmental Engineering and Key Laboratory of Hubei Province for Mineral Resources Processing and Environment, Wuhan University of Technology, Wuhan 430070, China

Mohamed E. I. Badawy
Department of Pesticide Chemistry and Technology, Faculty of Agriculture, Alexandria University El-Shatby, Alexandria 21545, Egypt

Yakubu Azeh and Gabriel Cheku Mohammed
Department of Chemistry, Ibrahim Badamasi Babangida University, Lapai, Nigeria

Ademola Olatunji
Department of Chemistry, University of Ilorin, Ilorin, Nigeria

Paul Andrew Mamza
Department of Chemistry, Ibrahim Badamasi Babangida University, Lapai, Nigeria
Department of Chemistry, Ahmadu Bello University, Zaria, Nigeria

WimMengerink, Jaap de Slegte and Ellen van den Heuvel
FrieslandCampina Research, Friesland Campina, 7400 AB Deventer, The Netherlands

Glenn Gibson and Robert Rastall
Department of Food and Nutritional Sciences, The University of Reading, White knights, P.O. Box 226, Berkshire, Reading RG6 6AP, UK

Eva Gietl
Friesland Campina Research, Friesland Campina, 7400 AB Deventer, The Netherlands
Department of Food and Nutritional Sciences, The University of Reading, White knights, P.O. Box 226, Berkshire, Reading RG6 6AP, UK

Venkanna Azmeera, P. Adhikary and S. Krishnamoorthi
Department of Chemistry, Banaras Hindu University, Varanasi 221005, India

A. Chandra Babu, B. Mallikarjuna and M. C. S. Subha
Department of Chemistry, Sri Krishnadevaraya University, Anantapur, Andhra Pardesh 515 003, India

M. N. Prabhakar and K. Chowdoji Rao
Department of Polymer Science and Technology, Sri Krishnadevaraya University, Anantapur, Andhra Pradesh 515 003, India

A. Suresh Babu
Department of Physics, Rayalaseema University, Kurnool, Andhra Pardesh 518 002, India

Ambily Chandran and Sunny Kuriakose
Research and Post Graduate Department of Chemistry, St. Thomas College, Mahatma Gandhi University, Kerala, Pala 686574, India

Tessymol Mathew
Research and Post Graduate Department of Chemistry, St. George College, Mahatma Gandhi University, Kerala, Aruvithura 686122, India

Yakubu Azeh and Paul AndrewMamza
Department of Chemistry, Ibrahim Badamasi Babangida University, Lapai, Nigeria

Gabriel Ademola Olatunji
Department of Chemistry, University of Ilorin, Ilorin, Nigeria

Gabriel Biringanine and Pierre Duez
Laboratoire de Pharmacognosie, de Bromatologie, et de Nutrition Humaine, Facult´e de Pharmacie, Universit´e Libre de Bruxelles (ULB), Campus Plaine CP 205/9, Boulevard du Triomphe, 1050 Bruxelles, Belgium

Moustapha Ouedraogo
Laboratoire de Pharmacologie et de Toxicologie, Unit´e de Formation et de Recherche en Sciences de la Sant´e, Universit´e de Ouagadougou, Ouagadougou 03, Burkina Faso

Bernard Vray
Laboratoire d'Immunologie Exp´erimentale, Facult´e de M´edecine, Universit´e Libre de Bruxelles (ULB), Campus Erasme, Route de Lennik, 1070 Bruxelles, Belgium

Anne Berit Samuelsen
Department of Pharmaceutical Chemistry-Pharmacognosy, School of Pharmacy, University of Oslo, P.O. Box 1068, Blindern, 0316 Oslo, Norway

Nadezhda Viktorovna Ivanova, Natalya Nikolaevna Trofimova, Lydmila Alexandrovna Es'kova and Vasilyi Anatolievich Babkin
A.E. Favorsky Institute of Chemistry, Russian Academy of Sciences, Siberian Branch, Irkutsk 664033, Russia

Meenu Kapoor, Dhriti Khandal, Ruchi Gupta, Pinklesh Arora, Geetha Seshadri and Saroj Aggarwal
Shriram Institute of Technology, Gautum Buddh Technical University, I.E.T. Campus, Sitapur Road, Lucknow 226 021, India

Rakesh Kumar Khandal
Gautum Buddh Technical University, I.E.T. Campus, Sitapur Road, Lucknow 226 021, India

Claudio Corradini, Antonella Cavazza and Chiara Bignardi
Dipartimento di Chimica Generale de Inorganica, Chimica Analitica, Chimica Fisica, Universit`a Degli Studi di Parma, Parco Area delle, Scienze 17/A, 43124 Parma, Italy

Leonel Pereira
IMAR-CMA, Department of Life Sciences, FCTUC, University of Coimbra, 3004-516 Coimbra, Portugal

Saly F. Gheda
Phycology Lab., Botany Department, Faculty of Science, Tanta University, Tanta 31527, Egypt

Paulo J. A. Ribeiro-Claro
Department of Chemistry, CICECO, University of Aveiro, 3810-193 Aveiro, Portugal

Issa M. A. IL Dueik and Gordon A. Morris
Department of Chemical and Biological Sciences, School of Applied Sciences, University of Huddersfield, Queensgate, Huddersfield, West Yorkshire HD1 3DH, UK

Jan Springer, Lambertus A. M. van den Broek and Gerrit Eggink
Wageningen UR Food & Biobased Research, P.O. Box 17, 6700 AAWageningen, The Netherlands

Carmen G. Boeriu
Wageningen UR Food & Biobased Research, P.O. Box 17, 6700 AAWageningen, The Netherlands
Universitatea "Aurel Vlaicu," Str. E. Dragoi nr. 2, 310330 Arad, Romania

Floor K. Kooy
Wageningen UR Food & Biobased Research, P.O. Box 17, 6700 AAWageningen, The Netherlands
Genencor International B.V., P.O. Box 218, 2300 AE Leiden, The Netherlands

Ziv Roth, Galit Yehezkel and Isam Khalaila
Avram and Stella Goldstein-Goren Department of Biotechnology Engineering, Ben-Gurion University of the Negev, P.O. Box 653, 84105 Beer Sheva, Israel

Zoran Herceg, Anet ReDek Jambrak and Tomislava Vukušić
Faculty of Food Technology and Biotechnology, University of Zagreb, Pierottijeva 6, Zagreb, Croatia

Verica Batur
High School of Food Technology, Gjure Prejca 2, Zagreb, Croatia

Ines Gmajnilki and Igor Špoljarić
Forensic Science Centre "Ivan Vuˇceti´c", Ilica 110, Zagreb, Croatia

Asmaa Farouk
Deutsches Textilforschungszentrum Nord-West gGmbH (DTNW), 47798 Krefeld, Germany
National Research Centre, Textile Division, Textile Chemistry and Technology, Department of Preparation and Finishing of Cellulosic Fibers, Tahrir Street, Dokki, P.O. Box 12622, Giza, Egypt

ShaabanMoussa
Genetic Engineering and Biotechnology Research Institute, Minoufiya University, P.O. Box 79/22857, El-Sadat City, Egypt

Mathias Ulbricht
Universit¨at Duisburg-Essen, Technical Chemistry II, Universit¨atsstrasse 7, 45141 Essen, Germany

Torsten Textor
Deutsches Textilforschungszentrum Nord-West gGmbH (DTNW), 47798 Krefeld, Germany
Center for Nanointegration Duisburg-Essen CeNIDE, 47048 Duisburg, Germany

N. Khebichat, K. Nekkaz and S. Ghalem
Laboratory of Natural Products and Bio Actives, Department of Chemistry, Faculty of Sciences, Aboubakr Belkaid University, Tlemcen 13000, Algeria

M. M. AbdElhady
Textile Research Division, National Research Centre, Dokki, Egypt

Naoko Igarashi, Atsuko Takeguchi, Kyohei Higashi and Toshihiko Toida
Graduate School of Pharmaceutical Sciences, Chiba University, 1-8-1 Inohana, Chuo, Chiba 260-8675, Japan

Shinobu Sakai and Hiroshi Akiyama
National Institute of Health Sciences, 1-18-1 Kamiyoga, Setagaya, Tokyo 158-8501, Japan

Arti Srivastava
Department of Chemistry, School of Physical Sciences, Guru Ghasidas Vishwavidyalaya, Bilaspur, Chhattisgarh 495009, India

Rajesh Kumar
Organic Polymer Laboratory, Department of Chemistry, Centre of Advanced Studies in Chemistry, Banaras Hindu University, Varanasi, Uttar Pradesh 221 005, India

K. Sudharsan Reddy, M. N. Prabhakar, P. Kumara Babu and G. Venkatesulu
Department of Polymer Science and Technology, Sri Krishnadevaraya University, Andhra Pradesh Anantapur 515003, India

U. Sajan Kumarji Rao, K. Chowdoji Rao and M. C. S. Subha
Department of Chemistry, Sri Krishnadevaraya University, Andhra Pradesh Anantapur 515003, India

Ananda S. Amarasekara and Bernard Wiredu
Department of Chemistry, Prairie View A&M University, Prairie View, TX 77446, USA